The Anthropocene: Politik—Economics—Society—Science

Volume 49

Series Editor

Hans Günter Brauch, Peace Research and European Security Studies (AFES-PRESS), Mosbach, Germany

http://www.afes-press-books.de/html/APESS.htm
http://www.afes-press-books.de/html/APESS_49.htm

Richard Marcantonio · Ellis Adjei Adams ·
Emmanuel Ojeifo · Emmanuel Katongole ·
Heather Foucault-Camm
Editors

Integral Ecology in the Anthropocene: Africa and Beyond

Springer

Editors
Richard Marcantonio
Keough School of Global Affairs
University of Notre Dame
Notre Dame, USA

Emmanuel Ojeifo
Department of Theology
University of Notre Dame
Notre Dame, IN, USA

Heather Foucault-Camm
McGrath Institute for Church Life
University of Notre Dame
Notre Dame, IN, USA

Ellis Adjei Adams
Keough School of Global Affairs
University of Notre Dame
Notre Dame, IN, USA

Emmanuel Katongole
Department of Theology, Keough School
of Global Affairs, College of Arts
and Letters
University of Notre Dame
Notre Dame, IN, USA

On This Book and Acknowledgements
This book emerged from the author's pursuit of understanding the operationalization in everyday life and beyond of integral ecology. The book is the culmination of a process of consultation, deliberation, and a convening at the Bethany Land Institute in Luwero District, Nandere Parish, Uganda. The resounding call from the conference and often found in this volume is for hope: Hope for a future where humans and nature simultaneously flourish.

ISSN 2367-4024 ISSN 2367-4032 (electronic)
The Anthropocene: Politik—Economics—Society—Science
ISBN 978-3-032-32124-4 ISBN 978-3-032-32125-1 (eBook)
https://doi.org/10.1007/978-3-032-32125-1

Language editing: Richard Marcantonio, Ellis Adjei Adams, Emmanuel Ojeifo, and Emmanuel Katongole
Copy editing and layout: PD Dr. Hans Günter Brauch, AFES-PRESS and HGBS, Alte Bergsteige 47, 74821 Mosbach, Germany

This Springer imprint is published by the registered company Springer Nature Switzerland AG
The registered company address is: Gewerbestrasse 11, 6330 Cham, Switzerland

This book is for all people of goodwill.
May we all find the internal and external
transformations so needed in our age.

Acknowledgements

As with any work, this volume would not have been possible without the generous support and labor of numerous people, organizations, and institutions. We are fortunate to have had so many and so steadfast a foundation. First and foremost, we must thank our contributors to this volume. We appreciate your talents and time and the amazing stories, experience, and research that you have gifted us.

We thank the Ford Family Program and the Kellogg Institute in the Keough School of Global Affairs at the University of Notre Dame for research grants and conference support that made this volume possible, as well as the Regenerating Ecologies and Economies for Livelihoods Lab. We also thank the Kroc Institute for International Peace Studies and the Ansari Institute for Global Engagement with Religion in the Keough School of Global Affairs at the University of Notre Dame for their conference support, as well as the Catholic Peacebuilding Network, the Just Transformations to Sustainability, the Environmental Change Initiative and the Institute for Social Concerns, all at the University of Notre Dame. Additionally, we thank Taproot Earth, a global organization focuses on liberation through climate justice, for their generous support to the conference.

We extend our sincere gratitude to the Bethany Land Institute in Nandere Parish, Luwero District, Uganda, and all of the tremendous people who are a part of it, from the staff to the Caretakers, for their inspiration for this volume and their immense support in bringing together the conference at their location. We hope that this volume can stand as our contribution to "starting small, starting well, and starting now".

Finally, we thank the wonderful crew that brought this volume to production. First, we thank the Hans Günter Brauch Foundation on Peace and Ecology in the Anthropocene, especially Dr. Hans Günter Brauch, for their vision to bring this into the volume series. We also thank our publishing partners at Springer, especially Dr. Christian Witschel and Birke Dalia with the Heidelberg team and Arulmurugan Venkatasalam in Chennai, Tamil Nadu, India, for this three-continental cooperation.

Introduction

Integral Ecology for the Anthropocene

The concept of the Anthropocene has emerged as a, if not the, primary descriptor of our modern moment. It signifies that humans have become the largest driver of geologic change on the planet (Steffen et al. 2007). From forest and grassland conversion to agriculture, the consumption of mineral resources for electronics, to the rate of fisheries extraction, our 'Great Acceleration' into the Anthropocene has seen mass change (Steffen et al. 2015)—geologically and ecologically, but importantly also societally and socially. In other words, our communities are co-evolving with our ecologies but with us in the driver's seat and stepping on the accelerator.

The Anthropocene does not on its own signal a positive or negative state, but rather a description of what is. It is how we interpret this change that assigns value to the unfolding processes and outcomes it describes. In many ways, the changes for humanity in the Anthropocene can be assessed positively—that is, as an improvement in the human condition. Humans today are living longer on average than ever in our evolutionary history. By many measures, the rate of global poverty is at historical lows—though notably progress has stalled. And humans are on average enjoying a larger share of material goods. We are more interconnected, communicating, and trading than ever. These are fair characterizations of the Anthropocene.

There are other ways to consider the Anthropocene and our modern moment that at best give pause and at worst are beyond alarming. The Planetary Boundaries framework is an Earth Systems approach that measures planetary change identifying nine key subsystems (Rockström et al. 2009a, 2009b). For each of the nine subsystems there is a safe operating space which allows for fluctuation within limits. But if the system sees too much change, it can move into the safe operating space which indicates potential cascade effects across the entire Earth System. When originally developed by Johan Rockstrom et al. (2009b), they estimated that four of the nine subsystems had transgressed the safe operating space. Today, it is estimated that seven of the subsystems are outside of the safe operating space.

While planetary systems outside of safe operating spaces might seem existential, there is growing clarity that human and planetary flourishing are intricately linked. In other words, planetary and human health are mutually dependent (Myers et al. 2025). Human systems, such as our local to global economies, are intricately tied to environmental realities whether that is the raw resources extracted, refined, value added, and sold or the extreme weather events that are increasingly disrupting these global supply chains. Earth Systems risks are thus cause and consequences of human activities. And these risks are not at all equally distributed, with the poor and most marginalized nations and communities most exposed to these risks—risks mostly not of their making (Landrigan et al. 2024; Pinho-Gomes et al. 2023; Rentschler and Leonova 2023).

The growing complexity of risks in the Anthropocene has been characterized as a polycrisis. Polycrisis occurs when "crises in multiple global systems become causally entangled in ways that significantly degrade humanity's prospects" (Lawrence et al. 2024). Such crises are simultaneously local and global because global impacts are the cumulative effect of on the ground changes in local economies, ecologies, and societies. The emerging idea of polycrisis is the simultaneity of stressors and triggers unfolding within and between natural and human systems. The interwoven complexity can carry the effect of one stress or trigger well beyond the localized bounds of its origin. Polycrisis is a concept to get at what all of the changes that characterize the Anthropocene can and are giving rise to—complex crises for humans and our planetary home.

Recognizing the integrated, co-evolving nature of earth and human systems, the late Pope Francis, former Pontiff of the Roman Catholic Church, called for a Care for Our Common Home that simultaneously addresses the cry of the earth and the cry of the poor. Pope Francis's charge was not simply that we find a way to be ecologically sustainable. He thoroughly warns that technocratic or technological solutions are insufficient to address the core issue. For Pope Francis, integral ecology starts with an intrapersonal conversion. Just as is signaled in the term the Anthropocene moniker of our moment, Pope Francis assessed our challenge as fundamentally a human challenge—an interwoven social and ecological problem from human roots. The causal pathways identified by Pope Francis are complex but shaped by expanding consumerism, a proliferating throw away culture (of both stuff and people), and disconnection—growing separation between people and between people and nature. To reconcile requires more than seeking greater efficiency in production processes or technologies that provide alternatives to more ecologically damaging ones in current use. Rather it requires an integral change of the heart to live and love each other and the earth differently such that 'sustainability' and 'planetary health' benefits unfold from a core conversion of individuals and societies, not new parts in the same system.

Descriptions of integral ecology vary, just as the deployment of it does—as will be seen in the chapters throughout this volume. Here we offer six emergent themes or pillars of integral ecology. These emergent ecologies tally with but build beyond traditional concepts of sustainability such as adaptation or resilience. The call to integral ecology is a call to a revolutionary transformation not just in how we are in the world and how we impact it, but also in how we see and experience the world—it starts with how we see and treat each other.

But first it is important to describe our own starting point. Specifically, why starting in Africa and then beyond for exploring and evaluating integral ecology for and in the Anthropocene. Our story, the story of the formation of this volume, begins at the Bethany Land Institute.

Born at Bethany

In 2012, Father Emmanuel Katongole, Father Cornelius Ssempala, and Father Anthony Rweza began discussing the various problems around their homes in rural Uganda. They were further inspired by Pope Francis' encyclical letter Laudato Si' which led them to realize that the major problems of deforestation, food insecurity, and poverty in Africa are all interconnected. Addressing them thus requires "an education program" to cultivate a "distinctive way of looking at things, a way of thinking, … a lifestyle and a spirituality" (LS #111).

With this idea, they decided to build BLI on a 95+ acre plot of land in Luwero, Uganda. It would serve as a demonstration of the kind of integral ecology that Pope Francis calls for—their farm would be one of the earliest operationalizations of integral ecology. Thus, Bethany Land Institute was formed as an educational program to form Ugandans in the practices, lifestyle, and spirituality of sustainable land care, food production, and economic expertise in rural Africa.

In 2016, the Bethany Land Institute established a new campus to expand its learning facilities, demonstration plots, outreach capacities and to build its own farm and cooperative store to develop financial self-sufficiency. BLI's location in Luweero District, Uganda, is strategic for two reasons. Firstly, Luweero was the site of some of the bloodiest clashes in Uganda's prolonged civil war (1980–1986). The effects of the civil war, which left tens of thousands dead, homes destroyed, and many people displaced, are still visible. Secondly, Luweero provides an archetypal lens into the problems of deforestation, food insecurity, and poverty which are experienced in much of rural Uganda, and broadly across Africa, and thus the research findings have broadly generalizable implications.

Uganda has the 4th fastest growing population on Earth, growing from 23 million people in 2000 to 44 million in 2019. This growing population, coupled with inadequate government ecological policies, has led to rapid deforestation. Trees are often cleared for firewood and unsustainable farming. Since 1990, 28,000 km^2 of forests have been lost. Global warming, reduced and irregular rainfall, and depleted topsoil cover have already created a crisis of reduced food production. Additionally, the lack of viable economic activities in rural Africa is a major challenge, especially for the youth. 57% of the country is under the age of 20 while 17% is under the age of 5. Educated youth often feel there is nothing for them to do in the villages and often end up in slums on the city edges. Those who remain in the village are forced to engage in ecologically disastrous tasks like charcoal burning and brick-making to make a living. This lack of viable economic options not only contributes to the ecological crisis, but undermines the dignity of millions of young Africans, who increasingly feel unwanted, excluded, and a failure in their own homelands. BLI addresses these challenges offering a three-pillared, two-year training program based on integral ecology for student 'Caretakers'. The three pillars are: (1) Mary' School—where regenerative agriculture and agribusiness skills are taught; (2) Martha's Market—a regenerative farm and cooperative store system facilitating collective bargaining for Caretakers and BLI; (3) Lazarus Forest—a native species forest serving as a base for ecological education, conservation and community outreach. While intrapersonal conversion and learning for Caretakers are central to BLIs foundation, it also reaches out providing technical assistance and training to communities within the district and beyond. BLI is carrying the message and methods of integral ecology far afield and further now through this volume that it instigated.

The editors of the volume, all involved with BLI in different capacities, recognized the need to evaluate integral ecology beyond BLI to support a community of practice and learning. This volume is but one part of disseminating, sharing, and reflecting on the work of BLI and integral ecology more broadly. All of the contributors to this volume convened at BLI to discuss integral ecology. Each came with varied, wide-ranging experiences, approaches, ideas, and observed outcomes. Most came from the African continent, but there was strong global representation. The composition of the volume speaks to the diversity and richness of understandings and deployments of integral ecology, but also persistent challenges and themes of the Anthropocene that replicated across these otherwise disparate settings and societies. This volume thus finds its origin and emerges from Bethany, a site and experiment examined in several chapters, but extends globally, offering insights for integral ecology across and well beyond Africa.

Emergent Framings of Integral Ecology: Six Ecologies

While scholars have interpreted what integral ecology means through the lens of *Laudato Si'*, there has not been a sufficient distillation of some of the key components of the concept. A close reading of the Encyclical yields at least six ecologies that are useful for thinking about the notion of integral ecology both in theory and in practice. These six ecologies are: (1) ecology of nature; (2) ecology of knowledge; (3) ecology of culture; (4) social ecology; (5) ecology of the spirit (6); intergenerational solidarity.

Ecology of Nature

We cannot talk about integral ecology without attentiveness to the physical environment. Early Christian thinkers are famous for developing the metaphor of the Book of Nature as one of two books (the other being the Book of Scripture) in and through which God reveals Godself and speaks to humans. They saw the natural world as a masterpiece of God's creative activity and a vehicle for God's self-disclosure. From a sociological point of view, the natural environment is the setting that makes human life and flourishing possible. According to the biblical creation accounts, God had to create the natural world first before creating the human person to dwell in the Garden (cf. Gen 1 and 2). But the natural environment is under serious threat today. In *Laudato Si'*, Pope Francis laments that "the earth, our home, is beginning to look more and more like an immense pile of filth… once beautiful landscapes are now covered with rubbish" (LS #21). This is due to human recklessness, irresponsible use, and abuse of creation. At the heart of this problem is an anthropological amnesia: "We have forgotten that we ourselves are dust of the earth (cf. Gen 2:7); our very bodies are made up of her elements, we breathe her air, and we receive life and refreshment from her waters" (LS #2). Pope Francis challenges us to recover love for nature and concern for its preservation. Placing before today's humanity the example of the patron saint of ecology, St. Francis of Assisi, he asks us to learn how to cultivate a mystical imagination of and an affective relationship with the natural world, which opens us to the horizon of joyful praise, infinite beauty, truth, and goodness: "If we approach nature and the environment without this openness to awe and wonder, if we no longer speak the language of fraternity and beauty in our relationship with the world, our attitude will be that of masters, consumers, ruthless exploiters, unable to set limits on their immediate needs" (LS #11). An integral ecology challenges us to see the natural environment as a common good, to foster a sense that we are co-partners and co-travelers with all other creatures each of which has an important and specific role in God's plan, and to be grateful for the life support system that nature is to us. As Pope Leo recently said, in our vocation to care for creation, we must never forget that "we are creatures among creatures, and not creators" (*Homily at the Inauguration and Blessing of the Borgo Laudato Si', Castel Gandolfo, 5 September 2025*).

Ecology of Knowledge

An informed understanding of integral ecology requires familiarity with the history and systems of thought that shape how humans relate to the environment. The imaginaries of these thought systems are found in politics, economics, science, technology, education, and culture. Knowing how these systems interact and are interconnected to one another can provide a more comprehensive picture of the many different facets of our present ecological crisis as well as potential building blocks of solutions.

What is central to the ecology of knowledge is how the parts are related to the whole and how the whole is related to the parts. In *Laudato Si'*, Pope Francis laments the fragmentation of knowledge into various specializations, which "though helpful for concrete applications… often leads to a loss of appreciation for the whole, for the relationship between things, and for the broader horizon" (LS #110). He argues that "the fragmentation of knowledge and the isolation of bits of information can actually become a form of ignorance unless they are integrated into a broader vision of reality" (LS #85). This is because the problems we are facing "cannot be dealt with from a single perspective or from a single set of interests" (LS #110). It all comes down to how we interpret reality: isolated, disconnected, or integrated? In addition, a truly integral ecology places value on all forms and sources of knowledge and advances an interdisciplinary dialogue that seeks to integrate science with the data generated by other branches of knowledge, including philosophy and social ethics. As Pope Francis forcefully argues, "no branch of the sciences, and no form of wisdom can be left out, and that includes religion and the language particular to it" (LS #63).

Ecology of Culture

Pope Francis tells us that it is not just the natural environment that is under serious threat today; the patrimony of culture is also endangered (cf. LS #145). Culture is part of the shared identity of each peoples and place and a foundation upon which to build a habitable society (LS #143). The destruction of indigenous cultures which has accompanied the destruction of the natural environment does not forebode well for humanity. If culture encompasses the totality of a people's way of life, to destroy culture is to cut humanity off from its life support system. While scientific reasoning seems to dominate current approaches to global ecology, it needs to be stated that the heritage of culture carries our ecological DNA through stories, myths, language, narratives, proverbs, songs, dance, and artefacts that helped our premodern ancestors to make sense of the world in which they lived and to cultivate the mental and spiritual habits to care for their places. While this is not an advocacy for a return to a Stone Age, the wisdom of the past needs to be reappropriated in a nuanced way today because it has so much to teach us. As Pope Francis states, "Culture is more than what we have inherited from the past; it is also, and above all, a living, dynamic, and participatory present reality, which cannot be excluded as we rethink the relationship between human beings and the environment" (LS #143).

Social Ecology

Social ecology studies how humans interact with and respond to the environment around them. It emphasizes that environmental problems are rooted in social systems, structures, and processes. Scholars working in social ecology believe that the domination of nature by humans is a symptom of and stems from the domination of humans by other humans. Thus, in order to achieve ecological sustainability, the transformation of social hierarchies is necessary. In Chap. 1 of *Laudato Si'*, Pope Francis highlights how disregard for human beings by other human beings (the "throwaway culture" LS #22) soon extends to the way humans treat the environment. His analysis shows how problems such as pollution, waste, water poverty, climate change, loss of biodiversity, decline in the quality of human life and the breakdown of society, and global inequality are linked to a reductionist anthropology and a defective sociology. This is why he argues that "we have to realize that a truly ecological approach *always* becomes a social approach; it must integrate questions of justice in debates on the environment" (LS #49; see also LS #139). In talking about the institutionalization of social ecology, Pope Francis speaks about how institutions from "the family, to the wider local, national and international communities" help "to regulate human relationships" (LS #142). Within each stratum, effort needs to be made to prioritize a culture of encounter and to build ecofriendly habitats (attention to "those who design buildings, neighborhoods, public spaces, and cities"—LS #150) that create spaces and opportunities for social fraternity to develop organically. At the heart of social ecology is social love, which, Pope Francis notes, "is the key to authentic development" (LS #231).

Ecology of the Spirit

The conviction shared by most world religions is that the universe is the work of a supreme creator, whom the Christians call "God." In thinking about the world, its origin, purpose, and end, it is impossible to overlook the supreme creator and the creator's plan for the world. Integral ecology thus needs a spirituality, "an interior impulse," to gird, nourish, and sustain it; for only "such a spirituality can motivate us to a more passionate concern for the protection of our world" (LS #216). In Chap. 2 of *Laudato Si'*, Pope Francis presents a gospel of creation that draws from the wisdom of biblical accounts. He believes that religious faith sheds a unique light on our present predicament, and that this requires respect for the interior life and spirituality of various peoples (LS #63. cf. also LS #111). In Chap. 6, he articulates what an ecological spirituality capable of responding to our ecological crisis looks like, summoning us to "profound interior conversion" (LS #217) and to (re)learn the virtues of sobriety, moderation, humility, simplicity, detachment from excessive possession, solidarity, and compassionate care (LS #210, #222), which were evident in "the sublime fraternity with creation which Saint Francis of Assisi so radiantly embodied" (LS #221). Spirituality ecology places "ecological conversion" at the heart of individual and communal activity, "which entails the recognition of our errors, sins, faults, and failures, and leads to heartfelt repentance and desire to change" (LS #218). One does not have to be a Christian to practice ecological spirituality. Pope Francis tells us that in whatever way we feel called to participate in the care of creation through little everyday gestures of love, "we should realize that this too is part of our spirituality" (LS #231). However, this spirituality has a transcendent dimension, which is made evident through sound education. "Environmental education," he writes, "should facilitate making the leap toward the transcendent which gives ecological ethics its deepest meaning" (LS #210).

Intergenerational Solidarity

Three times in *Laudato Si'*, Pope Francis uses the expression "intergenerational solidarity" (LS #159–160). He notes that, as a basic principle of justice, we must think about future generations when we talk about sustainable development "since the world we have received also belongs to those who will follow us" (LS #159). Thus, he asks: "What kind of world do we want to leave to those who come after us, to children who are now growing up?" (LS #160). Research shows that eco-anxiety and eco-grief are growing among children and young people today. They believe that the older generation has destroyed the planet and is leaving behind a host of catastrophic environmental challenges which they (the younger generation) have to deal with. Pope Francis had warned about this: "Doomsday predictions can no longer be met with irony or disdain. We may well be leaving to coming generations debris, desolation, and filth" (LS #161). The spiritual root of this problem lies in a crisis of eschatology that has come to define the modern age. We have lost a sense of the intrinsic meaning, direction, purpose, and goal of human life. Pope Francis invites us to wrestle with these deeper issues: "What is the purpose of our life in this world? Why are we here? What is the goal of our work and all our efforts? What need does the earth have of us?" (LS #160). The logic of intergenerational solidarity lies in the ethics of responsibility; it deconstructs the solipsism of the autonomous self. It tells us that we are not isolated monads bestriding the earthly firmament but that our lives are woven into a shared fabric, "a network of solidarity and belonging" (LS #148) with those who were here before us and those who will come after us. Each generation has received the environment as a loan, which must then be passed on to the next. "An integral ecology is marked by this broader vision" (LS #159).

We offer these six ecologies as a framework for understanding integral ecology as described in Laudato Si: Care for Our Common Home, but also to characterize how integral ecology is deployed and evaluated throughout this volume. The cultural, ecological, political and economic contexts of the cases and explorations consolidated in this volume vary, but the fundamental commitments of integral ecology persist. That said, the ways in which integral ecology are deployed and the resultant effects vary as well. The cases and considerations here put further flesh on the bones of our six ecologies, offering viewpoints into potentially generalizable understandings of engaging and deploying integral ecology while simultaneously highlighting areas for further exploration and evaluation. In other words, we offer a robust start to empirically and theoretically deploying and deciphering integral ecology, but it is a foundation for ongoing discursive analysis and experimentation.

Integral Ecology: Africa and Beyond in This Collection

The chapters assembled in this book provide both theoretical and applied insights on Integral Ecology (IE). The contributions center predominantly on three crosscutting parts: Operationalizing Integral Ecology Through Local Agency, Interconnected Ecological and Social Crises and Justice, and Theoretical Reflections of Integral Ecology. The chapters not only exemplify Integral Ecology as a useful normative framework but show the opportunities and challenges of implementing the framework's core principles. Taken together, the contributions' central synthesis is that although IE is a powerful optic and theological space for exposing intersecting social and environmental crises in the Anthropocene, its true promise lies in the ability to galvanize local participatory action with agency and ecological conversion.

The chapters under the thematic cluster on Operationalizing Integral Ecology through Local Agency use several case studies to illustrate how different actors use community-based initiatives and projects to illustrate the core underpinning principles of IE and Laudato Si. Driven by the question of how to move integral ecology from merely an aspiration to implementation, this set of chapters examines how the principles in Laudato Si' and IE translate into concrete, community-driven action and environmental management practices championed by individuals, community groups, and faith-based organizations. For example, de Llanes' chapter probes how women religious communities (*referred to as enablers*) in Francophone Africa operationalize *Laudato Si* through leadership, participatory governance, and planning. The chapter argues that IE offers deep insights on how faith-based actors can bridge spirituality and sustainable practices. Similarly, Kachungu investigates how the Young Emerging Farmers Initiative in Zambia mobilizes and empowers young people in regenerative agriculture, while Tumuhe et al., inspired by integral ecology, also analyze how civil society networks promote renewable energy solutions by empowering women and young people. Focusing on a decades-long journey of students at Holy Cross College, Notre Dame, Indiana over the past decade, Storer shares how integral ecology guides efforts to mold the hearts, minds, and hands of young students to respond to nature with faith, hope, and love. Comparable to but from a different context, Ojeifo relies on surveys with student interns and caretaker graduate of the

Bethany Land Institute (BLI)'s formation school in Uganda to demonstrates how IE can be anchored in a training program to nurture hope, spirituality, and care. Juskus' chapter addresses similar questions and uses Peruvian peace and justice organization *Paz y Esperanza* to show how an IE guided approach, relative to a human rights framework, offers much more possibility for ecological conversion, while Okoye's chapter foregrounds the role of the Catholic Church in Ghana in the fight against artisanal mining, highlighting the need for dialogue and collaboration with political leaders and calling for ecological conversion. Mugabi and Mpiima's chapter explores how smallholder farmers leverage mobile technology for regenerative agriculture in Uganda, citing IE as a foundational pillar for a sustainable nature-society relationship. And Liddick et al. provide what might be the first empirical evidence about the ecological impact of deploying, in name and nature, explicit integral ecology practices, with indications that the promise of integral ecology is plausible—integral ecology can be generative of lifeways and livelihoods.

Chapter contributions on Interconnected Ecological and Social Crises and Justice emphasize the inseparability of environmental degradation and the vulnerabilities of marginalized peoples and communities. Inspired primarily by the words of Pope Francis that we listen to *'both the cry of the earth and the cry of the poor' and "We are faced not with two separate crises, one environmental and the other social, but rather with one complex crisis which is both social and environmental.... Pope Francis, Laudato Si"*—these contributions acknowledge how different environmental crises disproportionately affect vulnerable communities and render them invisible. Further, the chapters hint at the role of external modern economic forces in undermining the lands and resources of marginalized people. A common thread in the other chapters under this theme is shining the spotlight on the lived experiences and challenges of marginalized people, including for example the Dayak People in Indonesia (Heeren) and energy injustice among refugee and host communities in northwestern Uganda (Bukenya et al.), the Oromos peoples of Ethiopia and the Sengwer peoples of Kenya experiences with disasters (Chigbu), how climate-induced loss and damage disproportionately affect Pacific communities (Barron), and how chemicals from artisanal mining affects women and children (Okoye). And Kinsey and Islanda explore how in the intersecting contexts of challenges, ecological and social, integral ecology can be a vehicle for promoting and sustaining non-violence in Kenya, while Bukenya et al. demonstrate how it foments agroecological and energy transitions for social and environmental justice in Uganda.

Mining was particularly instructive in showing these interconnected social and environmental crises. For example, Okoye uses the widespread effects of Artisanal Small-scale Gold Mining (ASGM) in Tarkwa, Ghana, to underscore how the exploitation of the environment inevitably leads to the exploitation of human beings. Considering the long-term threats of mining on indigenous communities, Offenheiser's uses Integral Human Development and IE as conceptual lenses to raise critical questions about the potential opportunities, challenges, and risks of new and emerging co-investment and benefit sharing partnerships between corporate mining entities and local communities. Shimave throws the spotlight on the consequences of exploitative mining practices in sub-Saharan Africa—including poisoned water, cultural disruption, and deepening poverty—and utilizes IE framework to not only examine the ethical implications of extractive mining but propose popular movements as a form resistance to extractive injustice in the region. Outside of mining but similarly in focus, Barron's analysis of loss and damage positions communities on the frontlines as core victims of climate change and invites leaders in such communities to respond to three major forms of climate loss—ecological, social, and spiritual loss, drawing lessons from integral ecology. The final contribution under this theme, by Ddungu and Shelton, directs attention to the role of structural poverty, historical conflicts, and ecological degradation in Uganda's escalating mental health crisis and presents BLI as a model for how land-based education, agroecology, and spiritual formation can promote mental health.

The chapters in the final thematic cluster, Theoretical Reflections of Integral Ecology, either provide exposés of the IE framework or bring it into conversation with other relevant frameworks. This set of papers also attempt to theorize new dimensions of IE that are largely absent in the literature. Chigbu also calls for the preferential protection of indigenous people from disasters by bringing Liberation Theology into dialogue with IE. Beasley also laments integral ecology's lack of focus on animals as a serious shortcoming of the framework. He uses "livestock revolution" occurring in many developing countries around the world to call for a new integral ecology that takes seriously animal welfare and animal agriculture. Consistent with these broad goals, Foucault-Camm uses the rise of digital technologies to question the limits of a technocratic mindset. She argues that integral ecology provides a useful interpretive framework for deducing whether technologies constitute true progress or threaten humanity's ultimate good. The chapter by Savino positions an "ecology of hospitality" as one antidote to the widespread "ecologies of neglect" that plague contemporary culture, as well as a framework more attentive both the cry of the earth and the cry of the poor. Jeremiah Anene Nwankwegu's chapter looks broadly at how Igbo cosmology and indigenous believes and practices can guide environmental policy in Africa, drawing inspiration from the hypothesis that environmental ethics expressed through sacredness and ecological taboos coheres with concept of *Integral Human Ecology* as articulated in Pope Francis's *Laudato Si.* In agreement with Nwankwegu's perspectives, Ikejimba and Alamezie also propose an alternative approach to environmental ethics that emphasizes sacredness and African cosmological tradition, while Berisso, situating his work in Ethiopia's pastoral lowlands with persistent droughts, considers the commodification of land as a theological-political phenomenon that undermines land's sacred and relational value and instead proposes eco-theopolitics as a counter alternative that rests on IE respect for the sacred and limits to human sovereignty over nature.

As has become clear, our volume is expansive in its geographical, methodological, empirical and theoretical, topical, and scalar breadth: with a set of contributors with matching diversity ranging from scholars to practitioners to religious workers and leaders. Such an assortment is required to effectively explore integral ecology for the Anthropocene.

Richard Marcantonio
Keough School of Global Affairs,
Mendoza College of Business
University of Notre Dame
Notre Dame, USA
rmarcant@nd.edu

Ellis Adjei Adams
Keough School of Global Affairs
University Notre Dame
Notre Dame, USA
eadams7@nd.edu

Emmanuel Ojeifo
Department of Theology
University Notre Dame
Notre Dame, USA
eojeifo@nd.edu

Emmanuel Katongole
Department of Theology, Keough
School of Global Affairs
University Notre Dame
Notre Dame, USA
ekatongo@nd.edu

References

Francis, P., 2015. *Laudato si: On care for our common home.* Our Sunday Visitor.

Landrigan, P.J., Britt, M., Fisher, S., Holmes, A., Kumar, M., Mu, J., Rizzo, I., Sather, A., Yousuf, A., Kumar, P., 2024. Assessing the Human Health Benefits of Climate Mitigation, Pollution Prevention, and Biodiversity Preservation. *Annals of Global Health* 90, 1. https://doi.org/10.5334/aogh.4161

Lawrence, M., Homer-Dixon, T., Janzwood, S., Rockstöm, J., Renn, O., Donges, J.F., 2024. Global polycrisis: the causal mechanisms of crisis entanglement. Global Sustainability 7, *e6*. https://doi.org/10.1017/sus.2024.1

Myers, S.S., Masztalerz, O., Ahdoot, S., Gabrysch, S., Gupta, J., Haines, A., Kleineberg-Massuthe, H., Lambrecht, N.J., Landrigan, P.J., Mahmood, J., Pörtner, L.M., Rohr, J., Traidl-Hoffmann, C., Wendt, A.S., Wray, B., Rockström, J., 2025. Connecting planetary boundaries and planetary health: a resilient and stable Earth system is crucial for human health. *The Lancet* 406, 315–319. https://doi.org/10.1016/S0140-6736(25)01256-5

Pinho-Gomes, A.-C., Roaf, E., Fuller, G., Fowler, D., Lewis, A., ApSimon, H., Noakes, C., Johnstone, P., Holgate, S., 2023. Air pollution and climate change. *The Lancet Planetary Health* 7, e727–e728. https://doi.org/10.1016/S2542-5196(23)00189-4

Rentschler, J., Leonova, N., 2023. Global air pollution exposure and poverty. *Nat Commun* 14, 4432. https://doi.org/10.1038/s41467-023-39797-4

Rockström, J., Steffen, W., Noone, K., Persson, Å., Chapin, F.S., Lambin, E., Lenton, T.M., Scheffer, M., Folke, C., Schellnhuber, H.J., Nykvist, B., de Wit, C.A., Hughes, T., van der Leeuw, S., Rodhe, H., Sörlin, S., Snyder, P.K., Costanza, R., Svedin, U., Falkenmark, M., Karlberg, L., Corell, R.W., Fabry, V.J., Hansen, J., Walker, B., Liverman, D., Richardson, K., Crutzen, P., Foley, J., 2009a. Planetary Boundaries: Exploring the Safe Operating Space for Humanity. *Ecology and Society* 14, 32.

Rockström, J., Steffen, W., Noone, K., Persson, Å., Chapin, F.S., Lambin, E.F., Lenton, T.M., Scheffer, M., Folke, C., Schellnhuber, H.J., Nykvist, B., de Wit, C.A., Hughes, T., van der Leeuw, S., Rodhe, H., Sörlin, S., Snyder, P.K., Costanza, R., Svedin, U., Falkenmark, M., Karlberg, L., Corell, R.W., Fabry, V.J., Hansen, J., Walker, B., Liverman, D., Richardson, K., Crutzen, P., Foley, J.A., 2009b. *A safe operating space for humanity. Nature 461, 472–475.* https://doi.org/10.1038/461472a

Steffen, W., Broadgate, W., Deutsch, L., Gaffney, O., Ludwig, C., 2015. The trajectory of the Anthropocene: the great acceleration. *The Anthropocene Review* 2, 81–98.

Steffen, W., Crutzen, P.J., McNeill, J.R., 2007. The Anthropocene: Are Humans Now Overwhelming the Great Forces of Nature. *AMBIO: A Journal of the Human Environment 36*, 614–621. https://doi.org/10.1579/0044-7447(2007)36[614:TAAHNO]2.0.CO;2

Contents

Abbreviations

ADF	African Defense Forces
AI	Artificial Intelligence
ASGM	Artisanal and small-scale gold mining
BLI	Bethany Land Institute
CAFO	Concentrated animal feeding operations
CCFD Terre Solidaire	French Catholic Committee against Hunger and for Development
CEDO	Community Enterprises Development Organization
CIAT	International Center for Tropical Agriculture
CLC	Customary land certificates
COP	Conference of Parties
CSO	Civil society organization
CST	Catholic Social Teaching
EPRDF	Ethiopian People's Revolutionary Democratic Front
FAO	United Nations Food and Agriculture Organization
FGD	Focus group discussions
FPIC	Free Prior and Informed Consent
HFIAS	Household Food Insecurity Access Scale
ICMM	International Council on Mining and Metals
ID	Institut Dayakologi
IE	Integral Ecology
IFC	International Finance Corporation
IFN	Integrity for Nonviolence
IHD	Integral human development
IRB	Institutional Review Board
JPDIC	Justice, Peace, Development, and the Integrity of Creation
KII	Key Impact Indicators
LCOY	Local Conference of Youth
LS	Laudato Si
MAAIF	Ministry of Agriculture, Animal Industry and Fisheries
MEMD	Ministry of Energy and Mineral Development

ND-GAIN	Notre Dame—Global Adaptation Index
NGO	Non-government organization
ODK	Open Data Kit
PCA	Principal component analysis
PCC	Pacific Conference of Churches
PELUM	Participatory Ecological Land Use Management
PERMANOVA	Permutational analysis of variance
PRODIMA	Program for Mestizo-Aguaruna Intercultural Development
RA	Regenerative agriculture
RCOY	Regional Conference of Youth
RICE	Rural initiative for community empowerment
SDG	Sustainable Development Goal
SGI	Small group interviews
SHF	Small-holder farmer
SSA	Sub-Saharan Africa
STP	Soil test phosphorous
TGCC	Tenure and global climate change
TN	Total nitrogen
TOC	Total organic carbon
UN	United Nations
UNDP	United Nations Development Program
UNDRIP	United Nations Declaration of Rights of Indigenous Peoples
UNFCCC	United Nations Framework Convention on Climate Change
UNICEF	United Nations Children's Fund
USAID	United States Agency for International Development
VSLA	Village Savings and Loan Associations
WHO	World Health Organization
WIM	Warsaw International Mechanism
WMPM	World Meeting of Popular Movements
YEFI	Young Emerging Farmers Initiative

List of Figures

List of Tables

Part I
Integral Ecology Through Local Agency

Chapter 1
Integrating Laudato Si' Goals into Community Development Projects in Africa: A Management Perspective

Alonso de Llanes

Abstract This chapter explores how the principles of Integral Human Ecology, as articulated in Laudato Si' (2015), are operationalized by women religious communities in Francophone Africa through management practices such as committed leadership, participatory governance, and strategic planning. The central research question is: How can management enablers foster the effective implementation of Laudato Si' in African community development projects?? The argument is that without integrating managerial tools into spiritual commitments, Integral Ecology risks remaining aspirational. Employing a qualitative methodology, the study draws on empirical evidence from five religious congregations in Cameroon, the Democratic Republic of Congo, and Benin. Data were collected through document analysis, semi-structured interviews (N = 15), and thematic coding of testimonies. The findings reveal a complex interplay between spiritual capital and structural fragility: while women religious from the Roman Catholic Church (used hereafter interchangeably as "sisters") possess high moral credibility and relational power, their projects face limitations in resources, training, and institutional resilience such that their capacity to sustain and scale ecological initiatives remains constrained. Nonetheless, small-scale initiatives, such as sustainable agriculture and ecological education, demonstrate credible pathways for lived ecological conversion. The chapter argues that Integral Ecology is not a fixed model but a dynamic praxis requiring contextual discernment, theological depth, and managerial adaptation. It offers insights into how faith-based actors can bridge spirituality and organizational strategy to foster just and sustainable transitions in the Anthropocene.

Keywords Community development · Participatory management · Gender · Transformational capacity

---

Alonso de Llane, Laudato Si Action Platform alonsodellanes@gmail.com

---

A. de Llanes (✉)
Laudato Si Action Platform, Washington, USA
e-mail: alonsodellanes@gmail.com

R. Marcantonio et al. (eds.), *Integral Ecology in the Anthropocene: Africa and Beyond*,
The Anthropocene: Politik—Economics—Society—Science 49,
https://doi.org/10.1007/978-3-032-32125-1_1

1.1 Introduction

The Anthropocene has ushered in an era marked by escalating ecological degradation, widening inequalities, and an erosion of the moral imagination necessary to sustain human and planetary flourishing. The term 'Anthropocene' refers to a proposed geological epoch in which human activity has become the dominant force shaping the Earth's systems (Crutzen/Stoermer 2000). In response to these interconnected crises, the Catholic Church, through Pope Francis's encyclical Laudato Si' (2015), has proposed a bold, integrative paradigm known as Integral Ecology. Far from being a niche theological position, Integral Ecology calls for a systemic rethinking of how humans relate to nature, to one another, and to God. As Pope John Paul II emphasized in Centesimus Annus (1991), the human vocation requires cultivating a "moral imagination" capable of safeguarding both creation and human dignity, a theme further developed by Pope Francis.

Yet, despite its ambitious scope and strong reception in theological and academic circles, Integral Ecology—understood as an integrative framework that weaves together environmental, social, economic, cultural, and spiritual dimensions of sustainability—remains difficult to implement on the ground. As Deneulin (2021) notes, the Roman Catholic Church often struggles to move from symbolic declarations to measurable institutional change Around the world, institutional inertia, political resistance, limited financial resources, and a lack of technical capacity have hindered the translation of this framework into actionable strategies. Many dioceses, religious communities, and Catholic institutions have expressed enthusiasm for the vision of Laudato Si', but empirical studies show that concrete, measurable, and sustained transformations remain rare and uneven (e.g., Taylor/Jones 2021; Conway 2022). Moreover, the global asymmetry in capacities—between the Global North and Global South, between urban and rural settings, between male-dominated leadership structures and the often-overlooked contribution of women—raises questions about who gets to lead, who decides, and who benefits from ecological transitions. understood here as processes that shift institutions, practices, and communities toward more sustainable modes of living

Africa represents both a challenge and an opportunity in this regard. On one hand, the continent faces acute ecological and socio-economic pressures, including desertification, water scarcity, energy insecurity, and food system fragility—often intensified by weak governance, guerrillas, colonial legacies, and global economic dependency, all of which affect the Church's ability to lead and scale ecological action (Mabhaudhi et al. 2019; Wilkins 2020). On the other hand, Africa is home to vibrant religious networks, deep communal traditions, and a growing ecosystem of grassroots initiatives that blend faith, ecology, and development in creative and resilient ways. Religious communities, and particularly women religious, are increasingly emerging as pioneers in applying Integral Ecology in concrete settings—from sustainable agriculture and clean water projects to educational initiatives and leadership development programs (Aikins 2021; UISG 2022; African Sisters Education Collaborative 2021).

This article examines the central research question: What are the management enablers that support the effective implementation of Integral Ecology in African community development projects inspired by Laudato Si'? The underlying assumption is that without robust management frameworks to implement integral ecology—specifically, committed leadership, participatory structures, and strategic planning—even the most compelling theological paradigms risk remaining aspirational. In other words, faith-driven ecological projects that lack integrated management tools in their design and implementation plans are more likely to struggle with systemic challenges, lack stakeholder ownership, and fail to generate sustainable and replicable models of change.

The article employs a qualitative, practice-oriented methodology. First, it presents a targeted overview on management theory in development contexts, with a particular focus on leadership, participation, and planning within faith-based organizations. This theoretical framing is complemented by a comparative analysis of five case studies drawn from religious congregations in Cameroon, Benin, and the Democratic Republic of Congo. While modest in scale, these projects offer valuable insights into how theological commitments can be operationalized through context-sensitive, community-led strategies.

By bridging theological ethics with development management, this article contributes to an emerging interdisciplinary field of practice that seeks to make faith-based ecological transformation both credible and actionable.

1.1.1 Management Insights for Implementing Integral Ecology in African Contexts

The implementation of Integral Ecology in Laudato Si' (2015) represents not only a moral and theological call but also a managerial challenge. Despite its global resonance, translating this vision into practice, especially in resource-constrained settings, faces significant barriers. Scholars note that success in the Church, particularly in Africa, depends on making theological commitments operational through leadership, participatory governance, and planning (Deneulin/Zampini-Davies 2017; Tavanti et al. 2017). This overview draws on management insights to examine three core themes—committed leadership, participatory management, and strategic planning—that shape the implementation of Laudato Si' principles in African contexts.

1.1.2 Committed Leadership as a Foundation for Transformation

Leadership, particularly in the form of value-driven and transformational leadership, is widely acknowledged as essential in initiating and sustaining ecological and social change (Friman et al. 2024; Ndlovu et al. 2018). The literature consistently emphasizes that without committed individuals who embody the values of Integral Ecology, structural reforms are unlikely to succeed and as Pope Francis reminds us in Laudato Si' (§13), "individual responsibility is essential if we are to bring about lasting change," and in Evangelii Gaudium (§183) he stresses that true transformation requires leaders who live what they preach. Without such witnesses, institutional reforms risk remaining merely technical or superficial.

In Laudato Si', such leadership requires more than competence; it demands spiritual and moral alignment with care for creation, justice for the poor, and intergenerational solidarity (Martins 2018; Lee 2025). Barron/Chou (2017) argue that faith-based sustainability efforts must be rooted in spirituality-based planning, led by ethical stewards, because only those leaders who integrate spiritual depth and organizational values can foster authentic stewardship, sustain collective commitment, and nurture transformation over time (Voynnet-Fourboul 2012)

In Africa, religious leadership plays a distinctive role in shaping norms and practices. Wakahiu/Keller (2017) highlight how the leadership training of women religious in Africa has led to greater institutional accountability and community engagement. Their study suggests that when sisters acquire leadership and organizational skills, they mobilize both Church and local resources to address environmental and social challenges effectively. Agadjanian (2015) reinforces this by pointing to how the increasing leadership of women in religious communities in sub-Saharan Africa is not only shifting traditional gender roles but also enhancing institutional responsiveness to community needs.

1.1.2.1 Participatory Management and Community Ownership

Integral Ecology insists on the interconnectedness of all life and the inclusion of the poor and marginalized in decision-making (Francis 2015). In management theory, this aligns with participatory and collaborative models of governance, which emphasize the co-creation of knowledge, empowerment of stakeholders, and distributed leadership (Burke 2012; Korff et al. 2012).

Participatory management has been found to increase the sustainability and relevance of development initiatives by fostering local ownership and accountability (Flacke et al. 2020). In African contexts, where top-down approaches have often failed, participatory strategies offer a corrective that resonates with both traditional communal values and the synodal ethos of the Catholic Church (Deneulin 2021).

Moreover, participatory management aligns with African ecclesial traditions of "small Christian communities," which prioritize localized discernment and horizontal collaboration. However, the literature also notes the challenges: true participation is often hindered by institutional inertia, gender biases, and a lack of training in facilitation techniques (Mengesha et al. 2023). Therefore, capacity-building remains a crucial condition for effective participatory management.

1.1.2.2 Strategic Planning and the Institutionalization of Ecological Vision

Beyond leadership and participation, effective implementation requires structured organizational processes. Strategic planning offers a managerial tool for articulating vision, setting goals, mobilizing resources, and monitoring progress (Zhovnirchyk et al. 2023; Martusenko et al. 2023). When rooted in sustainability ethics, strategic planning can translate theological commitments into measurable institutional behavior.

Research suggests that religious organizations often operate without clear planning frameworks, relying instead on charisma or informal practices (Tavanti et al. 2017; Cruz/Valdivieso 2023). While such approaches can generate enthusiasm and short-term mobilization, they frequently lack mechanisms for knowledge transfer, resource allocation, and accountability. As a result, initiatives risk depending too heavily on individual leaders, leading to discontinuity when leadership changes and limiting the scalability of ecological projects beyond local contexts.

Yet strategic planning is not a panacea. It must be context-specific, inclusive, and reflective of dynamic ecological and social realities. Scholars warn against over-reliance on technical plans divorced from local knowledge and spiritual values (Deneulin/Zampini-Davies 2017). For instance, empirical studies of Catholic agricultural cooperatives in East Africa show that plans succeed when they integrate community participation and spiritual practices alongside technical expertise (ASEC 2021). Therefore, successful planning requires a hybrid model that synthesizes managerial rigor with prophetic vision.

1.1.2.3 Gender, Power, and Transformational Capacity

An emerging theme across management and development literature is the centrality of gender in shaping the success of sustainability initiatives. Theologians and development scholars agree that Laudato Si' cannot be fully realized without confronting structural inequalities, including the underrepresentation of women in decision-making (Sánchez-Camacho/Moreno 2024; Jung 2021). From a management perspective, gender equity enhances organizational diversity, creativity, and resilience. For example, studies of women religious leading ecological agriculture projects in Kenya and Uganda show that their inclusion in governance structures improves project sustainability and community participation (ASEC 2021; UISG 2022).

Empirical studies in Africa highlight persistent gender gaps in Church structures. Women religious are frequently the driving force behind ecological initiatives, offering alternative models of Church-based entrepreneurship. Yet they remain excluded from financial planning, property ownership, and institutional leadership, which limits the scale of their impact (Burke 2012; Lynch et al. 2022).

1.1.2.4 Theological Groundings and Methodological Challenges

A critical strand in the literature concerns the methodological tension between theology and development studies, particularly regarding how religious worldviews interact with managerial logic. Deneulin/Zampini-Davies (2017) emphasize that the integration of religion into development praxis cannot be instrumentalized merely to enhance outcomes; rather, it must be approached as a dialogue between value systems, each with its epistemological and ethical foundations. This is particularly important in the context of *Laudato Si'*, which promotes not just ecological efficiency, but a spiritual and cultural conversion (*Laudato Si'*, 217).

Integral Ecology, by its nature, resists simplification. It critiques both consumerist capitalism and technocratic reductionism (Francis 2015, §§106–114). This poses challenges when applying traditional management tools, which often assume growth-oriented or efficiency-maximizing frameworks. As Tavanti et al. (2017) argue, management education for sustainable development must expand to include ethical reasoning, intercultural dialogue, and communal values. In African settings, where religion deeply informs identity and purpose, this integrative approach is not only viable but necessary.

However, some scholars caution that 'integral' language can become diluted if not critically anchored. For instance, Sasongko (2016) warns that ecclesial responses to ecological injustice sometimes risk spiritualizing structural problems without addressing their socio-economic roots. Similarly, Wilkins (2020) observes that Catholic clerical responses to climate change, while often well-intentioned, remain uneven and highly dependent on individual leaders' theological orientations. These critiques call for greater methodological clarity and institutional consistency in operationalizing Integral Ecology.

1.1.2.5 Education, Capacity Building, and Organizational Learning

Another recurring theme across management and organizational development literature is the role of education and organizational learning in sustaining Integral Ecology. Transformational change requires not just leadership or planning, but a culture of continuous reflection, learning, and adaptation (Malakoane et al. 2023). This is especially critical in African ecclesial settings, where many sisters and priests operate in environments with limited access to updated ecological and managerial knowledge.

Studies like those by Cruz/Valdivieso (2023) underscore the importance of competency-based training aligned with ecological ethics. Their analysis of *Laudato Si'* as a resource for business education proposes a shift from abstract moral appeals to practical skill development, such as environmental auditing, stakeholder analysis, and impact measurement.

Moreover, educational processes must be context-specific and co-created. As Mengesha et al. (2023) demonstrate in their work on maternal health in Ethiopia, social capital and trust networks are vital for knowledge transfer. In African Church settings, learning often occurs through storytelling, liturgical practices, and informal mentorship. Educational interventions, therefore, must balance technical content with culturally resonant methods.

Despite these insights, systemic challenges remain. Many congregations lack formal mechanisms for monitoring and evaluation or fail to document their ecological initiatives for broader learning (Tavanti et al. 2017). Building institutional memory and feedback loops is thus a pressing recommendation across the literature.

1.1.2.6 Critical Gaps and Future Research Directions

Despite these insights, important gaps remain. First, metrics for assessing the impact of Laudato Si' implementation are underdeveloped: without clear indicators, symbolic action risks substituting for structural change (Deneulin 2021). Second, most studies privilege successful cases, because failures are less documented and harder to access, yet analyzing them would reveal systemic obstacles. Third, comparative perspectives—across African regions or with other Global South contexts—are scarce, limiting understanding of how political structures or donor ecosystems shape outcomes. Finally, little is known about the roles of youth, male religious, and lay professionals, even though transformation requires the whole ecclesial community.

The literature thus suggests that the implementation of Laudato Si' in Africa depends on balancing theological integrity with managerial competence, while addressing persistent gaps in evidence, inclusivity, and evaluation (Deneulin/Zampini-Davies 2017; Tavanti et al. 2017)

1.1.3 General Organization and Project Overview

Religious communities across Africa are critical yet often under-recognized agents in implementing the principles of *Laudato Si'*. This section examines the organizational frameworks and project typologies of selected religious congregations, each illustrating the challenge of actualizing *Integral Ecology* amid socio-economic constraints and environmental vulnerability.

This study uses a qualitative case study methodology to explore how women religious communities in Francophone Africa enact *Laudato Si'* through management practices such as committed leadership, participatory governance, and strategic planning. The aim is not narrow quantitative measurement, but an interpretation of how Integral Ecology is translated, contested, and embodied within concrete community projects and institutional dynamics.

The research follows an interpretivist epistemology, recognizing the religious actors' motivations, practices, and resistances are shaped by culturally embedded worldviews and spiritual traditions (Deneulin/Zampini-Davies 2017). This aligns with *Laudato Si's*, calls for spiritual, ethical, and cultural conversion rather than purely technical solutions (Francis 2015). Accordingly, the study prioritizes depth over breadth, seeking to capture meanings and processes from within the lived experiences of communities rather than imposing external categories.

1.1.3.1 Selection of Cases

Five religious communities were selected for their active engagement in projects aligned with *Laudato Si'*, including sustainable agriculture, water and sanitation services, education for ecological awareness, and women's leadership development. Located in Cameroon, the Democratic Republic of Congo, and Benin, they include the Tertiary Sisters of St. Francis, the Sisters of the Divine Savior, the Benedictine Sisters of the Monastery of Emmaus, the Sisters of St. Marcellina, and the Daughters of St. Paul. The selection criteria included:

1. Clear reference to *Laudato Si'* principles in project objectives.
2. Participation in the Vatican's Laudato Si' Action Platform.
3. Demonstrable community impact (quantitative or qualitative).
4. Willingness to engage in reflective dialogue.
5. Geographical and thematic diversity within a Francophone African.

This purposive sampling strategy (Patton 2002) sought cases that, while not statistically representative, offered rich insight into diverse forms of religious ecological engagement.

In Cameroon, the Tertiary Sisters of St. Francis launched an organic farming project rejecting chemical fertilizers and drawing on traditional knowledge. The initiative improved soil health, cut costs, and fostered resilience. According to Sister Annette Tangwa, it brought "significant cost savings and improved health outcomes." The project directly engaged 200 sisters and indirectly benefitted 350 people, showing how collective commitment can catalyze local ecological transition.

In the Democratic Republic of Congo, the Sisters of the Divine Savior established a fruit tree plantation and agriculture education program that reached 6,719 students, indirectly impacting 12,000 people. This ambitious effort used participatory pedagogy and intergenerational formation, echoing Laudato Si''s call to cultivate "ecological virtues" (§211). The Benedictine Sisters of Emmaus Monastery addressed water scarcity through a sanitation project serving 140 people directly and 300 indirectly. As Sister Francine Manenga Samba put it, "Water is life, and now it flows abundantly," underscoring both the spiritual and material dimensions of ecology.

In Benin, the Sisters of Saint Marcellina promoted food sovereignty and environmental education through sustainable plantations that supported 250 families—about 800 people. This family-centered approach reflects Laudato Si''s vision that ecological conversion begins in domestic and communal life (§213).

Educational and leadership formation also play a central role. In the DRC, the Daughters of Saint Paul implemented a program to equip women religious with the knowledge and tools to integrate Laudato Si' into their ministries and missions. The initiative directly trained 120 participants and indirectly reached about 1,500 people, demonstrating how capacity building multiplies impact. The focus on training reflects Wakahiu/Keller's (2017) insight that leadership development enhances the transformative capacity of women religious, while also echoing Pope Francis's call for "ecological conversion," a change of mindset and worldview that precedes authentic ecological action (Laudato Si', §219).

1.1.3.2 Data Collection and Analysis

Data were collected through document analysis, semi-structured interviews, and reflective narratives from community members. Primary sources included project reports, monitoring documents, photo evidence, and testimonies from sisters involved in implementation, triangulated with internal documents and communications from congregational leadership.

In total, 15 semi-structured interviews (N=15) were conducted across the five congregations: three in each case, with (a) the general superior or local leader, (b) the sister responsible for Laudato Si'/ecology initiatives, and (c) a community member directly involved in the project. Interviews were conducted virtually or in writing when necessary, using open-ended questions to elicit narratives of transformation, struggle, and meaning making. Special attention was given to respondents' language, revealing how Laudato Si' vocabulary was adapted to local contexts and spiritual traditions.

An emergent thematic analysis was conducted, following the structured process outlined by Nowell et al. (2017) to ensure credibility and transparency. Categories emerged inductively, in line with grounded theory (Charmaz 2006), while also guided by sensitizing concepts from management literature—leadership, participation, planning, and resistance to change. Quotes were coded under themes such as "spiritual motivation," "project ownership," and "gendered leadership," interpreted considering the central question: how do these communities navigate the operational and cultural tensions of embodying Integral Ecology?

As a scholar connected to ecclesial networks and the Laudato Si' Action Platform, the researcher adopted a reflexive stance, balancing privileged access and trust with the need for analytical distance. Limitations include the small sample size, uneven data across projects, and the study's focus on Francophone Africa, which cannot represent the continent. Instead, it offers a contextualized window into how selected communities engage with Laudato Si', providing insights to inform future reflection and practice.

1.1.4 Discussion

1.1.4.1 Organizational Models and Capacity

The five congregations studied operate under hierarchical yet spiritually grounded leadership. Projects are typically initiated by a small group of sisters with training or external support, then expanded through assemblies, reflections, and workshops. Strategic planning is modest and informal, shaped by community life and liturgical rhythms, yet initiatives show high adaptability to local needs—an element often missing in large-scale NGO or government programs.

What sets these efforts apart is the integration of management practices with spiritual values. Strategic planning (Zhovnirchyk et al. 2023), participatory methods (Burke 2012), and committed leadership (Agadjanian 2015; Wakahiu/Keller 2017) are all evident in their organizational dynamics. Despite limited resources, these communities leverage moral authority, symbolic capital, and relational networks to offset logistical constraints, granting them access to the "moral infrastructure" (Deneulin/Zampini-Davies 2017) needed to promote sensitive changes like reducing pesticide use or shifting gender roles in agriculture.

Challenges remain: financial instability, aging membership, and lack of technical skills threaten long-term sustainability. Many lack access to digital tools, agricultural innovation, or policy platforms to scale their projects. Internal tensions also emerge as not all sisters share the same ecological urgency, slowing implementation.

Still, these congregations function as laboratories of faith-in-action. Their projects, though uneven, embody Integral Ecology beyond symbolic gestures. As one sister in Cameroon put it: "We are not experts, but our faith obliges us to act with what we have."

1.1.4.2 Laudato Si' Put to the Test

The field data gathered from the five communities shows how the theological and ethical principles of *Laudato Si'* are being translated—or sometimes distorted—by real-world complexity. While the encyclical provides an ambitious and inspiring vision of *Integral Ecology*, implementation is rarely straightforward. The projects analyzed display both creative fidelity to Pope Francis's message and friction between theological idealism and managerial or logistical constraints.

Laudato Si' calls for an ecological conversion that is personal and communal (Laudato Si', §219), rooted in contemplation, humility, and solidarity. Many projects embody this spirit, not as technical interventions but as spiritual acts grounded in liturgy, discernment, and vocational witness. In Cameroon and Benin, sustainable agriculture is framed as a "ministry of life" or "care for creation," linking soil health to divine grace and food security to Eucharistic values.

Yet here tensions arise. The encyclical's paradigm demands systemic change—economic reform, structural justice, policy transformation—while most communities operate at a micro-level. Their projects remain localized and lack the institutional capacity to confront agribusiness monopolies, subsidies, or deregulation. However, this bottom-up approach is precisely how Laudato Si' envisions ecological conversion: beginning with individuals and communities before scaling outward. The implication is that broader impact will depend on linking these grassroots efforts to networks, advocacy coalitions, and policy engagement

1.1.4.3 The Challenge of Operationalizing Holism

A central challenge in implementing Integral Ecology is its holistic nature. The encyclical calls for integrating environmental, social, cultural, economic, and spiritual dimensions in decision-making, yet the field study revealed fragmentation across all five communities. Spiritual motivation is consistently strong, but integration with technical, economic, and political strategies is uneven. For example, none of the communities. For example, none of the communities applied systematic monitoring or evaluation metrics to assess long-term sustainability—an absence noted in both the Tertiary Sisters' agriculture project in Cameroon and the Benedictine water initiative in the DRC. This gap, also highlighted in the literature (Zhovnirchyk et al. 2023), underscores the difficulty of translating holistic vision into operational practice.

This gap does not stem from negligence but from structural asymmetry: most sisters receive theological and pastoral training, not ecological or managerial. Even when committed to Laudato Si', limited tools and partnerships restrict durability. Projects often remain small and localized, an outcome not inherently problematic, since Integral Ecology values contextual responses. The risk, however, is isolation: without transferable learning or institutional support, innovations struggle to endure. As Tavanti et al. (2017) note, sustainability requires coupling ecology with management education and systems thinking.

1.1.4.4 Leadership Culture and Resistance in the African Church

A distinctive strength in the African context is the absence of ideological polarization around ecology. Unlike in Western countries, where Laudato Si' is politicized (Wilkins 2020), the sisters frame creation care through Christian responsibility, communal well-being, and ancestral reverence for nature. This consensus reduces ideological resistance and grounds ecological action in shared values.

Resistance persists in other forms. Generational divides are common: older members often regard ecology as secondary, while younger sisters embrace it as integral to vocation. As one interviewee from Benin explained, "Our superiors tell us to focus on prayer, but we see caring for creation as part of prayer." These tensions, combined with elder-dominated leadership, complicate implementation.

As Wakahiu/Keller (2017) and Friman et al. (2024) note, intergenerational dialogue and leadership development are essential.

Field evidence also shows that Laudato Si' functions as a diagnostic of leadership culture. Communities with participatory governance integrated ecological values more effectively, while top-down initiatives lacked ownership and continuity. The Salvatorian Sisters' agricultural workshops succeeded not only due to resources but because they fostered co-creation and discernment with local communities. This confirms findings by Burke (2012) and Shava/Heystek (2021) on the role of participatory and transformational leadership.

1.1.4.5 Resistance and Change: Navigating the Complexities of Integral Ecology

The implementation of Laudato Si' in African religious contexts is shaped by cultural worldviews and institutional inertia (Deneulin/Zampini-Davies 2017; Jung 2021). Long-standing traditions and norms often clash with the transformative demands of ecological conversion. While leadership and participatory management are promoted as enablers, their uptake is often constrained by hierarchical governance, limited resources, and logistical complexity.

Strategic planning, for instance, is widely recognized as crucial (Zhovnirchyk et al. 2023; Barron/Chou 2017), yet several congregations remain in early stages of articulating measurable goals or evaluation frameworks. The Tertiary Sisters of St. Francis, for instance, achieved benefits through organic farming, but lacked tools to monitor long-term effects (Malakoane et al. 2023; Cruz/Valdivieso 2023). Similarly, while participatory approaches are championed in theory (Burke 2012; Korff et al. 2012), their practice is uneven. Some communities, like the Salvatorian Sisters, who mobilized thousands through training; while more contemplative orders, remained insular, reflecting Deneulin's (2021) point on the difficulty of institutionalizing dialogue.

A further tension lies in the tacit spiritual dimension of Integral Ecology. Though most sisters in the study expressed faith-based motivations, few linked them explicitly to organizational strategy. This gap echoes Wilkins's (2020) concern that ecological responses remain fragmented unless spirituality and operations are integrated.

Resistance, however, is not merely an obstacle but a catalyst. Among the Sisters of the Divine Savior, initial skepticism toward ecological education was reframed through workshops, recasting teaching as care for creation—an example of the "hermeneutical flexibility" described by Deneulin/Zampini-Davies (2017). Likewise, planning tools initially dismissed as 'secular' were later embraced when reinterpreted as stewardship or discernment, echoing Barron/Chou's (2017) findings on spirituality-based planning.

Resistance also exposes intra-community dynamics. Generational divides were common, with younger sisters generally more open to ecological commitments than elders. This aligns with leadership studies stressing the need for renewal and intergenerational dialogue in African religious life (Wakahiu/Keller 2017; Friman et al. 2024).

Both literature and field data highlight that change is non-linear. Strategic planning and participatory management are not turnkey solutions but processes of trial, negotiation, and adaptation (Zhovnirchyk et al. 2023; Burke 2012). Many successful initiatives—such as fruit tree plantations or clean water systems—emerged not from flawless implementation, but from the willingness to experiment, collaborate, and learn from local experience. In this way, Integral Ecology is best understood not as a static model, but as a living process of relational engagement.

In conclusion, resistance—when approached with theological depth and managerial humility—can catalyze more authentic, context-sensitive forms of ecological action. The communities most open to wrestling with resistance, rather than avoiding it, are often those who succeed in rooting Laudato Si' not just in rhetoric, but in lived, credible transformation.

1.1.5 Spiritual Strength, Structural Fragility, and the Path Forward

The experience of African religious communities reveals a paradox: they hold deep spiritual capital yet remain structurally fragile. Sisters are recognized for their moral authority, relational depth, and rootedness in local realities—qualities that often allow them to reach where governments or NGOs cannot. This relational power serves as a unique enabler of Integral Ecology. Yet their projects are precarious, dependent on external donors and short-term grants, and vulnerable to shifts in leadership or national instability. Without sustained investment in institutional resilience and knowledge transfer, many initiatives risk dissolution.

This dual reality mirrors broader findings in the literature. As Deneulin (2021) notes, Integral Ecology cannot be implemented as a 'plug-and-play' model. It must be grounded in context, negotiated amid competing worldviews, and developed incrementally. The communities studied exemplify this slow and uneven process, not as a failure, but as part of what Laudato Si' describes as "the art of accompaniment" (Laudato Si', §223).

This article has examined how religious communities in Francophone Africa operationalize Laudato Si' through practical management strategies—spiritually grounded leadership, participatory structures, and strategic planning. While the theological and ethical framework of Integral Ecology offers a compelling vision of justice, implementation reveals a complex interplay of spiritual vitality, innovation, resistance, and cultural negotiation.

The literature emphasized that inclusive, spiritually anchored leadership acts as a catalyst for transformation. Participatory management ensures ownership and contextual relevance, while strategic planning provides structure and accountability. Field data, however, showed the limits of these approaches in contexts marked by scarce resources, traditional hierarchies, and limited technical support.

The case studies highlighted the impact of small-scale, locally led initiatives—organic farming, clean water, ecological education—as tangible expressions of Integral Ecology, often achieved with modest means. These projects show that Laudato Si' is not only a theological exhortation but also a living process of discernment, adaptation, and care for the vulnerable.

Resistance, whether from hierarchical rigidity, fear of change, or unfamiliarity with management tools, underscores the need for accompaniment, capacity-building, and pastoral reflection. Far from signaling failure, resistance can foster renewal when approached with humility and trust.

Ultimately, Integral Ecology is not a finished model but an evolving praxis. Its realization demands ongoing dialogue between spirituality and management, between global aspirations and local realities, between prophetic vision and organizational realism. The Holy Spirit may inspire, but implementation requires strategy, patience, and an "ecology of relationships." By embracing both promise and struggle, African religious communities show that credible, faith-rooted ecological conversion is not only possible but already unfolding

References

Agadjanian, V. (2015). Women's religious authority in a sub-Saharan setting. *Gender & Society*, 29(6), pp.982–1008. https://doi.org/10.1177/0891243215602106

Barron, K. and Chou, S. (2017). Toward a spirituality mode of firm sustainability strategic planning processes. *Society and Business Review*, 12(1), pp.46–62. https://doi.org/10.1108/sbr-01-2016-0008

Burke, K. (2012). Women's agency in gender-traditional religions: a review of four approaches. *Sociology Compass*, 6(2), pp.122–133. https://doi.org/10.1111/j.1751-9020.2011.00439.x

Charmaz, K. (2006). *Constructing Grounded Theory: A Practical Guide Through Qualitative Analysis*. London: SAGE Publications.

Cruz, C. and Valdivieso, R. (2023). Competencies for sustainability: insights from the encyclical letter Laudato Si'. Business Ethics: *The Environment & Responsibility*, 33(4), pp.606–616. https://doi.org/10.1111/beer.12628

Deneulin, S. and Zampini-Davies, A. (2017). Engaging development and religion: methodological groundings. *World Development*, 99, pp.110–121. https://doi.org/10.1016/j.worlddev.2017.07.014

Deneulin, S. (2021). Religion and development: integral ecology and the Catholic Church Amazon Synod. *Third World Quarterly*, 42(10), pp.2282–2299. https://doi.org/10.1080/01436597.2021.1948324

Flacke, J., Shrestha, R. and Aguilar, R. (2020). Strengthening participation using interactive planning support systems: a systematic review. *ISPRS International Journal of Geo-Information*, 9(1), p.49. https://doi.org/10.3390/ijgi9010049

Francis (2015). Laudato Si': On Care for Our Common Home. Vatican City: Libreria Editrice Vaticana.

Friman, S. et al. (2024). The characteristics of leadership and their effectiveness in quality management in healthcare – a systematic literature review and content analysis. *Health Services Management Research*, 38(1), pp.37–49. https://doi.org/10.1177/09514848241254921

Jung, J. (2021). Ecological challenges and injustice from a missiological perspective. *Transformation: An International Journal of Holistic Mission Studies*, 38(4), pp.279–295. https://doi.org/10.1177/02653788211038976

Korff, Y. et al. (2012). Implementing participatory water management: recent advances in theory, practice, and evaluation. *Ecology and Society*, 17(1). https://doi.org/10.5751/ES-04733-170130

Lee, Y. (2025). Integrating circular economy and Laudato Si': A Christian framework for sustainable development and environmental stewardship. *Religions*, 16(3), p.326. https://doi.org/10.3390/rel16030326

Lynch, R. et al. (2022). Religiosity is associated with greater size, kin density, and geographic dispersal of women's social networks in Bangladesh. *Scientific Reports*, 12(1). https://doi.org/10.1038/s41598-022-22972-w

Mabhaudhi, T. et al. (2019). The water–energy–food nexus as a tool to transform rural livelihoods and well-being in southern Africa. *International Journal of Environmental Research and Public Health*, 16(16), p.2970. https://doi.org/10.3390/ijerph16162970

Malakoane, B. et al. (2023). Implementing an intervention to improve leadership/management of public healthcare services in the Free State Province, South Africa: Lessons learned. *African Health Sciences*, 23(1), pp.469–482. https://doi.org/10.4314/ahs.v23i1.49

Martins, A. (2018). Laudato Si': Integral Ecology and Preferential Option for the Poor. *Journal of Religious Ethics*, 46(3), pp.410–424. https://doi.org/10.1111/jore.12224

Martusenko, I. et al. (2023). Strategic PR guidelines for marketing planning to ensure the sustainable development of socio-economic systems. *International Journal of Sustainable Development and Planning*, 18(4), pp.1113–1120. https://doi.org/10.18280/ijsdp.180414

Mengesha, E. et al. (2023). Social capital and its role to improve maternal and child health services in northwest Ethiopia: a qualitative study. *PLOS ONE*, 18(4), e0284592. https://doi.org/10.1371/journal.pone.0284592

Ndlovu, W. et al. (2018). Transformational leadership and employee organisational commitment in a rural-based higher education institution in South Africa. SA *Journal of Human Resource Management*, 16. https://doi.org/10.4102/sajhrm.v16i0.984

Okunade, B. et al. (2024). Community development programs in rural Africa: An effectiveness review. *International Journal of Science and Research Archive*, 11(1), pp.1217–1226. https://doi.org/10.30574/ijsra.2024.11.1.0185

Sánchez-Camacho, J. and Moreno, J. (2024). Foundations and implications of the integral ecology and sustainable development goals in Catholic university education. *Religions*, 15(4), p.480. https://doi.org/10.3390/rel15040480

Shava, G. and Heystek, J. (2021). Managing teaching and learning: Integrating instructional and transformational leadership in South African schools context. *International Journal of Educational Management*, 35(5), pp.1048–1062. https://doi.org/10.1108/IJEM-11-2020-0533

Tavanti, M., Brennan, M. and Helgeson, S. (2017). Management education for sustainable development: Integrating ecology and community values in social impact practicums Capacity building: A phenomenological. *Journal of Management for Global Sustainability*. https://doi.org/10.13185/jm2016.04106

Wakahiu, J. and Keller, D. (2017). Study of the African women perceptions and experiences in the leadership training program. *African Women in Leadership Journal*, 31. https://doi.org/10.21423/awlj-v31.a85

Wilkins, D. (2020). Catholic clerical responses to climate change and Pope Francis's Laudato Si'. Environment and Planning E: *Nature and Space*, 5(1), pp.146–168. https://doi.org/10.1177/2514848620974029

Zhovnirchyk, Y. et al. (2023). Strategic planning for the sustainable development of territorial communities. *International Journal of Sustainable Development and Planning*, 18(7), pp.2097–2105. https://doi.org/10.18280/ijsdp.180712

Chapter 2
Regenerative Agriculture Among Young People in Zambia: The Frontline Work of Young Emerging Farmers Initiative

Richard Kachungu

Abstract This paper explores the role of regenerative agriculture in empowering young people in Zambia, focusing on the work of the Young Emerging Farmers Initiative (YEFI). Over the past decade, YEFI has mobilized youth to adopt sustainable agricultural practices such as agroecology. This research outlines the challenges, progress, impacts, and future direction of youth-led regenerative agriculture in Zambia. The study relies on primary insights from YEFI's operational data and secondary sources, to test under what conditions of engaging youth in climate-resilient, sustainable farming can best facilitate Zambia's food security, ecological balance, and socio-economic development.

Keywords Youth · Livelihoods · Regenerative farming · Empowerment · Zambia

2.1 Introduction

Regenerative agriculture has emerged as a beacon of hope in the face of climate change, land degradation, and youth unemployment, 26.4% (Zamstat, P75). In Zambia, where over 77.89% (15,338,802 people) of the population is under the age of 35, agriculture remains a central livelihood activity (CSO 2022). However, youth engagement in agriculture has been limited due to systemic barriers, including lack of land, poor access to finance, inadequate skills, and sociocultural perceptions of farming as unprofitable or outdated (FAO 2014). The Young Emerging Farmers Initiative (YEFI) is working to reverse this trend by placing youth at the center of regenerative agricultural innovation.

Richard Kachungu, Youth Empowerment Farmers Initiative Zambia, richard.kachungu@gmail.com.

R. Kachungu (✉)
Youth Empowerment Farmers Initiative Zambia, Lusaka, Zambia
e-mail: richard.kachungu@gmail.com

R. Marcantonio et al. (eds.), *Integral Ecology in the Anthropocene: Africa and Beyond*, The Anthropocene: Politik—Economics—Society—Science 49,
https://doi.org/10.1007/978-3-032-32125-1_2

This paper documents the pioneering role of YEFI in promoting regenerative agriculture, providing a model for youth engagement, rural development, and ecological restoration. It analyzes the data generated from YEFI's fieldwork and broader academic insights to offer an evidence-based map for Zambia's journey toward sustainable agriculture through the lens of youth empowerment.

2.2 Conceptual Framework: Regenerative Agriculture and Youth Empowerment

Regenerative agriculture refers to a system of farming principles and practices that restores and enhances the resilience of ecosystems, increases biodiversity, improves water cycles, and strengthens the health of soils (Rhodes 2017).

Regenerative agriculture, as applied by the Young Emerging Farmers Initiative (YEFI), is both a philosophy and a practice that restores balance between people, soil, and nature. It goes beyond sustainability by actively regenerating ecosystems and improving community wellbeing. YEFI defines regenerative agriculture as any farming system that enhances soil fertility, biodiversity, water retention, and carbon capture while improving food security and livelihoods.

YEFI's training model integrates practical modules drawn from agroecology, permaculture, and indigenous knowledge systems. The curriculum includes: (1) *Soil regeneration techniques* such as composting, mulching, green manures, and minimal tillage; (2) *Agroforestry and crop diversification*, focusing on integrating trees, legumes, and cover crops; (3) *Organic input production*—biofertilizers, compost tea, and natural pest repellents; (4) *Water management* through contouring, swales, and rainwater harvesting; (5) *Community nutrition gardens* promoting local food sovereignty; and (6) **Agro-enterprise** *and cooperative management* for economic empowerment.

This integrated training approach helps young farmers restore degraded land, build resilient agro-enterprises, and strengthen food systems. By connecting ecological regeneration with income generation and civic engagement, YEFI's model empowers youth not only to cultivate the land but also to lead the transformation toward a regenerative and equitable food future.

In the context of youth empowerment, regenerative agriculture offers a twofold advantage. It allows young people to create sustainable livelihoods while participating in climate action. Regenerative agriculture, as championed by the Young Emerging Farmers Initiative (YEFI), provides a tangible pathway for youth empowerment through both economic resilience and environmental stewardship. Far from being a theoretical model, YEFI's 2024 Impact Report provides clear data showing how this approach results in sustainable livelihoods. Between 2022 and 2024, over 3000 indigenous youth and community members across 26 districts in Zambia reported an average income increase of 40%, a shift directly linked to their adoption of regenerative practices such as agroecology, agroforestry, and conservation farming (YEFI 2024). These gains were not abstract; they reflected in their ability to pay school fees, invest in assets, and save collectively through group enterprises.

YEFI-trained young farmers transitioned from subsistence to income-generating agro-enterprises. For instance, the ZYCALA project supported the creation of over 20 youth-led cooperatives/clubs, generating more than 1000 green jobs. Through diversification, such as honey production, poultry, and organic vegetable farming, these young entrepreneurs reduced economic risk while restoring degraded ecosystems. In Serenje and Chitambo districts, youths like Speancer and Maxiwell used regenerative techniques not only to improve soil health but also to scale small businesses, creating both personal income and community employment (YEFI 2024).

Moreover, YEFI's peer-to-peer training model enabled 2200 households to replicate these techniques, amplifying impact. Further benefits determined from these shifts included improved child nutrition and food security, reinforcing the link between ecological farming and broader wellbeing. By integrating environmental care with economic opportunity, regenerative agriculture has moved beyond promise into measurable, life-changing outcomes for some Zambia's youth. This approach aligns with the goals of Zambia's National Youth Policy and the African Union's Agenda 2063, which emphasize youth as drivers of inclusive economic growth and environmental stewardship (African Union Commission 2015).

2.2.1 The YEFI Youth Empowerment Model

The YEFI Youth Empowerment Model is a holistic framework that situates youth at the center of Zambia's food system transformation. It integrates four interdependent pillars: *Capacity Building, Enterprise Development, Policy Advocacy, and Climate Resilience*, reinforced by continuous learning and community ownership (See Fig. 2.1). YEFIs Theory of Change (Fig. 2.2) builds on this core framework to specifically address pathways to impact. YEFI believes that true empowerment (See Fig. 2.3) occurs when youth have access to **knowledge, capital, and voice** to lead inclusive and regenerative agricultural systems.

2.3 The Zambian Context: Youth and Agriculture

Agriculture remains a cornerstone of Zambia's economy, contributing 19% to the national GDP and employing more than half the population (World Bank 2021). Yet, youth involvement in the sector is alarmingly low (Average age for a farmer in Zambia is above 50 years), despite young people making up over 70% of Zambia's population. This disconnect stems largely from systemic barriers: many young people lack access to land titles, cannot secure affordable loans, and are excluded from input distribution programs like Farmer Input Support Programme (FISP). From YEFI's experience engaging young indigenous farmers, market access remains another hurdle, with

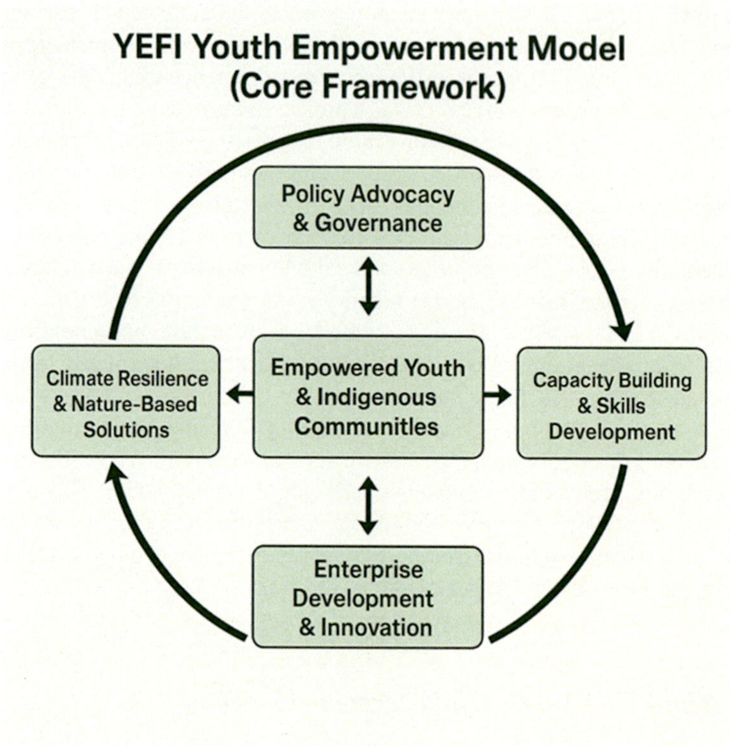

Fig. 2.1 YEFI Youth Empowerment Model (Core Framework): Each pillar interacts dynamically to strengthen youth leadership. *Source* Author created

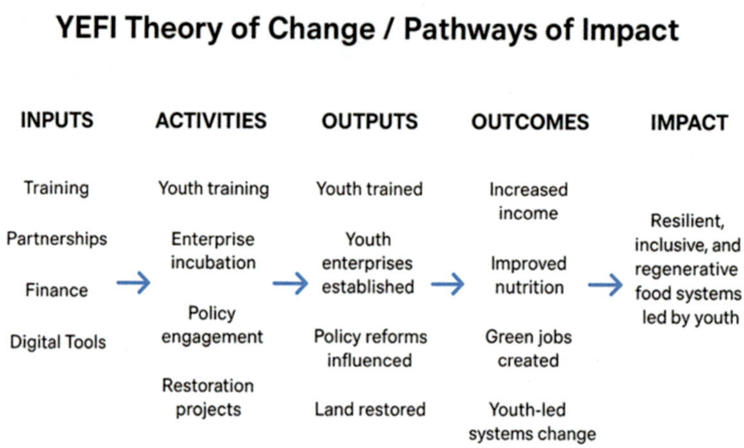

Fig. 2.2 YEFI Theory of Change/Pathways of Impact: YEFIs participatorily developed theory of change with identified pathways of impact for their programming. *Source* Author created

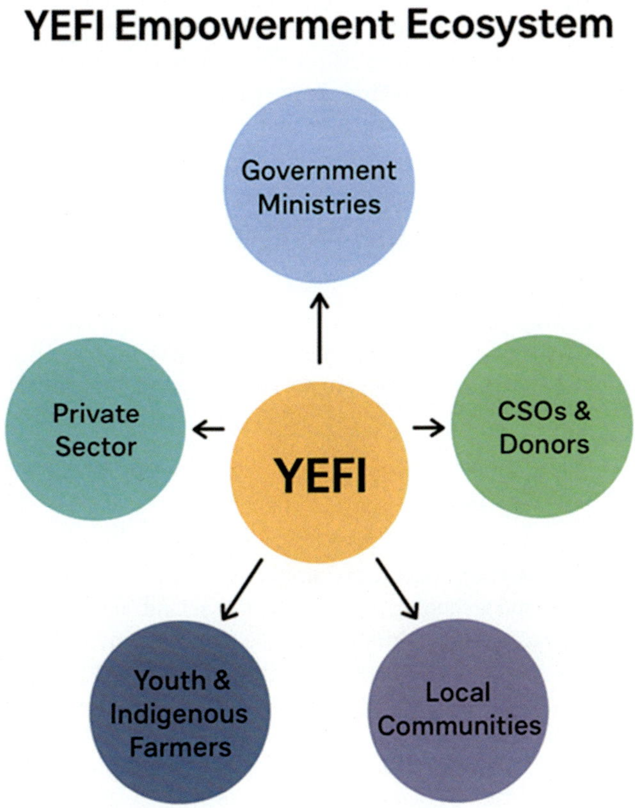

Fig. 2.3 YEFI Empowerment Ecosystem: YEFI operates as a *connector and enabler*, linking grassroots youth initiatives with institutional, market, and policy-level systems for sustainability and scale. *Source* Author created

poor infrastructure, volatile pricing, and weak cooperative systems limiting returns on their produce.

Additionally, climate change continues to disrupt traditional farming cycles. Zambia has experienced increasingly severe droughts, floods, and unpredictable rainfall over the past decade, severely affecting smallholder farmers who depend on rain-fed agriculture. These climatic shocks have worsened food insecurity and discouraged youth from pursuing agriculture as a viable livelihood.

However, programs like those run by the Young Emerging Farmers Initiative (YEFI) have shown that regenerative agriculture can offer a new path. Through practices such as Agroecology; agroforestry, composting, crop diversification, and organic pest control, youth are beginning to reclaim degraded land and producing food sustainably (2000 hectares). According to YEFI's 2024 Impact Report, young farmers who adopted these techniques experienced a 50% improvement in food security. Youth also reported greater pride and interest in farming, viewing it as a respected, modern livelihood so long as profits or incomes were being generated. In this context, regenerative agriculture is not just an ecological response, it is a practical, empowering solution to structural and environmental challenges.

2.4 Challenges Facing Young People in Agriculture in Zambia

Despite its successes, YEFI recognizes that young people face persistent barriers. Primary identified barriers include but are not limited to the following.

2.4.1 Access to Resources

A significant number of young people in Zambia struggle to access critical resources such as land, finance, and technology. Most do not own land, and those who do often lack security of tenure, making it difficult to make long-term investments in agriculture. For example, USAID's TGCC project in Chipata District (2014–2018) mapped land and issued 17,871 customary land certificates (CLCs) across five chiefdoms, documentation recognized locally but not a formal state deed. Of those parcels, 52% were held or co-held by women; the public report does not provide a youth (15–35) breakdown. This illustrates that project documentation exists, but youth-specific titling data were not reported (YEFI 2024). NORC.

UN-Habitat also notes that only 13% of formal title deeds in Zambia (2000–2019) were issued to women, again not youth-specific, but it underscores how formal deeds are limited and skewed, while most landholding (including that accessed by youth) is under customary tenure.

Access to affordable credit is also limited, preventing them from purchasing farming inputs, machinery, and adopting modern, climate-resilient technologies. This lack of investment hinders productivity, discourages innovation, and makes it harder for youth to transition to sustainable, market-oriented farming systems. The impact of climate change further worsens this situation, as many young farmers lack the tools and knowledge to adapt effectively.

2.4.2 Policy and Institutional Gaps

Although youth represent a large portion of Zambia's population, their specific needs and potential in the agricultural sector are often overlooked in national policy frameworks. This lack of intentional focus on youth contributes to gaps in policy coherence around regenerative agriculture. For example, the need to have guarantee quotas for youths in accessing Land, Insurance and finance. As a result, programs designed to support sustainable agriculture or rural development often operate in silos, not necessarily because they are *incoherent* themselves, but because there is limited coordination and alignment across ministries, donor agencies, and civil society organizations to jointly prioritize and integrate youth engagement.

2.4.3 Market Limitations

Young farmers face major hurdles when it comes to accessing markets. Poor road networks and limited storage and processing facilities make it difficult to transport and preserve produce, resulting in significant post-harvest losses for all farmers especially young farmers. Additionally, the absence of consistent pricing structures and weak bargaining power means youth often earn less than their produce is worth. A lack of access to digital tools and platforms also isolates many from wider markets and information on demand trends, pricing, and potential buyers. For example, young people YEFI is working with in Gwembe, Sopani area, have network connectivity issues most of the time, thus, find it difficult to utilize digital platforms to access markets. These constraints reduce motivation and lead to low returns on investment.

2.4.4 Cultural Barriers

In many communities, agriculture is still perceived as a low-status occupation associated with poverty and hard manual labor. This perception discourages young people from pursuing farming as a viable livelihood option, especially in comparison to more urban, white-collar professions. Without efforts to reframe agriculture as a modern, profitable, and respectable career, it remains challenging to attract and retain youth in the sector.

2.5 The Genesis and Mission of the Young Emerging Farmers Initiative

The Young Emerging Farmers Initiative (YEFI) was founded in 2014 as a grassroots response to the persistent exclusion of youth from Zambia's agricultural transformation agenda. At a time when youth unemployment rates were soaring and young people were migrating from rural to urban areas in search of opportunity, YEFI recognized that the solution lay not in abandoning agriculture but in reimagining it. Over the past decade, the organization has grown into one of Zambia's most influential youth-driven movements in the food system, now operating in over 50 districts and directly reaching more than 500,000 young people across all 10 provinces.

YEFI's mission is to empower indigenous youth aged 15–35 to become agents of change in agriculture through regenerative and climate-resilient practices. Its work is centered on equipping youth with practical skills in agroecology, conservation agriculture, and agroforestry, while also addressing critical barriers such as limited market access, weak policy support, and the lack of affordable financing. By promoting locally rooted solutions that harmonize with the environment, YEFI offers a model that blends traditional knowledge with modern innovation.

The organization deploys a decentralized model driven by District Coordinators (subunits of Provinces), who lead youth mobilization, training, and mentorship within their communities. YEFI also uses demonstration plots, peer-to-peer learning, and participatory planning to foster ownership and sustainability. Over the years, the model has adapted to include digital tools, social enterprise development, and advocacy campaigns that give youth a stronger voice in local and national policy.

In recognition of its impact, YEFI has received multiple awards, including international recognition (2023 Equator Prize under the UNDP Equator Initiative) for innovation in regenerative agriculture and youth empowerment. Today, it stands as a global award-winning organization, influencing both local policy and international development discourse on youth and sustainable agriculture. Its journey is a testament to what is possible when communities are trusted to lead their own transformation.

2.6 Impact and Achievements

YEFI has made substantial contributions to Zambia's agriculture and youth development landscape over the past decade. Notable achievements include.

2.6.1 Training and Engagement

YEFI has directly trained and empowered over 50,000 young people across all 10 provinces of Zambia, since 2014. Through hands-on learning and community-based mentorship, youth have been equipped with knowledge and skills in agroecology; conservation agriculture, agroforestry, integrated pest management, and other climate-smart farming practices. These capacity-building initiatives have created a strong network of young producers committed to sustainable agriculture. 60 percent of these young people are females and over 70 percent are in rural areas of Zambia.

2.6.2 Income Growth

Farmers engaged by YEFI report a consistent increase in their household incomes, with an average growth of 40% (Impact Report 2024). According to YEFI's 2024 Impact Report, the reported 40% average increase in household income among participating farmers is based on a five-year longitudinal impact assessment conducted across 26 districts, involving 3000 indigenous youths and community members. The data was collected through YEFI's Monitoring and Evaluation (M&E) system, which includes baseline and endline surveys, income tracking tools, and focus group discussions led by District Coordinators. Income improvements were measured by comparing household earnings before and after participation in YEFI programs, using indicators such as revenue from crop and livestock sales, savings group records, and investment in household needs like school fees, farming inputs, and construction materials.

Farmers transitioned from subsistence farming where production was largely for home consumption, to small-scale commercial operations characterized by surplus production for local markets, group marketing, and value addition. Models that integrated multiple income streams, such as combining poultry, vegetable farming, and honey production, proved most resilient. Youth who adopted agroecological methods alongside entrepreneurship training were better able to scale their farms sustainably. This rise is largely attributed to the adoption of diversified farming techniques, improved productivity, better post-harvest handling, and expanded access to markets. Many youth-led farming enterprises are transitioning from subsistence to small-scale commercial operations, creating a ripple effect of economic improvement in their communities.

Challenges emerged in cases where youth lacked collective cohesion or business management skills. For instance, some poultry ventures struggled due to poor planning or unclear ownership. These experiences highlighted the need for stronger cooperative management and follow-up support, which YEFI has since integrated into its evolving training model.

2.6.3 Food and Nutritional Security

Between 2022 and 2024, YEFI conducted a comprehensive impact assessment across 26 districts involving 3000 indigenous youth-led households. Data was collected through household surveys, food diaries, and collaboration with 15 local health centers. Food security was assessed using the Household Food Insecurity Access Scale (HFIAS), while child nutrition outcomes, specifically for children under six, were measured by reviewing clinic-based growth monitoring records and conducting interviews with community health personnel.

The reported 50% improvement in food security reflects increased dietary diversity, meal frequency, and year-round food access. This shift was largely due to training in permaculture, compost use, and organic pest management, enabling households to produce a range of vegetables such as rape, tomatoes, cowpeas, and indigenous greens. Youth-led implementation programs reached over 2200 additional households, spreading these practices community-wide.

Nutritional gains, over 40% improvement in under-six child health, were noted in reductions in underweight and stunted growth cases, corroborated by clinic records. Interventions such as kitchen gardens, compost-fed nutrition beds, and seasonal food planning made substantial differences. These results demonstrate that YEFI's model goes beyond advocacy, it produces measurable health and food system outcomes rooted in youth-led, regenerative agriculture.

2.6.4 Green Jobs

YEFI tracked the creation of 32 youth-led agro-enterprises across 26 districts, verified through cooperative registrations, business licenses, and field visits by District Coordinators. These enterprises employed approximately 1038 youth in roles such as compost production, poultry rearing, organic vegetable sales, agro-processing, and community tree nurseries. In Serenje and Chitambo, youth groups formed processing hubs for dried vegetables, mushrooms and honey, generating monthly incomes between ZMW 1500–4500 per member. YEFI's enterprise support included business planning, market linkages, and startup capital through savings groups. While enterprises with strong leadership and collective ownership thrived, loosely coordinated groups struggled to scale.

2.6.5 Land Restoration

YEFI's land restoration achievements since 2014 is backed by data from field records and community land-use surveys conducted in collaboration with district agricultural offices. The 2113 hectares restored included areas in Chitambo, Serenje, Mkushi and Kasempa where youth adopted agroecological methods, agroforestry (planted 30,000+ trees, mostly fruit trees), cover cropping, and compost application. Rotational grazing practices were introduced in four districts to rehabilitate pastureland. The preservation of 10,380 hectares of forest was linked to youth-led honey production activities, which incentivized forest protection. Beekeeping groups documented hive placements and forest use agreements that discouraged charcoal production and illegal logging.

2.6.6 Policy Influence

YEFI's policy influence extends beyond community-level organizing to active participation in national and international policy processes. At the national level, YEFI has consistently engaged with the Ministry of Agriculture and the Ministry of Green Economy and Environment, contributing to dialogues on youth inclusion, climate-smart agriculture, and land access. Since the first quarter of 2025, YEFI has been a formal member of the Technical Working Group drafting Zambia's National Agroecology Strategy, chaired by the Ministry of Agriculture and co-chaired by the Ministry of Green Economy and Environment, where it provides direct input from youth-led agroecological field models. YEFI's data and field-based insights have helped shape language around youth participation and indigenous knowledge systems.

Internationally, YEFI has co-hosted Local Conference of Youth (LCOY) Zambia and the 2024 SADC Regional Conference of Youth (RCOY), serving as a launchpad for elevating rural youth voices into global platforms like the UNFCCC COP processes. These convenings have informed regional declarations calling for stronger policy support for youth in food system transitions. YEFI also liaises with global networks such as Taproot Earth to align grassroots experiences with global climate advocacy. This multi-level engagement positions YEFI as both a grassroots and policy actor in agroecological transformation.

2.7 Lessons Learned from YEFI's Model

Several key insights emerge from YEFI's decade-long journey, offering guidance on how to initiate and develop youth empowerment programming for regenerative agriculture and sustainable livelihoods.

2.7.1 Economic Incentives Drive Adoption

One of the most concrete takeaways from YEFI's programming over the past five years is that youth engagement in agriculture rises sharply when economic returns are immediate and measurable. In early phases of the ZYCALA project, participation hovered at moderate levels, with many young people still skeptical about agriculture's viability. However, this changed significantly when YEFI introduced value chains tied to export markets. For example, in Serenje, Chitambo and Mkushi, youth trained in beekeeping were linked to LUSH, a UK-based ethical cosmetics company, through a field-level honey export partnership. This model provided not only training in hive management but also structured routes to formal markets. Within a year, 185 hives were operational across 12 demonstration sites, and youth began earning ZMW 2000–5000 per harvest, an income level far above local casual labor or subsistence farming.

The financial outcomes spurred a surge in demand for participation, including from youth who had initially opted out. Those who earned from early cycles reinvested in additional hives, poultry, or vegetable beds. The spillover effect was equally notable: new cooperatives formed independently, some pooling funds to scale operations. This experience confirms that without real and predictable income opportunities, youth interest remains limited. Economic incentives are not just helpful, they are essential for sustained engagement.

2.7.2 Community Ownership is Essential

YEFI's experience across over 50 districts has demonstrated that lasting change depends on deep community ownership, especially among youth. In districts like Chama, Chitambo, and Gwembe, project success directly correlated with the strength of local leadership. YEFI invests in identifying and training local youth, particularly young women, to become District Coordinators who are embedded in their communities. These coordinators lead training sessions, monitor implementation, and provide mentorship to their peers.

In Serenje, for instance, it was local youth who organized follow-up visits, launched farmer field schools, and facilitated collective marketing. Before any intervention begins, YEFI undertakes participatory needs assessments through community meetings, mapping exercises, and interviews with local leaders and youth groups. This groundwork ensures that the design reflects real needs and builds community trust. In Serenje and Lundazi, this approach led to spontaneous formation of new farming groups after the initial intervention phase, without external prompting. The community-led model has proven far more resilient and adaptive, particularly during periods of drought or economic stress.

2.7.3 Policy Advocacy Matters

YEFI's policy engagement has evolved from informal advocacy to formal participation in Zambia's national policy-making structures. Recognizing that practical, field-level impact must be matched with institutional change, YEFI has positioned itself within key decision-making spaces. Since the first quarter of 2025, YEFI has been a core member of the Technical Working Group for Zambia's National Agroecology Strategy, chaired by the Ministry of Agriculture and co-chaired by the Ministry of Green Economy and Environment. In this role, YEFI provides field-based data and youth perspectives that inform provisions around land access, seed sovereignty, and youth financing mechanisms.

Additionally, YEFI has played a leadership role in coordinating Zambia's 2022, 2023 Local Conference of Youth (LCOY) and co-hosted the SADC Regional Conference of Youth on Climate Change (RCOY) in 2024. These platforms generated youth-led declarations calling for climate-resilient food systems and youth access to decision-making. Domestically, YEFI has participated in ministerial dialogues on land tenure reforms and green financing mechanisms, contributing case studies and testimonies from beneficiaries.

Through these engagements, YEFI ensures that youth are not seen as passive recipients but as co-authors of Zambia's development agenda, embedding grassroots innovations into national strategies that affect food security, environmental conservation, and climate resilience.

2.7.4 Digital Technology Expands Reach

Embracing digital innovation has significantly enhanced YEFI's impact, especially in reaching rural and underserved youth. Over the past decade, the organization has invested in developing and scaling digital tools that provide access to agricultural training, market linkages, early warning systems, and precision farming techniques. By building the digital capacity of young people, YEFI has enabled them to develop tech-driven solutions tailored to local agricultural challenges, for instance in 2017 YEFI hosted a Youth in Agribusiness Conference where some youths participated in an Hackathon to enhance the food system especially for small-scale farmers.

During that event, one of the winners of the Hackathon, ended up being recruited by a market digital innovation called LIMA links that provides real time market information for farmers in some regional markets in Zambia. This includes mobile platforms for weather forecasting, pest alerts, soil health monitoring, and market price tracking. As more young farmers gain access to digital tools, they are better equipped to make informed decisions, increase productivity, and connect with broader value chains. The integration of technology has not only modernized farming practices but has also made agriculture more appealing to tech-savvy youth.

2.8 Future Directions and Scaling Strategies

To scale regenerative agriculture among youth, YEFI is pursuing several strategies.

2.8.1 *Establishment of the Youth Agriculture Hub*

One of YEFI's most ambitious steps forward is the creation of the Youth Agriculture Hub. This space will serve as both a center of innovation and a business incubator, designed to empower young agro-entrepreneurs. The hub will operate as a social enterprise, providing hands-on training in sustainable agricultural practices, business development, and access to financing. It will serve as a launchpad for youth-led start-ups, equipping them with skills and mentorship to build viable, environmentally friendly agribusinesses. Beyond technical skills, the hub will foster a culture of collaboration and peer learning, helping young people grow their enterprises in ways that are sustainable and rooted in local realities. YEFI's planned Youth Agriculture Hub is driven by findings from its works since 2014 and confirmed by its 2024 impact assessment, which revealed that 68% of youth trained lacked follow-up support to scale their agribusinesses. Similar models in Kenya (Egerton's AgriHub) and Nigeria (IITA Youth Agripreneur Program) have shown increased youth enterprise success and employment (IITA 2018), validating YEFI's localized incubation approach.

2.8.2 *Operationalizing a Digital Platform*

To extend its reach and efficiency, YEFI is moving forward with a digital platform (USSD code access platform) that will connect young farmers to essential services and opportunities. This platform will provide digital extension services, real-time market information, and learning resources. It will also serve as a space for youth to share their innovations and stories. Importantly, YEFI is not working in isolation, it aims to collaborate with existing digital tools and platforms to avoid duplication and ensure broader reach. This approach will make it easier for rural youth, often left out of mainstream systems, to stay informed, access new markets, and receive timely support.

2.8.3 *Launching the YEFI Fund*

To ensure long-term support for youth-driven initiatives, YEFI has initiated the YEFI Fund. This youth-led financing facility is designed to support nature-based solutions and scale models that have already shown success, such as the ZYCALA project.

Though launched, the fund is currently in its fundraising phase, with efforts focused on mobilizing resources from philanthropic foundations, social investors, and development partners. Once fully operational, the fund will offer flexible, targeted support to young innovators working in agriculture and environmental stewardship.

Since 2014, YEFI work with youths has reviews that finance is one of the biggest challenges young people face in the food system and de-risking the agricultural financing is one of YEFI goals for a sustainable, just, fair and transformative food system. Since 2014, 98 percent of the young people in rural areas state that limited access to agriculture financing is their biggest challenge.

2.8.4 Expanding Strategic Partnerships

Building strong, diverse partnerships is central to YEFI's approach. The organization continues to work closely with government departments, local NGOs, community groups, private sector actors, and international bodies. These partnerships help amplify reach, share resources, and align efforts for greater collective impact.

2.8.5 Participating in Global Networks

YEFI recognizes the value of global exposure and cross-learning. By engaging with international platforms such as Taproot Earth, faith-based networks, and philanthropic foundations, YEFI aims to strengthen its visibility, attract new forms of support, and exchange ideas that can refine and enhance its model.

2.9 Conclusion

The story of YEFI illustrates that young people can be powerful agents of environmental and agricultural transformation. Through regenerative agriculture, they are restoring landscapes, feeding communities, creating jobs, and influencing policy. However, to sustain and expand these gains, investment in youth-led models, digital tools, inclusive financing, and policy reform is essential for replication and sacle. Zambia, and indeed Africa, stands at a crossroads. The future of food systems must be both regenerative and youth-led.

References

African Union Commission. (2015). *Agenda 2063: The Africa We Want*. Addis Ababa: AUC.

Central Statistical Office. (2022). *Zambia Demographic and Health Survey 2021–2022*. Lusaka: CSO.

Food and Agriculture Organization of the United Nations. (2014). *Youth and Agriculture: Key Challenges and Concrete Solutions*. Rome: FAO. (https://www.fao.org/3/i3947e/i3947e.pdf)

IITA. (2018). *Youth initiative in agribusiness helping transform African agriculture*. Retrieved from https://blogs.iita.org/youth-initiative-in-agribusiness-helping-transform-african-agriculture/

Rhodes, C. J. (2017). The Imperative for Regenerative Agriculture. *Science Progress*, 100(1), 80–129. https://doi.org/10.3184/003685017X14876775256165

UN-Habitat (2020). *Land Governance Assessment Framework: Zambia Country Report*. Nairobi: United Nations Human Settlements Programme. https://unhabitat.org/news/05-mar-2024/empowering-women-through-land-rights-in-zambia?utm_source=chatgpt.com.

USAID (2018). *Tenure and Global Climate Change: Land Tenure and Resource Governance Research in Zambia's Eastern Province—Final Report*. Washington, DC: United States Agency for International Development. https://www.norc.org/content/dam/norc-org/pdf2023/cel-zambia-land-final-report.pdf?utm_source=chatgpt.com.

YEFI. (2024). *Internal Impact Report*. Lusaka: Young Emerging Farmers Initiative. https://drive.google.com/file/d/1YRACiHQzM9aEvI9WwsEQVc342XOQUPwP/view?usp=drive_link

Zambia Statistics Agency (ZamStats). (2022). *2022 Labour Force Survey Report*. Lusaka: Central Statistical Office (CSO). https://www.zamstats.gov.zm/

Zambia Statistics Agency, Ministry of Labour and Social Security. (2023). *Labour Force Survey*. Lusaka.

Chapter 3
Bittersweet Relationships: Renewable Energy, Integral Agroecology and Climate Action in Uganda

Charles L. Tumuhe, Denis Katusiime, David Ssekamatte, Vincent F. Bukenya, and Landing Diedhiou

Abstract This study examines the integration of renewable energy solutions into agroecological initiatives to enhance climate adaptation and mitigation in Uganda. Using a qualitative case study conducted across 17 districts, data were collected through focus groups, interviews, and document reviews. Findings show that organizations within the civil societies' network of Participatory Ecological Land Use Management (PELUM) in Uganda promote renewable energy solutions, empowering women and youth as key actors. These innovations aim to reduce deforestation and strengthen local resilience, although barriers such as policy gaps and financing persist. By aligning ecological sustainability with social equity and community agency, the study reflects the principles of integral ecology, emphasizing the interdependence of environmental, social, and economic systems. The study recommends that scaling renewable energy within agroecological systems requires strengthened community leadership, enhanced institutional support, andcoherent policy alignment to achieve equitable, sustainable, and climate-resilient rural development.

Charles L. Tumuhe, Uganda Martyrs University, tumchaz34@gmail.com; Denis Katusiime, Uganda Martyrs University, dkatusiime@unik.ac.ug; David Ssekamatte, Uganda Management Institute, dmssekamatte@gmail.com; Vincent F. Bukenya Uganda Martyrs University, bukenya.vincent@stud.umu.ac.ug; and Landing Diedhiou, Assane Seck University of Ziguinchor, l.d7@zig.univ.sn.

C. L. Tumuhe (✉) · D. Katusiime · V. F. Bukenya
Uganda Martyrs University, Kampala, Uganda

D. Katusiime
e-mail: dkatusiime@unik.ac.ug

V. F. Bukenya
e-mail: bukenya.vincent@stud.umu.ac.ug

D. Ssekamatte
Uganda Management Institute, Kampala, Uganda

L. Diedhiou
Assane Seck University of Ziguinchor, Ziguinchor, Senegal
e-mail: l.d7@zig.univ.sn

R. Marcantonio et al. (eds.), *Integral Ecology in the Anthropocene: Africa and Beyond*, The Anthropocene: Politik—Economics—Society—Science 49,
https://doi.org/10.1007/978-3-032-32125-1_3

Keywords Agroecology · Renewable energy · Climate adaptation · Community empowerment · Uganda

3.1 Introduction

Rural communities in Uganda face mounting environmental degradation driven by their overreliance on traditional biomass energy sources, such as firewood and charcoal. This dependence accelerates deforestation and exacerbates climate vulnerability in ecosystems already stressed by erratic rainfall and resource scarcity (Isgren 2018). Women and youth, often tasked with sourcing energy, bear a disproportionate burden that limits their participation in economic and agroecological activities (FAO 2021). While agroecology offers a holistic framework that encompasses sustainable resource use and community empowerment (Gliessman 2020), the integration of renewable energy solutions into agroecological interventions remains under-documented in Africa. Local innovations, such as biogas from cow dung, charcoal briquettes from recycled waste, and solar irrigation, introduced by agroecology network organizations, demonstrate promising results (Kung 2013; Surtia et al. 2016). However, the systemic barriers to scaling these technologies, especially in marginalized and displaced populations, persist. There is thus an urgent need to explore how renewable energy can be mainstreamed within agroecological transitions, reduce ecological pressure, and strengthen climate resilience among vulnerable rural populations (Tittonell 2020; Bottazzi/Boillat 2021).

The introduction of renewable energy technologies into agroecological initiatives reflects a vital shift toward holistic and locally driven climate adaptation strategies. Participatory Ecological Land Use management, Uganda Country Chapter (PELUM Uganda) member organizations have embraced diverse clean energy solutions for food preservation and efficient cooking. These innovations are not isolated technical fixes but are embedded in broader agroecological frameworks emphasizing ecological regeneration, gender equity, and participatory learning (Altieri et al. 2015; Wezel et al. 2020). Women and youth have emerged as key agents in this transition, receiving training to produce and disseminate clean energy alternatives such as charcoal briquettes and solar technologies. These efforts are believed to contribute to reduced deforestation, improved public health, and enhanced food sovereignty. For instance, the recycling of cow dung into biogas addresses waste management while providing an accessible energy source, illustrating the integration of agroecological farming, energy production, and environmental stewardship (Gliessman 2013; CIDSE 2018). Yet, despite their ecological and social benefits, such initiatives remain at the margins of mainstream agricultural policy frameworks in many countries, including Uganda, that continue to prioritize industrial, input-intensive models (de Brauw/Bulte 2021; van Beusekom 2021). The lack of policy support, financing, and institutional backing limits the replicability of these solutions (UNDP 2024). Consequently, further investigation into the scaling, sustainability, and governance of renewable energy within agroecological framing is critical to advancing equitable climate adaptation strategies across Uganda and similar contexts (Isgren/Ness 2017; Sseguya et al. 2019).

The promotion of renewable energy within agroecological systems is not merely a technical solution but a moral and spiritual commitment to a more just and sustainable world. By empowering rural women and youth to lead in the adoption of clean energy technologies, PELUM member organizations operationalize this vision, fostering community resilience while respecting the integrity of creation. The transformation of waste into energy, the shift from exploitative extraction to regenerative practices, and the participatory governance of resources reflect the core tenets of integral ecology (Francis 2015).

This means that when supported by decentralized renewable energy, agroecology can empower smallholders, reduce vulnerability, and align with Uganda's broader goals of inclusive development and ecological stewardship. That builds climate justice and food sovereignty (Altieri et al. 2015; Francis 2015). Put differently, if renewable energy is deployed to power agroecological systems, like solar-powered irrigation, cold storage, or processing, rural communities, such as in Uganda, could unlock a resilient, low-carbon rural economy. This is vital synergistic integration (Ouko et al. 2024). Bridging these dimensions requires reframing climate action not just as emissions reduction, but as a holistic strategy for livelihoods, ecosystems, and intergenerational equity. This provides an operational clarity, at least, for policymaking and policy uptake.

3.2 Methods

This study was conducted in Uganda, an East African country characterized by diverse geographical landscapes and climatic conditions. The climate is predominantly tropical, with the central, eastern, and southern regions experiencing a tropical rainforest climate marked by two main rainy seasons (Adem 2024): the first season (March to May with a peak in April) and the second season (October to December with a peak in November). The mean annual rainfall is 1614 mm, with the temperature averaged between 18.14 and 26.53 °C (Dagoudo 2024).

This paper is part of a broader study that employed a qualitative, phenomenological case study design to explore how renewable energy technologies are integrated into agroecological initiatives by PELUM Uganda member organizations. Conducted across 17 districts (Fig. 3.1), the study engaged participants through 26 Focus Group Discussions (FGD), six Small Group Interviews (SGI), and 11 in-depth interviews (IDI), targeting women, youth, farmer leaders, and organizational staff involved in renewable energy solutions such as biogas, solar dryers, and energy-saving stoves. Data collection was supplemented by document reviews and analysed using thematic content analysis (Braun et al. 2019), supported by NVivo software, alongside network analysis to map collaborations. Out of the 25 PELUM member organizations that participated in this study, seven were promoting renewable energy solutions. The study engaged a diverse group of participants with a median age of 35, ranging from

22 to 72 years, reflecting a mix of youthful energy and seasoned experience. Participants spanned a wide array of roles across civil society, development, academia, and farmer institutions, ranging from field-based practitioners, such as Program Officers and Farm Managers, to senior decision-makers, including Executive Directors and CEOs. Technical experts, including Monitoring and evaluation and Agribusiness Officers, as well as academic representatives and student leaders, highlighted the sector's interdisciplinary nature. Additionally, leaders from 17 farmer institutions, including cooperatives and farmer groups, held roles like chairpersons, facilitators, and coordinators.

Figure 3.1 shows the locations of PELUM member organizations that participated in the study. The study covered seventeen districts. PELUM member organizations are denoted by the prefix 'MOS' followed by a number. The map also shows the districts where PELUM member organizations operate. PELUM Uganda, part of a 14-country African civil society (CSO) network, promotes agroecology through community-driven, gender-responsive, and climate-resilient practices. Guided by values of equity, innovation, and sustainability, it works with 74 civil society members across 122 districts in Uganda. Through partnerships, advocacy, and farmer-led training, PELUM advances its vision of just, inclusive, and ecologically sustainable food systems (PELUM 2020).

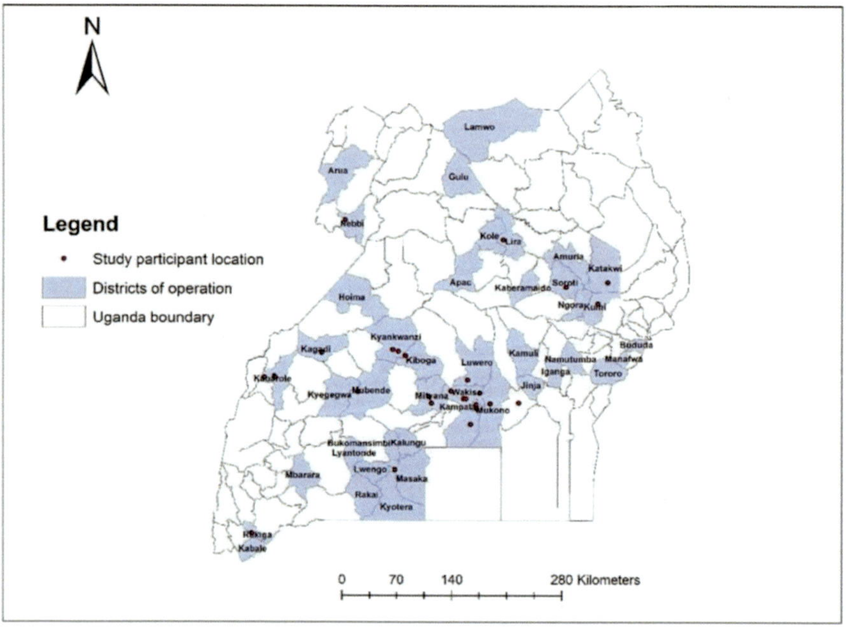

Fig. 3.1 Map of the study area. *Source* Author created

3.3 Findings

PELUM Uganda member organizations are promoting efficient and eco-friendly energy solutions among farming communities. These include energy-saving stoves, solar dryers, biogas systems, and solar energy systems for cooking, lighting, and irrigation. The PELUM Uganda member organizations train women and youth in these technologies, who later spread them to other members of the community. The training programs aim to equip young people with the skills to recycle waste materials into valuable products, such as charcoal briquettes, for use as an environmentally friendly fuel alternative to traditional fuels. Besides preventing deforestation, this training makes a significant contribution to environmental conservation, and it has economic implications, thereby fostering a sense of responsibility and innovation among the youth. Participants further explained that this kind of training promotes environmental stewardship. For example, one member organization staff was quoted as saying:

> "They [an NGO] brought in the solar irrigation kit … for [crop] irrigation, […]" (FGD, MOS1)

This quote demonstrates the introduction of solar-powered irrigation into farming practices, which signals a shift toward integrating renewable energy within agroecological systems. While the quote does not provide direct evidence of reduced fossil fuel use or quantitative improvements in water access, it shows that solar technology is being adopted as an alternative to conventional methods, which are often diesel- or petrol-powered. It reflects a shift toward sustainable, climate-resilient agriculture, empowering farmers with low-cost, clean technologies that support ecological balance and improve productivity in water-scarce environments. Other organizations didn't just bring the finished product but helped farmers make them using local materials. See quote below.

> "And where we come in with renewable technologies, the women themselves have been told how to produce safe energy [technologies], for example, producing charcoal briquettes, biogas, the integration between farming, biogas, and producing energy, and the solar that we are also using. […] to help them have access to safe energy, […]. […], there's also a lot of awareness, mass awareness, like where we have the women themselves trained as […] ambassadors of agroecology and energy, where they go and do programs to sensitize others, to let them know the effects and how they can adapt, […]." (IDI, MOS23)

A report accessed from MOS23 confirmed that information on renewable energy is disseminated to farmers.

> "Over 1000 persons reached with information on improved energy solutions and better agriculture practices" (MOS23 Annual Report, p. 22)

These two quotes from MOS23 exemplify how renewable energy technologies are integrated into agroecological initiatives to empower women as both users and educators. The quote below confirms this assertion.

Fig. 3.2 Solar dryer technology at MOS23 in central Uganda. *Source* Author created

"And also, we have the biogas. As MOS23, we promote the use of renewable energy and encourage our farmers to tap into this source of energy. We recycle all the by-products from the cows, such as the dung, and use them to produce biogas. This system supports our training in the use of renewable energy. We have those systems at the farm specifically for training our farmers. We have also established biogas systems in some of the homes of our farmers, to show them that you can tap into the resources available at their homes." (IDI, MOS23). (See example in Fig. 3.2)

By training women as ambassadors of agroecology and clean energy, the approach fosters local leadership, environmental awareness, and community-wide adoption of sustainable energy practices. See quote below:

"We are trying also to build women-led energy cooperatives… and these women are also part of our community resource persons. We are trying to build women as ambassadors of agroecology and energy so that they can be able to sensitise their communities."—IDI, MOS23

Farmers are also part of the movement. They are even engaging with local government authorities, e.g., sub-counties, as shown in the quote below.

Farmer group leaders mentioned that "… we try to sensitize the community to recycle [plastics]. They turn them into other materials like…" (FGD, MOF2). Through […], we got a solar dryer. Through this dryer, we have become known even at the Sub-County level. We made a market field day, and we sold … (FGD, MOF14).

This collection of testimonies from the PELUM network illustrates a vibrant transformation where agroecology and renewable energy converge to empower rural communities and foster ecological sustainability. Solar irrigation kits introduced by PELUM member organizations like MOS1 enabled smallholder farmers to irrigate their crops more efficiently during dry spells, reducing dependence on unpredictable rainfall and manual labour. Through support from local initiatives, such as the solar drier received by a farmer group leader in MOF14, communities are preserving food, including fruits and vegetables, for longer shelf life. This allows them to participate in events like sub-county market field days, where they gain visibility and income from their preserved foods".

Demonstration farms offer hands-on, context-specific learning where farmers can engage with agroecological techniques in a real-world setting. In addition to training, these demonstration farms allow practitioners and organizations to observe learning outcomes in real time. By seeing how farmers respond, adapt, or replicate the practices at their own farms, staff can informally assess the retention, comprehension, and implementation of fidelity of training efforts. As a staff member of MOS23 noted:

> The main purpose there is to learn some agroecological practices and learn how to sustainably improve their agricultural business

This statement reflects how learning outcomes are not abstract—they are observed through how farmers interact with the demonstration components, ask questions, and replicate practices in their own contexts.

The data we obtained revealed a lack of formal evaluation mechanisms to assess learning outcomes or training effectiveness. There is no mention of structured evaluation frameworks such as pre- and post-tests, or baseline and endline surveys that would measure knowledge or behavior change over time. Additionally, there is no evidence of systematic tracking of individual learner progress, nor the use of defined impact indicators aligned with specific learning objectives. This suggests that while training activities were widespread, their effectiveness may not have been rigorously measured or documented.

Across multiple sites, participants framed renewable energy not merely as a fuel replacement but as a pathway to dignity, environmental care, and collective self-reliance. For instance, at MOS23's agroecological demonstration farm, biogas systems were introduced to transform waste into clean energy, rooted in values of ecological regeneration and learning:

> "We recycle all the by-products from the cows, like the dung… to produce biogas… we have those systems at the farm purposely for training our farmers."—IDI, MOS23

This training-oriented approach reveals how renewable energy is embedded within participatory education and sustainability. Biogas systems were also extended into homes, guided by the value of local ownership:

> We have also established the biogas systems in some of the homes of our farmers… to show them that you can tap into the resources available at your home

Similarly, communities described charcoal briquettes and cookstoves as part of climate-conscious and socially empowering innovations, often emphasizing women's agency. A member from MOF17's farmer group shared:

> "They come and train us on different things like for example making of Lorena cookstoves… we also get training on making pesticides, local pesticide rather than buying these ones from the shop."—FGD, MOF17

The emphasis on *training* and *making* rather than *receiving* reflects deep-rooted values of self-sufficiency, appropriate technology, and adaptation. One participant described the broader impact of such initiatives:

> "MOS17 has done for us good, the knowledge MOS17 has given to us as a group… it has given us visibility in this area… the visitors come with different ideas that develop us."—FGD, MOF17

These statements demonstrate how renewable energy is tied not only to ecological goals, but to social recognition, visibility, and empowerment, particularly for marginalized rural groups. As such, these initiatives reflect a bottom-up, justice-oriented model of energy transition that is deeply values-driven and embedded in agroecological thinking.

At the heart of this clean energy movement is women's empowerment: women supported by MOS23 (also a PELUM member organization) are not only using but also producing clean energy solutions such as charcoal briquettes and biogas systems, the latter derived from recycled cow dung sourced from their livestock. These biogas systems are not just prototypes; they are fully operational and in use. They are established at both training farms and in individual homes, showing that renewable energy can be accessible and replicable at the household level. Furthermore, farmers are integrating farming with biogas production, closing the loop between agriculture and energy generation. Women trained by MOS23 are also becoming ambassadors of agroecology and renewable energy, leading community sensitization campaigns to spread awareness about climate change, energy access, and adaptive farming practices like the use of bio-slurry from biogas systems and solar-powered irrigation. Their (women's) role extends beyond education; they are mobilizers for change, showing others how to implement what they've learned. Beyond energy, environmental stewardship is gaining momentum; for example, farmer groups like MOF2 are advocating for plastic recycling, transforming waste materials into reusable products. These interconnected actions demonstrate how agroecological systems, when supported by renewable energy and grassroots leadership, can foster sustainable livelihoods, restore ecosystems, and promote social equity from the grassroots level.

Refugee settlements and their host communities often face significant environmental challenges, particularly deforestation caused by the high demand for firewood and charcoal. To mitigate this impact, PELUM member organizations, e.g., MOS17 in Soroti and Kumi, MOS23 in Kiboga and Kyankwanzi districts, are promoting energy-saving cookstoves, which enhance fuel efficiency and reduce pressure on forest resources. By improving biomass utilization, these interventions help conserve forests while ensuring that both refugees and host community members have sustainable cooking solutions. As highlighted in the quote below, promoting clean energy technologies is a crucial strategy for addressing environmental degradation and supporting the livelihoods of vulnerable communities.

> "… to reduce the impact of deforestation on tree cutting of forests, we also promote the use of energy-saving stoves at the household level, both in refugee settlements and in host community households, so that the little biomass in terms of firewood or even charcoal that they get can be utilised well and the impact on forests is reduced. So, promotion of cookstoves is one of the interventions that we have been doing …" (IDI, MOS20)

The cookstove initiative promoted by MOS20 embodies agroecology values by ensuring that scarce biomass resources are used more efficiently, thus easing tensions over firewood collection and reducing the ecological footprint of displacement. It fosters a shared responsibility for environmental stewardship among refugees and host populations alike, creating a practical basis for coexistence, cooperation, and healing. This approach reflects agroecology's broader commitment to restorative justice, where ecological regeneration and social inclusion are inseparable. In doing so, it reimagines humanitarian response not as a temporary relief effort, but as a regenerative, community-led process rooted in sustainability, dignity, and peace (Table 3.1).

3.4 Discussion

The findings of this study reveal that renewable energy technologies, such as biogas systems, solar dryers, and energy-saving stoves, are being actively promoted by agroecology organizations within the PELUM Uganda network as part of climate adaptation strategies. These technologies are not implemented in isolation but are integrated into broader agroecological practices that emphasize recycling, local resource use, and ecological sustainability (Wezel et al. 2020). For instance, biogas systems that convert livestock waste into clean energy exemplify the principle of nutrient cycling, thereby reducing dependence on firewood and mitigating deforestation (Altieri et al. 2015; CIDSE 2018). The study also highlights how women and youth are central to these transitions, not only as beneficiaries but as agents of change, leading training sessions and community sensitization on clean energy use. This aligns with the agroecological principle of co-creation of knowledge and reinforces the CIDSE framework's emphasis on socio-political and gender-inclusive transformation (Gliessman 2020; Bottazzi/Boillat 2021).

Despite these successes, systemic barriers persist in scaling renewable energy innovations. Limited institutional support, fragmented policy frameworks, and constrained financing continue to restrict the widespread adoption of these technologies (Isgren 2018; Tittonell 2020). Renewable energy initiatives in Uganda are mainly promoted by NGOs with limited government support. The Uganda Renewable Energy Policy (2007), developed under the Ministry of Energy and Mineral Development (MEMD), promotes some renewable energy initiatives aimed at increasing the share of renewable sources in the national energy mix. Key initiatives include the

Table 3.1 Renewable energy technologies promoted by PELUM Uganda member organizations. *Source* Author created

Technology type	Quotes
Solar energy	They brought in the solar irrigation kit that they are using for irrigation, [...]—FGD—MOS1 Through [...], we got a solar dryer. Through this drier, we have become known even at Sub-County level. We made a market field day, and we sold ...—FGD—MOF14 And where we come in with renewable technologies, the women themselves have been told how to produce safe energy, for example, producing charcoal briquettes, biogas, the integration between farming, biogas and producing energy, and the solar that we are also using. [...]—IDI—MOS23
Biogas	And, we have the biogas. Because as MOS23, we promote renewable energy use, we encourage our farmers to tap into renewable energy. [...]—IDI—MOS23 So, at the farm, we do biointensive farming... Box gardens, Mandela gardens, and we have the irrigation systems... We use that dung to produce biogas... All purposely for training our farmers—IDI—MOS22
Charcoal briquettes	The women themselves have been told how to produce safe energy, for example, producing charcoal briquettes, biogas...—IDI—MOS23 So, the other thing is the challenge of these alternative livelihoods within the environment. [...] the briquette-making machine itself is expensive, [...]—IDI—MOS25
Energy-saving stoves	... To reduce the impact of deforestation on tree cutting in forests, we also promote the use of energy-saving stoves at the household level [...]—IDI—MOS20
Irrigation systems	... We train farmers how to use a cost-effective irrigation system and support the domestic work—IDI—MOS23
Plastic recycling	... We try to sensitize the community to recycle [plastics]. They turn them into other materials like [what? Complete the code] ...—FGD—MOF2

expansion of biomass energy technologies such as improved cookstoves and biogas digesters, solar energy systems for lighting, water pumping, and refrigeration, mini-hydro power development for off-grid rural electrification, and wind and geothermal exploration. The policy supports private sector participation, feed-in tariffs, and regulatory frameworks to accelerate the adoption of clean, affordable, and sustainable energy solutions across Uganda.

Although organizations have piloted renewable energy systems in both household and community contexts, including refugee settlements, their efforts often remain under-recognised in national adaptation planning. Moreover, prevailing agricultural policies continue to favour industrial inputs over locally appropriate, low-cost solutions, such as biogas and solar technologies (de Brauw/Bulte 2021; van Beusekom 2021). This disconnect underscores the need for stronger policy integration, capacity-building, and long-term investment in community-led renewable energy as a pillar of agroecological transformation. Ultimately, scaling these initiatives requires recognizing that renewable energy is not only a technical fix but a socially embedded tool for climate justice, ecological restoration, and livelihood empowerment.

From the interviews, renewable energy solutions such as biogas systems, charcoal briquettes, and solar-powered irrigation are implemented not merely to reduce carbon emissions, but to transform social realities in rural Uganda. These technologies are embedded in local agroecological knowledge systems, serve marginalized communities, and are driven by collective agency rather than top-down imposition.

Moreover, these results resonate with recent critiques of the technocratic bioeconomy model advanced in existing literature.

In contrast to these extractivist trajectories of the global bioeconomy, our findings reveal that community-led renewable energy systems embedded in agroecology can foster equity, ecological regeneration, and climate resilience. While bioeconomy agendas frame biomass-based innovations as green growth opportunities, literature highlights how large-scale, industry-driven schemes often perpetuate land grabbing, displace smallholders, and divert fertile soils toward export-oriented fuel or bioplastic crops, thus eroding the right to food and deepening ecological debt in the Global South. For example, in Pakistan, smallholder farmers are increasingly marginalized by bioeconomy initiatives promoting biotech seed systems that consolidate control among industrial actors (Zuberi 2024), while similar patterns of hybrid seed proliferation in Pakistan and Tajikistan reflect a broader neoliberalization of seed regimes (Spies 2025). In Africa, large-scale sugarcane expansion serves the interests of state-capital alliances. It reproduces colonial patterns of dispossession (Martiniello 2023), while in the Amazon, agribusiness-driven models increasingly dominate policy, leading to the displacement of indigenous communities and accelerating ecosystem degradation (Costa et al. 2024). More broadly, conceptual critiques demonstrate how the global soy complex exemplifies the bioeconomy's entanglement with global inequalities and extractivist trade structures (Backhouse et al. 2022) and how biofuels exacerbate distant socio-environmental harms, thereby undermining the goals of sustainable energy transitions (Singh 2024). Additionally, new agri-tech and renewable energy deployments have been shown to create regimes of dispossession in both the Global North and South through enclosures facilitated by neoliberal governance models (Stock and Gardezi 2022). Finally, research from Tanzania indicates a disconnect between smallholder worldviews and the narratives embedded in environmental policy, which contributes to their ongoing marginalization (Schopp 2023).

The findings of this study align closely with the concept of integral ecology in *Laudato Si'*, which calls for an integrated approach that links environmental care with social justice, economic equity, and cultural renewal. Pope Francis emphasizes that responses to climate change must restore dignity to the excluded while protecting nature (§139), a vision reflected in the community-led renewable energy practices documented here that not only reduce deforestation but also empower women, youth, and smallholder farmers. These initiatives go beyond technological fixes; they represent a more profound cultural and ethical shift toward stewardship, reciprocity, and ecological citizenship (§211), grounded in local knowledge and participatory governance. As *Laudato Si'* urges respect for diverse ecological cultures and inclusion of marginalized voices (§143), the farmer-led innovations in this study model a real-world form of "ecological conversion"(Francis 2015; Tilche/Nociti 2015).

To align with global climate justice commitments and the principles of *integral ecology* (Francis 2015), the World Bank, International Monetary Fund and International Fund for Agricultural Development should redirect funding and related resources from outdated, extractive energy paradigms to regenerative energy transitions rooted in local agency, circular economies, and food sovereignty. Not all "renewables" are ecologically or socially just. The bioeconomy model, where biomass is utilised as a renewable resource in industrial agriculture, perpetuates extractivist, carbon-intensive, and unequal trade relations, particularly between the Global South and the North. Agroecology activists have a profound critique of the bioeconomy mode (Faucon et al. 2023; Fiore et al. 2024). This critique calls for a more context-sensitive and justice-oriented approach to renewable energy transitions. The bioeconomy as currently structured fails to embody holistic sustainability (Backhouse et al. 2022). Large-scale renewable energy development projects are framed as sustainable and climate-friendly, yet in practice, they reinforce neoliberal, technocratic governance structures that marginalize local communities and undermine democratic control over resources (Stock/Gardezi 2022). Stock and Gardezi show how emerging technologies, even when branded as green or sustainable, can entrench inequality and ecological harm if governance does not prioritize local agency and social inclusion.

By contrast, the renewable-energy interventions documented in this study, such as the solar dryers that preserve household harvests, biogas systems that recycle livestock waste, and energy-saving cookstoves that curb deforestation, do embody the principles of agroecology, solidarity economy, and food sovereignty They are farmer-designed, locally governed, and embedded in circular nutrient flows that regenerate soils while reducing the labour burdens of women and youth. In so doing, they demonstrate that climate-resilient energy transitions need not follow extractive, growth-oriented logics; instead, they can advance agroecology-aligned pathways that redistribute power, centre community agency, and uphold human-rights commitments within agroecological food systems.

Agroecology cannot flourish without community-controlled, affordable, and regenerative energy systems that reinforce food production and rural livelihoods. This is affirmed by this study showing that decentralized renewable energy, like solar irrigation, biogas, and solar dryers, supports not only ecological balance but also farmer agency and gender inclusion in Uganda. Moreover, technocratic energy transitions risk reproducing extractive dynamics unless they are rooted in justice, local knowledge, and democratic governance. Studies have shown that "green" projects, such as large-scale solar farms or lithium mining, often replicate the same displacements and inequalities as fossil-based models (Stock/Gardezi 2022). In contrast, community-centred alternatives can empower local actors and restore ecological integrity (Backhouse et al. 2022). To realise this vision, policy and financing must shift from market-led energy infrastructure to locally governed, regenerative systems that align with agroecological values.

3.5 Conclusion

What's at stake in Uganda's "bittersweet relationships" between renewable energy, integral agroecology, and climate action is the potentially productive tension that lies in how these dimensions interact; sometimes synergistically, sometimes in friction. That is, bridging these domains requires reframing climate action not just as emissions reduction, but as a holistic strategy for livelihoods, ecosystems, and intergenerational equity. This provides an operational clarity, at least, for policymaking and policy uptake.

As such, affordable and clean energy (SDG7) and climate action (SDG13) are operational pathways toward realizing integral ecology, with integral agroecology as a grounded subsect within that vision and enduring commitment. To ensure integral adoption of global transitions toward agroecological and low-carbon systems is not optional; it is foundational to sustainable development.

References

Altieri, M. A., Nicholls, C. I., Henao, A., & Lana, M. A. (2015). Agroecology and the design of climate change-resilient farming systems. *Agronomy for Sustainable Development*, 35(3), 869–890.

Backhouse, M., Lühmann, M., & Tittor, A. (2022). Global Inequalities in the Bioeconomy: Thinking Continuity and Change in View of the Global Soy Complex. *Sustainability (Switzerland), 14*(9). https://doi.org/10.3390/su14095481

Bottazzi, P., & Boillat, S. (2021). Agroecology as a political process: The case of food sovereignty in Africa. *The Journal of Peasant Studies*, 48(1), 20–45.

Braun, V., Clarke, V., Hayfield, N., & Terry, G. (2019). Thematic analysis. In *Handbook of Research Methods in Health Social Sciences*. https://doi.org/10.1007/978-981-10-5251-4_103

CIDSE. (2018). *The Principles of Agroecology: Towards just, resilient and sustainable food systems.* Brussels: International Cooperation for Development and Solidarity.

de Brauw, A., & Bulte, E. (2021). Colonial legacy and agricultural productivity in Sub-Saharan Africa. *World Development*, 138, 105254.

FAO. (2021). *The State of Food and Agriculture 2021: Making agrifood systems more resilient to shocks and stresses.* Rome: Food and Agriculture Organization of the United Nations.

Faucon, Michel-Pierre, Thierry Aussenac, Romain Debref, Stéphane Firmin, David Houben, Elisa Marraccini, Loïc Sauvée, Isabelle Trinsoutrot-Gattin, and Romain Gloaguen. Combining agroecology and bioeconomy to meet the societal challenges of agriculture. Plant and Soil 492, no. 1 (2023): 61–78.

Fiore, Vincenzo, Massimiliano Borrello, Domenico Carlucci, Giacomo Giannoccaro, Simone Russo, Sarah Stempfle, and Luigi Roselli. The socio-economic issues of agroecology: a scoping review. *Agricultural and Food Economics* 12, no. 1 (2024): 16.

Francis. (2015). *Laudato Si': On care for our common home [Encyclical letter]. The Holy See.* https://www.vatican.va/content/francesco/en/encyclicals/documents/papa-francesco_20150524_enciclica-laudato-si.html

Gliessman, S. R. (2013). *Agroecology: The ecology of sustainable food systems* (3rd ed.). Boca Raton, FL: CRC Press.

Gliessman, S. R. (2020). A voice for African agroecology. *Agroecology and Sustainable Food Systems*, 44(7), 825–826.

Isgren, E. (2018). Towards just and sustainable food systems? Smallholder farmer perceptions and responses to agricultural transformation in Uganda. *Journal of Rural Studies*, 63, 128–136.

Isgren, E., & Ness, B. (2017). Agroecology to promote just sustainability transitions: Analysis of a civil society network in Uganda. *Sustainability*, 9(8), 1357.

Kung, K.S. (2013) Low-Cost Production of Charcoal Briquettes from Organic Waste CONTACT. Available at: https://api.semanticscholar.org/CorpusID:165162253.

Stock, R., & Gardezi, M. (2022). Arrays and algorithms: Emerging regimes of dispossession at the frontiers of agrarian technological governance. *Earth System Governance*, *12*(April 2021), 100137. https://doi.org/10.1016/j.esg.2022.100137

Surtia, G. et al. (2016) Improving livelihood initiatives through environmentally friendly solutions derived from livestock byproduct. At: https://api.semanticscholar.org/CorpusID:211102042.

Tilche, A., & Nociti, A. (2015). Laudato Si': The Beauty of Pope Francis' Vision. *FEEM Briefs & FEEM Reports*. https://api.semanticscholar.org/CorpusID:73600242

Tittonell, P. (2020). Assessing resilience and adaptability in agroecological transitions. *Agricultural Systems*, 184(May), 102862.

UNDP. (2024). *Financial Aggregation for Distributed Renewable Energy in Uganda: Market Assessment and Action Plan*. New York, USA. At: https://africaminigrids.org/wp-content/uploads/2024/02/undp-financial-aggregation-for-distributed-renewable-energy-in-uganda.pdf?utm_source=chatgpt.com.

van Beusekom, M. (2021). The history of export agriculture in Africa: A review of key debates. *Journal of Agrarian Change*, 21(1), 3–24.

Wezel, A., Herren, B. G., Kerr, R. B., Barrios, E., Gonçalves, A. L. R., & Sinclair, F. (2020). Agroecological principles and elements and their implications for transitioning to sustainable food systems. *Agronomy for Sustainable Development*, 40(6), 1–13.

Chapter 4
Teaching Radical Hope: Growing Young People with Roots Who Live *Laudato Si'*

Shawn T. Storer

Abstract This piece presents and explores the journey of students at Holy Cross College, Notre Dame, Indiana over the past decade as their hearts, minds and hands are trained and formed to listen and respond in an embodied way to God's call through the school's *Laudato Si'*-informed curriculum that seeks to educate students into an integral ecology and the theological virtues of faith, hope and love and the ecological virtues of that encyclical letter read in continuity with the broader social magisterium of the Catholic Church. Theological portraiture and reflection drawing on the writings of Pope Francis, Wendell Berry, Emmanuel Katongole, Alasdair MacIntyre, Stanley Hauerwas, L. Gregory, Jones, Ellen F. Davis, William T. Cavanaugh others are employed to frame and approach how the question: "Where is Africa then years after *Laudato Si'*?" prompts, inspires and challenges a diverse range of students in this particular context. In the final section, the regenerative work and instructive witness of Bethany Land Institute is given special consideration situated within an exegetical treatment of the story of Jesus' raising of Lazarus in the Gospel according to John.

Keywords Hope · Regenerative transformation · Place · Care

4.1 On Having Imaginations Lighted by the God's Revelation of a New Creation

> "*Found your hope, then, on the ground under your feet.*
> *Your hope of Heaven, let it rest on the ground*
> *underfoot. Be it lighted by the light that falls*

Shawn T. Storer, Holy Cross College, sstorer@hcc-nd.edu.

S. T. Storer (✉)
Holy Cross College, Notre Dame, Indiana, USA
e-mail: sstorer@hcc-nd.edu

R. Marcantonio et al. (eds.), *Integral Ecology in the Anthropocene: Africa and Beyond*, The Anthropocene: Politik—Economics—Society—Science 49,
https://doi.org/10.1007/978-3-032-32125-1_4

freely upon it after…the darkness of our ignorance and madness.
Let it be lighted also by the light that is within you,
which is the light of imagination. By it you see
the likeness of people in other places to yourself
in your place. It lights invariably the need for care
toward other people, other creatures, in other places
as you would ask them for care toward your place and you."
—Wendell Berry[1]

In his remarkable book *Radical Hope: Ethics in the Face of Cultural Devastation*, philosopher Jonathan Lear explores ways in which members of various Native tribes in the Great Plains of North America considered questions of how to practice virtues integral to the character of their ways of life that were vanishing amidst the dramatic social upheaval and ecological destruction of the late 1800s in the face of American expansion and industrialization. Lear focuses most intensely on Plenty Coups, a chief of the Apsáalooke (Crow) tribe as they struggled to persevere as a people into a new way of life on a reservation in present-day Montana.

Lear profiles how some of these native leaders of great integrity decided to 'make' virtues that were crucial to their previous way of life 'small' in order that these could hopefully become seeds that would contribute to the regeneration of new forms of communal life that could only be discerned, discovered and developed incrementally through prudence practiced across generations. In the face of cultural devastation, the tribal members had to consider how virtues essential to who their identity as a people could take alternative forms and differ in both relative and absolute importance due to the dramatically uncontrollable circumstances that they were attempting to navigate together. Lear explores how Plenty Coups and his people seek to courageously—and creatively—live out a vision he is given which requires persevering in the practice of a radical hope that was only possible through the practice of faith. Lear's work is not written simply as a historical explanation; it is a meaningful philosophical exercise undertaken for all who read the book to consider what he recounts and explores can teach us about founding hope and practicing the virtues related to it in these times.

For many years, my wife Stephanie and I lived at a rural, Catholic Jesuit School in a community with the Oglala Lakota descendants of some of those profiled in Lear's book. Many members of these families co-administered, taught at or were otherwise involved in the life of the school. Economically, the region of the Dakotas where we lived was the poorest in the U.S. but was and is remarkable in a multitude of ways, when considered in its wholeness. One large part of this is because of the moral discernment practiced by the people there over the past few generations, people who striving to be, in the language of L. Gregory Jones, "traditioned innovators." (For more on this see Jones' work in, *Christian Social Imagination* and *Navigating the Future* and for more on the illuminating moral practices and virtues of the Lakota way of life, see Alasdair MacIntyre, *Dependent Rational Animals*).

[1] All section epigraphs are extracts from Wendell Berry's longer "Sabbaths 2007, VI", *This Day*, 305–306.

In Pine Ridge, Stephanie and I witnessed first-hand, how—over decades—God has been providentially drawing on elements of the African context to 'water' these 'seeds' to bring to about reconciliation, healing and restoration in the Great Plains of North America. Our work as educators (and bus drivers) with Mahpiya Luta-Red Cloud School touched on many common aspects of secondary school education while also addressing questions of inculturation, Truth & Reconciliation efforts related to the boarding school, as well as community health, economics and food sovereignty questions. (Providentially, it was during this time that some of working at the school began reading Father Emmanuel Katongole's books *Beyond Universal Reason*, *African Theology Today* and *A Future for Africa* to inform our discernment about the next chapter of the life of the school). Some of this work brought us into direct cooperation with an organization called Inter-Tribal Buffalo Council which started in the 1990s. Through a host of moves in the nineteenth Century, the U.S. Government attempted to weaken or eradicate Native peoples and tribal ways of life by destroying—almost to the point of extinction—the keystone species upon which those people, ecosystems and patterns of life depended—the Buffalo (or North American Bison). Our humble interactions with the work of this council consisted of working with the school in its relations with the tribal buffalo herd and with some families restoring bison to considerable family lands alongside the tribal herd.

As a child in the late twentieth Century, one of the founders of that council, Jason Baldes of the neighboring Eastern Shoshone tribe, lamented to his father that his people could not hunt the buffalo anymore. "Why was this so?" he asked. (Julianna Brannum, *Homecoming*). Rather than simply tell him stories about the past, Baldes' father, a trained biologist, decided to take his young son with him on a research trip to study ecosystems in Uganda as well as to track elephants in Kenya and Tanzania. While doing so, they witnessed the beauty of the great wildebeest migration through the Masai Mara plains. Baldes and his father saw over 1.5 million of these animals on the move over hundreds of miles. Beholding this, Baldes realized that this multitude was a mere fraction of the number of bison that once resided across the middle of North America. Baldes' father ennobled his son to become a traditioned innovator. Baldes said this was an 'epiphany' for him and that he returned to Wind River grateful—"with a newfound appreciation for his home, his people and community." Baldes decided to study science to begin focusing on devoting the rest of his life to the slow work of buffalo restoration.

Baldes' imagination of space and time was expanded, inspired and disciplined by this 'epiphany,' and it calls to mind Pope Francis' observation, "The world sings of an infinite Love: how can we fail to care for it?" (*LD*, #65). Baldes began to envision how to practically work for common goods that he did not yet see or that did not seem possible before these encounters. Baldes also began to envision the possibility of a different kind of intergenerational and intertribal politics and his role as a traditioned innovator in his place. He began to imagine the art of the possible and to practice hope and other virtues ordered toward what we have been calling "integral ecology" in a new way.

Now, the Black Hills-based council, includes 86 member tribes (many of whom were, at one time, historical enemies) caretaking 20 million bison and 32 million acres across 20 different vast U.S. states—from large intact grasslands of Montana to small desert herds of New Mexico. Its members are responsible for the caretaking of more than 32 million acres of Tribal lands and have restored and protected more than 20,000 buffalo on nearly one million of those acres and Numerous Community Health and Development and Food Sovereignty programs have developed across many parts of Indian Country because of and alongside this work across various tribal homelands. These tribal members are building on the radical hope and virtues made small by their predecessors and often historically supported by parts of the Church so that these could regenerate in new ways in a different social reality that might at one time have seemed unimaginable to most.

A Christian reading of history sees Divine Providence at work here both in the broad strokes that stretch across generations and around the earth but also in the fine details of all the particular lives that find a part in this story. It is a wonder to behold God's reconciling, restoring and regenerating activity across time and continents here, but also in doing so in and through the particular lives of people and creatures. As Emmanuel Katongole and Chris Rice note, God has ordained it that "Creation takes time." (Katongole and Rice, 58) An example like this shows us that it pleases God that the New Creation should take time as well. What is recounted here is a but a brief preface to the cooperation with God of which we are now a part at Holy Cross College in Notre Dame, Indiana. While we have certainly "saved seeds of hope" from this chapter of life and brought them to sow in our current context, we have also come to a greater understanding of the ways God is tending his Garden across time and how to train students to learn to see this.

4.2 On Sowing Hope and Cultivating Young People with Roots

"It is hard to have hope…
for hope must not depend on feeling good…
…withdrawn belief in the present reality
of the future…
…hope is harder when it cannot come by prediction
any more than by wishing…stop dithering.
The young ask the old to hope. What will you tell them?
Tell them at least what you say to yourself."
—Wendell Berry

Over the past decade, Holy Cross College—the small, teaching college where Stephanie and I now work—has orchestrated much of its integrated curriculum as a response to Pope Francis' teachings, especially *Laudato Si'*, locating it in continuity with the Church's broader magisterium across the ages with special attention to Vatican II and papal teachings since that ecumenical council. The entire faculty of the college has been engaged in a conversation about how to strive as educators in the faith to begin to answer the summons of Chapter Six of *Laudato Si'* as well as the spiritualities of *Dilexit Nos* and our founding religious order: the Congregation of Holy Cross. Our goal is to form and integrate the students who come to us so that we are growing what Pope Francis called "young people with roots" (Francis, *Christus Vivit*, Chap. 6). We follow John Henry Cardinal Newman in understanding that a college is not simply for intellectual instruction for a profession or a credential, but exists as a durable, nurturing, dynamic place intended to provide stability for the shaping and transformation of character for the students' life-long flourishing (See Newman, *Rise and Progress of Universities).* This sort of edification is one that is concerned with cultivating and synthesizing the mind, the heart and the hands of our students (as Pope Francis so often put it echoing the sentiments of the Congregation of Holy Cross founder Blessed Basil Moreau). Our college's approach takes seriously that students need to be mentored and trained in the acquisition of *techne* for a profession but also a deep and profound engagement with *sophia* that will enable them to learn to listen for, discern their vocation from God and respond to this personal call. (For more on this see Christopher J. Thompson's *The Joyful Mystery*). This is especially crucial as so many are increasingly cut off from traditional sources of wisdom both through the breakdown of intergenerational relationships and the rise of addictive attention-capturing devices and platforms that tend to focus only on the shallow and the immediate, distracting or cutting the young off from more shared, embodied relationship in and across time as well as the deep wells of traditional human wisdom handed on from generation to generation. The social teachings of Saint Pope John Paul II, Pope Benedict XVI and Pope Francis all repeatedly sound great concern for and lament this alarming trend. More specifically, Pope Francis begins "Ecological Education and Spirituality"—the chapter of *Laudato Si'* that is foundational and orienting for our entire college curriculum—by underscoring that:

Many things have to change course, but it is we human beings above all who need to change. We lack an awareness of our common origin, of our mutual belonging, and of a future to be shared with everyone. This basic awareness would enable the development of new convictions, attitudes and forms of life. A great cultural, spiritual and educational challenge stands before us, and it will demand that we set out on the long path of renewal. (*LS*, #202).

My classes at Holy Cross, and an annual co-taught, shared integral ecology course with our larger Congregation of Holy Cross sister schools—the University of Notre Dame and Saint Mary's College—are often comprised of a mixture of local students from the geographic region and community our college is in—the largely post-industrial cities and towns and strained rural communities of the Midwestern/Great Lakes of North America—*and* students from many other places around the world including communities across Africa and other parts of the Global South. A considerable number of our students are migrants or the children of migrants and many of our students are first-generation college students. Only a select few of our incoming students have heard of, let alone, read parts of *Laudato Si'* before they join us each year, usually because they have attended Catholic Secondary Schools or because they have participated in summer programming with our college as high school students which introduced them to Pope Francis' letter. For most of our students, before we can consider the question at the heart of this conference: "Where is Africa Ten Years After *Laudato Si'*?", it must first be acknowledged that most of our incoming students would likely first ask, "What is *Laudato Si'*? Just as when we begin reading a book like Lear's *Radical Hope* or consider examples from places like Pine Ridge, students will commonly ask, at least initially, "What does this have to do with me?" While some of the less individualistic students might ask, "Why are we reading this in a theology course?" Such questions reveal how little sense most have of how God moves and acts in and through history and their own lives.

Our approach to reading *Laudato Si'* at Holy Cross is informed by theologian William Cavanaugh's guidance on reading the Church's social tradition and is aimed at anticipating and addressing such questions. At this stage, when most students read *Caritas in Veritate, Laudato Si, Laudate Deum, Fratelli Tutti* or other such encyclicals, they rightfully have their doubts about these somehow being applied to and transforming the idolatrous structures and exploitative and extractive systems challenged in those documents. Cavanaugh's noting that it is "easy to fault papal encyclicals for their generality," helps the students to begin to consider what the telos of this letter actually is. Encyclical letters often contain beautiful and important exhortations and diagnoses but are written so generally in hopes of communicating with every member of the Church, or, in the case of *Laudato Si'*—every person on the planet -that when a host of theologians, scholars, journalists, lobbyists and pundits poised and ready to distill down each letter into a short list of principles, pillars, key takeaways or action points the day the letter is shared this only exacerbates this problem. Francis often lamented that *Laudato Si'* was reduced to or dismissed as a "green encyclical" rather than being considered for the "social encyclical" it was intended to be. Cavanaugh posits that encyclicals "are not meant to be blueprints for the reconstruction of the world order"; rather, these "open up different ways of imagining the world" (Cavanaugh, *Field Hospital*, 121) and help us to seriously consider the principle of subsidiarity and conceive of new forms of "dispersed political authority" that might offer meaningful alternatives to the systems within the fusion of the modern state and capitalist markets that such these documents often suggest need subverting. (Cavanaugh, *Field Hospital*, 121–126). The Church's social tradition is first about the formation of people's consciences in local places around God's world—it is a learning to "know with" God and God's Church and to begin to discern how to respond in our particular communities in God's Creation. It is an exercise in what Cavanaugh calls the shaping of "disciplined imaginations of time and space" (Cavanaugh, *Theopolitical Imagination*, 1)—a learning to become traditioned-innovators imagining communities together with God. In the examples of the Crow, Lakota and Eastern Shoshone since the 1800s, the students begin to discover a human connection between their striving to faithfully discern how to live ethically in the wake cultural devastation and our struggle now after assessing Francis' assessment of the past two hundred years of industrialization and the rise of godless technocracy. They also begin realize they are being called to attend to a place and be a part of a community.

Since *Laudato Si's* promulgation in 2015, every student at our college has been presented with at least parts of it along with many other of Francis' writings (not something to be taken for granted in the U.S. context, even in Catholic settings), with many our students reading the encyclical in its entirety, some multiple times. It is striking how *Laudato Si'* and many of Francis' other writings intimately articulate the half-formed questions and concerns in the hearts of so many of our students. When I began teaching at Holy Cross in 2017, most of the students in my classes were somewhat concerned with ecological degradation and the prospects of a shared, livable future for everyone, with a vocal minority denying that this, especially climate instability, was much of an issue. Most, at that time, however, were rather enamored with the promises of new digital technologies. Now, those in our classrooms have lived much of their lives thus far with these concerns, at least in the background of their daily existence, with others arriving at our college with a more pronounced anxiety or sense of fear that often arcs toward despair about such matters. They are also coming now with a much more critical view of emerging technologies than their predecessors tended to hold—these new students having lived through what Jonathan Haidt calls the "Great Rewiring" of childhood. (Haidt, *The Anxious Generation*). It is not uncommon to hear students realize and lament having 'lost' their childhood to their phones and describe the isolation of feeling "scrolled past" by others who often chose to give their attention to a device rather than another human being in their presence much of the time. Conversation in the classroom—an increasingly disappearing art that must also be taught at this stage—usually resonates profoundly with Pope Francis' assessment that "[y]oung people demand change. They wonder how anyone can claim to be building a better future without thinking of the environmental crisis and the sufferings of the excluded," (*LS*, #13) and again when Francis notes that so many people "no longer seem to believe in a happy future; they no longer have blind trust in a better tomorrow based on the present state of the world and our technical abilities. (*LS*, #113). As our students read *Laudato Si'*, they begin to wrestle with fundamental questions, particularly around what Pope Francis calls the ecological virtues—especially the theological virtue of hope. Many students ask, "What is hope?" and "Show us why we should hope?" While others ask, "Why bother?"-succumbing to the deadening indifference that Pope Francis so often warned is being globalized throughout the world in our time. They begin to understand that the Enlightenment and capitalist language of 'values' will not be enough to address their concerns and that many of the systems and institutions in which we are embedded are not ordered toward God. The students awaken to the stark reality that, as Pope Francis observes, "Humanity has changed profoundly, and the accumulation of constant novelties exalts a superficiality which pulls us in one direction. It becomes difficult to pause and recover depth in life." (*LS*, #113). Most of the students begin to see that seeking such depth will require more than claiming to vaguely assent to some theoretical values and that, for almost all of them, they have only lived after this change. The majority of our students begin to realize that an education into the ecological virtues of *Laudato Si'* is necessarily embodied and will require sacrifice, discipline, courage, creativity, experimentation, perseverance and community. There will be no progress in integral ecology without this. Sadly, some see this as too daunting to take on and live into at this stage of their lives—often citing that it is the widespread, entrenched and uninspiring example of older generations that contributes to this decision.[2] (There have been cases of some who return years after graduation who say their minds were changed about this often, interestingly enough, after having children). Many of our students discover a turning point when we study farmer and writer Wendell Berry's dismissal of asking if we have time to address the sorts of problems that Pope Francis names in *Laudato Si*. Berry warns, "We don't have a right to ask that question. We have to ask what's the right thing to do and go ahead and do it and take no thought for the morrow." (Moyers & Company, *Wendell Berry: Poet & Prophet*). This insight "clears the brush" for the students in a way akin to Ludwig Wittengenstein's exhortation: "Don't think, but look" (Wittgenstein, *Philosophical Investigations*, #66). Berry's changing of the question and opening another path often leads students toward a virtue ethic that inspires some to persevere further. Rather than languishing and succumbing to despair, the students begin to resist the temptation to the vice of acedia amidst what seems like impossibly large problems and find a deeper wellspring to draw upon. Berry's words help them immensely with this:

> "[W]e've acknowledged that the problems are big, now where's the big solution? When you ask the question: 'what is the big answer?', then you're implying that we can impose the answer. But that's the problem we're in to start with, we've tried to impose the answers. The answers will come not from walking up to your farm and saying, 'This is what I want ...' and 'This is what I expect from you...' You walk up and you say, 'What do you need?' And you commit yourself to say, 'All right, I'm not going to do any extensive damage here until I know what it is that you are asking of me.' And this can't be hurried. This is the dreadful situation that young people are in. I think of them, and I say well, the situation you're in now is a situation that's going to call for a lot of patience. And to be patient in an emergency is a terrible trial. (Moyers & Company, *Wendell Berry: Poet & Prophet*)

Like a modern-day Evagrius of Pontus, Berry's charge to practice zeal patiently helps the students shake off temptations to acedia and discover a new dimension of what it means to practice peace.

A smaller strand of students begins to realize that this sort of questioning leads to a deeper and more radical form of *metanoia*. They begin to ponder Pope Francis' very Augustinian insights in *Laudato Si'* and *Laudate Deum* that we must embrace a humility that acknowledges God as the center of the Cosmos and Lord of All Creation if we are to begin to begin to order the chaos parts of humanity have unleashed especially over these past two centuries. These students discover that, as Pope Francis notes at the end of *Dilexit Nos*, that the teaching of the social encyclical *Laudato Si'* "is not unrelated to our encounter with the love of Jesus Christ. For it is by drinking of that same love that we become capable of forging bonds of fraternity, of recognizing the dignity of each human being, and of working together to care for our common home." (*DN*, #217). These students begin to hear the words of Jesus,

---

[2] A non-exhaustive list of ecological virtues in *Laudato Si'* includes praise, faith, hope, love, awe and wonder, respect, attentiveness, generosity, gratitude, intergenerational justice and solidarity, perseverance, sobriety, humility, patience, delight, compassion, simplicity, musicality; joy, courage, and prudence (practical wisdom).

Genesis, the Psalmist, the Prophets, St. Paul and St. John the Revelator with new ears and even consider our college's oft-repeated motto "Ave Crux Spes Unica (Hail the Cross—Our Only Hope)" in a new light. They begin to wrestle with and understand hope not as another 'feeling' in an emotivist culture or as a kind of naïve optimism, but as a necessary discipline and practice for making our way together through God's world in light of Eternity. If we look to the Scriptures for insights on this we find that practice of authentic hope is challenging and marked by a tension of being drawn (the Hebrew evokes the language of being pulled taut like a measuring line), into a position of asserting a boundless hope in God who rescues and reconciles even when we are faced in danger, surrounded by the threat of rapacious, beast-like enemies (see Psalm 27), the practice of the biblical/theological virtue of hope trains us in a constellation of other virtues.[3] In light of what God reveals to His people, we are led to learn to practice a peace and patience and to: "Wait for the LORD, to take courage; be stouthearted, wait for the LORD!" In this psalm and in the practice of the Christian way of life, this becomes intimately tied up with the practice of trust and faith in God.

Here we encounter serious obstacles, as so many of the students from North America and Europe have consistently breathed the air of unbelief and have imbibed the modern assumption that real and meaningful transformation is not really possible (see Charles Taylor, *A Secular Age*). Even though most of our students have been around elements of Christianity often they have been formed either by what Charles Wesley, first, and Stanley Hauerwas, William Willimon and L. Gregory Jones, in our time, call "practical atheism." Others are adherents to what Christian Smith dubs "moral therapeutic deism" (Smith & Denton, *Soul Searching*). What is decried in these is a crisis of genuine, practiced faith. This is not far from what Pope Francis says in *Laudate Deum* when he observes, with the passage of eight years of time since *Laudato Si'*, "our responses have not been adequate, while the world in which we live is collapsing and may be nearing the breaking point." (*LD*, #2). Dan DiLeo notes that the *Catechism of the Catholic Church* describes faith as the "adequate response" to God's love (*CCC*, 142) and that Pope Francis is naming a crisis or lack of practice of faith on the part of so many of us in responding to God's Love, the cry of the earth and the cry of the poor. (DiLeo, "Earth Day 2024").

When we read *Laudato Si'* in common at Holy Cross College asking the question: "Where is Africa in relation to this encyclical letter?" also prompts and lands in a myriad of ways. While some students are striving to tell stories from different parts of the African context, others are attempting to simply begin to imagine life in African places, let alone realize the significance these examples might have to offer. However, when our students consider the theological portraiture in Emmanuel Katongole's work they begin to believe in and appreciate more how God works in our lives. When they hear Angelina Atayam, proclaim even amidst real pain and struggle that, "God wants us to be practical Christians," (Bynasse, "Perilous Presence") or Maggy Barankitse say in the midst of daunting undertakings, "God is God. He will

[3] I am indebted to Scripture scholar and agrarian theologian, Ellen F. Davis for sharing this insight with me.

help me," (Katongole, "Who are my People?" [Lecture]) an awakening takes place. Generally, our students are challenged and impressed by the joy and seriousness they find in these and other witnesses. The students begin to see the ways in which they have much to learn from the practical holiness of such figures.

This also contributes to the ongoing growth of our African students in a notable way during their time in studies with us. When considering *Laudato Si'*, one particular student, Nicholet Zabuna, comes to mind. Nicholet hails from Uganda. Active in the Laudato Youth Initiative as a student at a Holy Cross High School, she is one of the few students I have taught who had some developed sense of integral ecology upon starting at our college. In considering such theological portraits and reading Pope Francis' teachings in class, Nicholet has consistently voiced her reflections on *Laudato Si'* in ways that expand her fellow students' understanding of the strengths and difficulties of her homeland in a measured and honest way that help students from the U.S. and other places to better understand how the forces of modernity and the technocratic paradigm shape their lives and the cultures of our campus and their home places. Nicholet's emergence as a leader in the classroom, has contributed to her becoming a mentor of secondary school students studying integral ecology at our college's summer theology institute. Nicholet came to us considering the nursing profession but as her sense of care has broadened and she has helped to found our Common Home student group and with time has assumed leadership of it. She will also be the first to complete our college's integral ecology academic track. With confident humility, Nicholet has become a transformative speaker giving impassioned addresses to students on our campus and at our neighboring sister schools. She has allowed *Laudato Si'* to shape her heart and has become a sort of missionary to her peers and others. With directness, humor and even music, she patiently, courageously, and joyfully challenges her classmates to begin to see, or to see more deeply, the ways that their lives and communities are so often malformed and distorted by the problems that Pope Francis names in his letter. Nicholet has invited her fellow students to begin to unlearn some of the ways they have been shaped by the abstractions of global and national thinking and to begin disciplining their imaginations to begin to think more 'interlocally.' She helps them to imagine a bond between South Bend and Jinja. At the conclusion of Emmanuel Katongole's lecture, "Where are My People?" he shares the witness of several holy exemplars from different parts of Africa, concluded by directing people to do the work of finding stories and communities of love, compassion and tenderness in their places that can allow them to share in the holy work God is doing in their places. Similarly, in the way Nicholet shares her life, stories and insights with others, she invite her peers to ponder the different ways "Where is Africa?" can be answered but also to move beyond that question to consider how the life of their family, neighborhood, town or city is both like and unlike the life of her village and what we can learn from this exercise. She also challenges people to wonder about how we can better think with the Church and *Laudato Si'* specifically about making common cause with the place where we all live and work each day instead of simply approaching it with such a thoughtless, extractive and wasteful mindset.

As the students move together beyond the realm of optimism and pessimism and begin to realize that the seeds of faith and hope and the other ecological virtues that have been sown in their hearts by God will ask for a different sort of response and require an openness to allowing God to transform their imaginations about the places in the world that they inhabit, a different set of questions arises for them. They begin to discover that the words of *Laudato Si'* #202 are words written directly to each and every one of us reading those together in our classroom and they want to know: what are the next steps for *us*, *here*, on this "long path of renewal?" They ask: how do we practice the ecological virtues of *Laudato Si'* in a manner akin to the Crow example in *Radical Hope* in order to heal our common home and regenerate society?

4.3 On Learning to Belong and to Care for a Place

"*Because we have not made our lives to fit*
our places, the forests are ruined, the fields eroded,
the streams polluted, the mountains overturned. Hope
then to belong to your place by your own knowledge
of what it is that no other place is, and by
your caring for it as you care for no other place, this
place that you belong to though it is not yours,
for it was from the beginning and will be to the end.

Belong to your place by knowledge of the others who are
your neighbors in it..."
—*Wendell Berry*

With each year, the students who come to us are increasingly suspicious of and isolated by illusions of limitlessness and destructive ideologies that seem inescapable. It used to be, in the late 2010s, that this was much more of an educational concern for professors, as the students were, so-often, trapped by these. In recent years, though, comments by a broad cross section of our students from differing parts of the world, more readily echo what Pope Francis characterizes in *Dilexit Nos*, #218 as the captivity of a demeaning world system that distracts our gaze and stifles our personal and collective imaginations. From the incoming first-year students to our graduating seniors, there is a near-universal frustration with this. Our students are willing to admit and clearly name this in a way that their predecessors were largely unable to in the past. While some are still not yet willing or ready to accept that even in a fallen world, as Emmanuel Katongole and Chris Rice write, "the way things are is not the way things have to be." (Katongole & Rice, *Reconciling All Things*), there is, when we read *Laudato Si'* together, an unprecedented consensus about Pope Francis' assessment of the way things. At Holy Cross, our professors and other formators seek ways to *show* the students (and not simply tell them) that in the face of the ways of the worldly, that as Francis continues in *Dilexit Nos*:" The love of Christ has no place in this perverse mechanism, yet only that love can set us free from a mad pursuit that no longer has room for a gratuitous love. Christ's love can give a heart to our world and revive love wherever we think that the ability to love has been definitively lost. (*DN*, #218).

This 'showing' involves the training of attention with an emphasis on its relationship to awe, wonder and ultimately generosity. This involves understanding both study and work beyond the classroom, as Simone Weil puts it, "with a view toward the love of God." (see Weil, "On the Right Use of School Studies").

As the students progress through the early stages of our integrated Catholic curriculum, they are presented with the Greek origins of the word 'Catholic' (*kat'holon*) and its meaning of attending to the whole and "home economy/household" (*oikos*). The language of "care for our common home" and "integral ecology" ring differently as they are led beyond the classroom and the campus on excursions to the distinct and diverse ecosystems of our region—the area around the southern part of Lake Michigan—where in the late 19th and early 20th centuries concern and appreciation for this biodiversity amidst rapid and drastic industrialization gave rise to the pioneering of the contemporary academic discipline of ecology. They encounter both the Gospel and Grammar of God's Creation on these site visits while also discover the cry of the earth and the cry of the poor (*LS*, #49) in and around our local region at farms, urban slums, lakes, dunes, nursing homes, rivers, and landfills, schools, food pantries and rescue centers, refugee/migrant education organizations and Catholic Worker apostolates around Saint Joseph County. Many of these become practicum host sites where our students apprentice under mentors and are practically educated into some of the virtues. These site visits and apprenticeships are often crucial in building on the witness of those aforementioned exemplars we study in class and capacitating our students for finding authentic grounds for practicing hope. For as Berry says, "If you can find one good example, then you've got the grounds for hope. If you can change yourself, if you can make certain requirements of yourself that you are then able to fulfill, you have a reason for hope." (Moyers & Company, "Wendell Berry: Poet & Prophet").

With each year that passes, the way digital technologies have drawn people's attention away from the places they inhabit and the people and creatures of those places—it is the case that whether the students born and raised in the region of our college or whose families have recently migrated to this region or others in the U.S., or those who travel from half the world away—all require an introduction to the history of this community and landscapes of this place due to the dislocation and destructive hyper-mobility that has not only become a hallmark of life in this society and many others, but that is also, unfortunately exacerbated by the institutions of higher education. This has prompted our college to take seriously Berry's invitation to ponder what might "an education for homecoming" look like? How do we awaken a desire and sense of responsibility to becoming local to whatever place one is given and instill in them sense of concern as well as the habits and capacities to seek to make what Berry calls "common cause" with whatever place they settle in after they are sent forth from the training grounds of our campus be that where they were raised, here or a new community. Put differently, how do we together cultivate "virtues of place"? (See Baker and Bilbro, *Wendell Berry and Higher Education*). Students are often surprised when they encounter such questions in the classroom and appreciate the space that is made to consider what it means to discover, receive and co-author their life with God rather than simply ignore those questions only to have it haunt them later.

Often, the students who are most responsive to Pope Francis' words have been inundated and overwhelmed by information for much of their lives (frequently removed and detached from the framing and guidance of personal relationships), but their hearts are longing for true wisdom and mentorship. Many of our students arrive at our college prepared to ask and even attempt to define for themselves the answer to the question: "Who am I?" but as they come to know better that the Love that is God is creating, sustaining and reconciling them each is communicating to them through God's Creation and that the question to which they are really seeking an answer to is "Whose am I?" It is when they realize they are also yearning for what Frederick Christian Bauerschmidt distinguishes as the crucified "Love that is God" (Bauerschmidt, *The Love that is God*), that what seems most daunting about the challenging messages of *Laudato Si'* and *Laudate Deum* seems to become less of an insurmountable and impossible wall and more of an arduous path, but a path to be journeyed upon with God and others. Once the students begin to understand that our ultimate common good is God, then they become liberated and empowered to *discover* a diversity of common goods of which they are called by God to devote themselves in the drama of their day-to-day lives in their home places.[4] The witness of the saints and virtuous exemplars as well as the loving accompaniment and guidance provided by teachers and local mentors inspire them to take the next steps in radically practicing faith, hope and love—where we live.

4.4 On Believing and Seeing the Glory of God

> "*…The world*
> *is no better than its places. Its places at last*
> *are no better than their people while their people*
> *continue in them. When the people make*
> *dark the light within them, the world darkens.*"
> —*Wendell Berry*

In my classes, we begin with developing attention, awe and wonder, before the ultimate and uncreated Common Good—God. Then we explore common goods all the way down to towns, villages, neighborhoods, families, schools, watersheds, parishes,

[4] The mystic, St. Thomas Aquinas writes on God as our ultimate telos and common good and divine goodness as the end for all things in *Summa Theologiae* I, q. 44, a. 4.

workplaces, athletic teams, musical ensembles and even roommate arrangements—where and with whom students share their lives. We seek understanding about the God to Whom we belong, how the world is and how our lives and local places might be reconciled if we respond together to God's summons and grace. The students often need exemplars from the cloud of witnesses to encourage this imaginative growth and make the integral ecology of *Laudato Si'* and what they study in other parts of the Church's social tradition come alive.

Certainly, Bethany Land Institute is one such embodied example of educating people into the ecological virtues that inspires and helps them to understand that such virtues are lived by people in particular places and in relationships in concert with God. The students are impressed by the mission, the practice of integral ecology and the fruits of BLI, but many begin to see that it is its religious rooting that is the 'seedbed' for this. Even the name is charged with this. In the Gospel, when Lazarus is deathly ill, Jesus' disciples do not want to face difficulty and the prospect of death. They plead to go back to Judea rather than to Bethany. Jesus says to his followers, "Are there not twelve hours in a day? If one walks during the day, he does not stumble, because he sees the light of this world. But if one walks at night, he stumbles, because the light is not in him." (Jn 11:9–10). Jesus tells his fearful followers, "Our friend Lazarus is asleep, but I am going to awaken him." (Jn 11:11) and assures Lazarus' sister Martha, grieving and frustrated that Jesus had not come sooner, "Your brother will rise." (Jn 11:23). In this story, we encounter Jesus' weeping in sorrow, snorting in anger at the evil of death, and most importantly his love for Lazarus and those mourning him. In his intimate conversation with Martha, we behold the Lord stretching Martha and capacitating her for greater faith. "I know he will rise, in the resurrection on the last day," Martha speaks through her grief. (Jn 11:24). Jesus then begins to prepare her to have a faith capable of practicing an even more radical hope. Jesus says: "I am the resurrection and the life; whoever believes in me, even if he dies, will live, and everyone who lives and believes in me will never die." He then asks Martha, "Do you believe this?" (Jn 11: 25–26). Though Martha is able to reply, "Yes, Lord. I have come to believe that you are the Messiah, the Son of God, the one who is coming into the world …," (Jn 11:27), she cannot *yet* imagine what Jesus will do when he has for the stone from her brother's tomb to be taken away and raises concerns about doing so. Before calling out to his father in a voice that will eventually sound out to all tombs, Jesus asks Martha, "Did I not tell you that if you believe you will see the glory of God?" (Jn 11:40). This 'glory' is not a mere fleeting splendor; it is rather a communication, an in-breaking, an epiphany of divine presence, mercy and love.

Jesus cryptic words about light could be taken by some as an example of ancient peoples not understanding how light and the human eyes. Berry's poetic language invites a more theological and spiritual reading of this. Elsewhere in the Gospel According to John, Jesus says to his followers, "I am the light of the world. Whoever follows me will not walk in darkness, but will have the light of life." (Jn 8:12) and that one day His followers will come to "realize that I am in my Father and you are in me and I in you." (Jn 14:20). Jesus exhorts his followers/friends not to fear or fall prey to the illusory lie that we are orphaned, instead he promises the consolation and steadfast advocacy of the Holy Spirit until the heavens and the earth are renewed and allows His light to live in each of us and be a part of the slow, reconciling work of God's making all creation new in Christ, in and through each *local* home place we belong to, inhabit and care for—for as the Lakota Catechist the Servant of God Nicholas Black Elk often spoke: "Everywhere is the center of the world."

In his book *The Need to Be Whole*, Wendell Berry narrates the many ways in which those in government and electoral politics, journalism, and educational systems from Higher Education to local school districts seem to be extinguishing the light within them to the detriment of everyone. He eloquently and plainly explains his reasons for frustration with the failure of the food movement the U.S. to be able to bring about significant cultural transformation and problems inherent in our educational systems that do not suggest the capacity to take seriously what is really involved with caring for the whole of creation and the fostering of genuinely communities that can hope to hand on a livable future to their descendants. Berry highlights the need for fostering "local adaptation" and the handing on of this to be built upon in distinct places and regions so as to "resettle America" regenerating local culture and agriculture in a variety of local settings over generations. Berry argues that such a project could not be done in of the "great universities" of this time because of where their concerns would lie, he proposes the creation of a sort of "small, poor, humble university, a ramped-up old teachers' college in one of the underpopulated states of rural America." (Berry, 425). What Berry sketches out imaginatively bears a strong resemblance to Bethany Land Institute embodies and is cultivating in practice in Nandere. The witness of those with Bethany Land Institute raise important questions for those not only in contemporary Uganda but also in the United States. When my students study BLI they frequently sound like a young Jason Baldes posing questions to his father. A chorus of questions cascades after each class reads about BLI: "Why not here? What happened here that makes this seem so far out of reach? What could it look like here? Where do I find others who care enough to begin to try?" At the Ten Years After *Laudato Si'*: Where is Africa? Conference at BLI participants encountered the kind of education that Bl. Basil Moreau, CSC called a "work of resurrection" which offered a beautiful glimpse of the glory of God. I heard many other participants and fellow pilgrims from around Africa and other parts of our common home inspired to ask the question similar to those posed by my students, as they prayed at Mass, shared meals together, hiked through Lazarus' Forest, toured Mary's Farm and heard word of Martha's Market Cooperative, guided by BLI Faculty and Caretakers who in founding their hope of heaven on the ground underfoot have shared a light that allows others to find the grounds on which they can begin to further imagine and practice integral ecology and a more radical and authentic form of hope in their home places.

References

Bauerschmidt, F.C. (2020) *The Love that is God: An Invitation to Christian Faith.* 1st edn. Grand Rapids: Wm. B. Eerdmans Publishing Company.

Berry, W.E. (2013) *This Day: Collected and New Sabbath Poems.* 1st edn. San Francisco: Counterpoint Press.

Berry, W.E. (2022) *The Need to be Whole: Patriotism and the History of Prejudice.* 1st edn. Berkeley: Shoemaker & Company.

Baker, J. R. and Bilbro, J. (2017) *Wendell Berry and Higher Education: Cultivating Virtues of Place (Culture Of The Land)*: 1st edn. Lexington: University Press of Kentucky.

Brannum, J. (2023) *Homecoming.* Florentine Films. https://www.pbs.org/kenburns/the-american-buffalo/homecoming.

Byasee, J. (2009) "Perilous Presence: Christians in Uganda." *The Christian Century.* February 10th Issue. https://www.christiancentury.org/article/2009-02/perilous-presence.

Catechism of the Catholic Church (1993). 1st. edn. Citta del Vaticano, Libreria Editrice Vaticana.

Cavanaugh, W.T. (2003) *Theopolitical Imagination: Christian Practices of Space and Time: Discovering Liturgy as a Political Act in an Age of Global Consumerism.* 1st edn. London: T & T Clark.

Cavanaugh, W.T. (2016) *Field Hospital: The Church's Engagement with a Wounded World.* 1st edn. Grand Rapids: Wm. B. Eerdmans Publishing Company.

DiLeo, D. (2024) Earth Day 2024: It's time for the Catholic Church to give up fossil fuels for good." *America Magazine.* April 19, 2024. https://www.americamagazine.org/politics-society/2024/04/19/earthday-2024-diocese-decarbonization-laudato-si-247709/

Francis (2015) *Laudato Si'*

Francis (2019) *Christus Vivit*

Francis (2023) *Laudate Deum*

Francis (2024) *Dilexit Nos*

Haidt, J. (2024) *The Anxious Generation: How the Great Rewiring of Childhood Is Causing an Epidemic of Mental Illness.* 1st edn. New York: Penguin Press.

Hauerwas, S and Willimon, W. (2014) *Resident Aliens.* 2nd edn. Nashville: Abingdon Press

Jones, L.G. (2016) *Christian Social Innovation.*1st edn. Nashville: Abingdon Press

Jones, L.G. and Hogue, A.P. (2021) *Navigating the Future: Traditioned Innovation for Wilder Seas.* 1st edn. Nashville: Abingdon Press.

Katongole, E. (2000) *Beyond Universal Reason: The Relation between Religion and Ethics in the Work of Stanley Hauerwas.* 1st Edition. Notre Dame: University of Notre Dame Press.

Katongole, E., ed. (2002) *African Theology Today.* 1st edn. Scranton: University of Scranton Press.

Katongole, E. (2005) *A Future for Africa: Critical Essays in Christian Social Imagination.* 1st edn. Scranton: University of Scranton Press.

Katongole, E. and Rice C. (2008) *Reconciling All Things: A Christian Vision for Justice, Peace and Healing.* 1st edn. Lisle, IL: Intervarsity Press.

Katongole, E. (2017) Who are My People?: Christianity, Violence, and Identity Politics in Sub-Saharan Africa. (Lecture); https://www.youtube.com/watch?v=3GtUXbxHSMc.

Lear, J. (2006/2008), *Radical Hope: Ethics in the Face of Cultural Devastation.* 1st paperback edn. Cambridge: Harvard University Press.

MacIntyre, A.C. (1994) *Three Rival Versions of Moral Enquiry: Encyclopaedia, Genealogy, and Tradition.* 1st edn. Notre Dame: University of Notre Dame Press.

MacIntyre A.C. (2001) *Dependent Rational Animals: Why Human Beings Need the Virtues.* 1st edn. Notre Dame: University of Notre Dame Press.

Moyers & Company (2013) "Wendell Berry: Poet & Prophet" October 4, 2013. https://billmoyers.com/episode/wendell-berry-poet-prophet.

Newman, J.H. (2001) *Rise and Progress of Universities and Benedictine Essays (Works of Cardinal Newman: Birmingham Oratory Millennium Edition) (Works of Cardinal Newman: Birmingham Oratory Millennium Edition, III):* 1st edn. Notre Dame: University of Notre Dame Press.

Roche, A. (2023) *The Gardens of God*. 1st edn. Washington, D.C.: Word on Fire Press.
Smith, C. and Denton, M.L (2005) *Soul Searching: The Religious and Spiritual Lives of American Teenagers*. 1st edn. Oxford. Oxford University Press.
Taylor, C. (2007) *A Secular Age*. 1st edn. Cambridge: The Belknap Press of Harvard University.
Thompson, C.J. (2017) *The Joyful Mystery: Field Notes Toward a Green Thomism*. 1st edn. Steubenville: Emmaus Road Publishing.
United States Conference of Catholic Bishops, (2011) *New American Bible-Revised Edition*. 1st edn. Oxford: Oxford University Press.
Weil, S. (2009) "On the Right use of School Studies with a View Toward the Love of God" from *Waiting for God*. 1st edn. New York: Harper Perennial Publishing.
Wittgenstein, L. (1973) *Philosophical Investigations*. 3rd edn. G.E.M Anscombe, Translator. London: Pearson Press.

Chapter 5
Leveraging Mobile Technologies for Regenerative Agricultural Information and Extension Services Delivery to Men and Women Smallholder Farmers in Central Uganda

Nicholas Mugabi and David Mugambe Mpiima

Abstract The adverse effects of climate change have trickled down to men and women smallholder farmers, worsening their precarious living and working conditions. To moderate the climate crisis, regenerative agriculture is vital to improve soil health, protect the climate and biodiversity, and increase farm productivity in harmony with nature. This paper underscores the *Laudato Si* manuscript that nuances integral ecology efforts for a sustainable relationship between nature and the society that lives in it. In this paper, we assess how smallholder farmers have leveraged mobile technology tools to access and utilize regenerative agricultural information and extension services in central Uganda. The chapter is structured around two research questions: Which mobile technology platforms are used by men and women smallholder farmers? Which regenerative agricultural information is disseminated through mobile technology platforms? Primary data was collected using survey questionnaires with 390 farmers: 52% and 48% were women and men, respectively. Seven focus group discussions (three were mixed men and women, two for men, and two for women alone), and nine key informant interviews (six were men and three women). Study findings revealed that several uncoordinated mobile technology platforms were used to disseminate regenerative agricultural information. The majority of platforms were mainly project-based; driven by start-ups, donor-funded through civil society organizations, and at different levels of ideation, trials, and not mainstreamed within the government extension systems. Using personal, social, and environmental resources, Village Enterprise Agents, as lay extension workers, leveraged the *Kulima* mobile platform to disseminate agronomic and climate change mitigation information. The predominant regenerative agricultural information was: use of good planting materials, soil conservation, and weather information. However, with a limited scope on the regenerative agricultural content package. The proliferation of uncoordinated

N. Mugabi (✉) · D. M. Mpiima
Makerere University, Kampala, Uganda
e-mail: mugabinicoxs@gmail.com

R. Marcantonio et al. (eds.), *Integral Ecology in the Anthropocene: Africa and Beyond*, The Anthropocene: Politik—Economics—Society—Science 49,
https://doi.org/10.1007/978-3-032-32125-1_5

mobile platforms and technological malfunctioning were the main challenges. Thus, to nuance the call of *Laudato Si*, the regenerative agricultural information content package should be mainstreamed in government extension systems, translated into indigenous languages, and ensure quality assurance.

Keywords Extension services · Technical assistance · Rural transformation · Information for change

5.1 Introduction

5.1.1 Background to Climate Change

Uganda and indeed most developing countries in Sub-Saharan Africa (SSA) are primarily agricultural-based and dependent on natural conditions. Changes in nature, which are mainly induced by climate change, grossly affect agricultural production. This is because agricultural activities are heavily dependent on climate and meteorological conditions; this vulnerability makes agriculture the most significant and susceptible sector influenced by climate change (Ndhlovu 2025). Climate change has been exhibited by variation in temperature and precipitation, which affect soil moisture, extreme temperatures, and higher concentrations of carbon dioxide (Ibid). The temperatures continue to increase while rainfall amounts have been decreasing in the region (Ndhlovu 2025). In addition, the Food and Agriculture Organization (FAO) observed that a lot of cultivated soils throughout the world have lost 50–70% of their original carbon, and excess carbon dioxide in the atmosphere comes from destroying the soil (Moodley et al. 2024). In warmer regions like SSA, the temperatures are soaring, which has affected agriculture with reductions ranging from 26% to 34% in productivity and losses in agricultural revenue (Dabalen et al. 2024). These occurrences directly affect Africa's agricultural production. The adverse effects of climate change have trickled down to smallholder farmers (SHFs), worsening their precarious living and working conditions. The impacts of climate change are gendered, mediated by incomes, productive resources, power relations, land ownership, farm inputs, labor, access to extension services, and prevailing social and gender norms to food and nutritional security (Duff et al. 2017; FAO, IFAD, UNICEF, WFP, and WHO, 2023).

As a result, at global, regional, and national levels, there is an urgent need to address and moderate the adverse effects of climate change. In this case, climate change adaptations imply "the processes of adjustment to actual or expected climate and its effects; in human systems, adaptation seeks to moderate harm or exploit beneficial opportunities; in natural systems, human intervention may facilitate adjustment to expected climate and its effects" (Ndhlovu 2025:195). Addressing the impact of climate change requires an urgent and concerted commitment, which is underscored in the *Laudato Si* manuscript. The manuscript nuances integral ecology efforts for a sustainable relationship between nature and the society that lives in it (Laudato Si 2015). While various climate change mitigations are proposed, for example, agricultural insurance, climate change information, resilient cultivators, and water management techniques (Ndhlovu 2025). In our view, integral ecology through regenerative agricultural practices is more sustainable. Regenerative agriculture is vital to improve soil health, protect the climate and biodiversity, and increase farm productivity in harmony with nature. Thus, in this attempt to answer two research questions: Which mobile technology platforms are used by men and women smallholder farmers? And which regenerative agricultural information is disseminated through mobile technology platforms?

5.1.2 The Laudato Si and Integral Ecology Model

In the Encyclical letter to the Catholic Church and global audience, Saint Pope Francis echoed a dialogue with all people about our common home. Aligned with FAO, the pope underscored the potential for an "ecological catastrophe under the effective explosion of industrial civilization", and stressed "the urgent need for a radical change in the conduct of humanity" (Francis 2015). The current destruction of global biological diversity due to irresponsible social and economic gains poses a great danger to hitherto beautiful Mother Earth. Thus, causing changes in its climate, by stripping the earth of its natural forests or destroying its wetlands; contamination of waters, land, air, and life is a commit a crime against the natural world (Francis 2015). One of the mitigations to the human destruction of Mother Earth calls for integral ecology, which is conscious of the environmental, social, and economic dimensions of sustainable life on earth.

Before the *Laudato Si* call, there have been efforts for example, 1992 Earth Summit in Rio de Janeiro (concerns for sustainable development), 1972 Stockholm Declaration (care for the ecosystem of entire earth), the Basel Convention on hazardous wastes, the Vienna Convention for the protection of the ozone layer, United Nations on Sustainable Development, and "Rio+20" (Rio de Janeiro 2012) among others. The United Nations (UN) 23rd Convention of the Parties (COP23) in Bonn, Germany, in November 2017 also underscored Carbon Capture and Storage technologies, reforestation efforts, and soil erosion prevention, as key climate change mitigation (Codur/Watson 2018). Nevertheless, the Laudato Si invokes Mother Nature and the vulnerable sister in reference to nature, mirroring the threats women face but also highlighting their caring and life-giving roles (Kanayankal 2016; Brazal 2021). Thus, Laudato Si positions women as vulnerable, taps into feminine descriptors like caring and emotions, feelings of compassion [no 210] and tenderness [no 20] (Brazal 2021). Thus, integral ecological practices such as regenerative agriculture are paramount to contribute to restoring the climate conditions.

5.1.3 Regenerative Agricultural Practices

The origins of the term "regenerative agriculture" date back to the late 1970s, gaining wider recognition during the early 1980s when it was stumbled upon by the Rodale Institute in the United States (Moodley et al. 2024). In the 1980s, the concept of regenerative agriculture was used in response to the growing concerns about the negative impacts of traditional agricultural practices, characterized by soil degradation, loss of biodiversity, and over-reliance on synthetic inputs (Rodale 1983). According to Rodale (1983), regenerative agriculture is an agroecosystem that is free of biocides, to increase levels of productivity, improving soil biology. Thus, regenerative agriculture is a farming approach that focuses on soil health, ecosystems (to protect the climate and biodiversity in harmony with nature), an essential element of sustainable agriculture amidst the adverse effects of climate change (Dabalen et al. 2024). Regenerative agriculture is particularly key for women, who form a substantial part of the agricultural workforce in many regions (Business Fights Poverty, 2023). According to Dabalen et al. (2024), regenerative agriculture has a bit of relation with other practices like climate-smart agriculture and conservation agriculture (with the overriding goal to maintain soil from erosion and degradation). Nonetheless, the use of the climate-smart agriculture concept gained momentum in the 2000s as an approach to increase agricultural productivity; adapt and build resilience to climate change; and reduce and remove greenhouse gas emissions (Dabalen et al. 2024). Other scholars use the concepts of climate-smart agriculture (CSA) and regenerative agriculture synonymously with Agroecology (Codur/Watson 2018). Below is a matrix showing the relationship between the three concepts of regenerative agriculture, climate-smart agriculture, and conservation agriculture.

In practice, and within the context of smallholder farmers, there are no distinctions among these three agricultural practices. The practices are more complementary in purpose and function, though they only differ in nomenclature and with emphasis on particular practices. Also, beyond the conventional farming practices, regenerative practices focus more on maximizing yields with a priority on soil health and preserving biodiversity. Some of the common regenerative agricultural practices include reduced-tillage or zero-tillage, vegetation cover, intercropping, applying compost, integrated grazing, and organic waste instead of chemical fertilizers, and livestock (Codur/Watson 2018). This is intended to restore soil health (reduce soil disturbance), increase crop yields, and improve resilience. Other regenerative agricultural practices associated with smallholder farmers are crop rotations, cover cropping, plant spacing, ecological pest management, and integrating livestock (Giller et al. 2015; Moodley et al. 2024). In essence, regenerative agriculture is hinged on six principles of (1) Minimal Soil Disturbance, (2) Diversity, (3) Protect Soil Surface, (4) Maintain Living Roots, (5) Livestock Integration, and (6) Understanding the Context of the Farm Operation (Moodley et al. 2024). These practices can provide essential long-term benefits to millions of resource-poor small-scale, subsistence, and commercial farmers worldwide (Moodley et al. 2024).

However, there is a gender dimension to how men and women embrace these practices. This is because gender is part of the social structure that determines how men and women interact, including participation in agricultural production (Mpiima & Mugabi, 2025). In addition, rural masculine and feminine identities have been largely concerned with farming; for example, women mostly engage in activities such as planting, weeding, and men do the marketing of the products (Ibid). Studies have also shown that once women are exposed to the same information, knowledge, and skills for regenerative agriculture, they can significantly increase farm yields, contribute to food security, and climate resilience (Business Fights Poverty, 2023). Thus, in this paper, we interrogate the kind of mobile technology platforms that are used to disseminate regenerative agricultural information to men and women, and whether the disseminated information is intentional to promote regenerative agricultural practices.

5.1.4 Mobile Technologies for Agricultural Information Dissemination

The fourth industrial revolution emerged with massive technological innovations and changes, influencing various sectors, including agriculture. In the agricultural sector, technological innovations are sought to increase prospects for rural and agricultural development through access to knowledge, increase efficiency, and productivity (Ndhlovu 2025). For example, different digital platforms have been deployed to acquire real-time and accurate information on farming activities (Ibid). There is a shift from traditional extension service delivery to an e-extension scenario by fast-tracking and offering real-time information to farmers (Mirembe/Lubega 2019; Mugabi 2019; Mugabi et al. 2018). Studies have also revealed several digital platforms used by smallholder farmers for agriculture production systems (Mugabi 2019; Ndhlovu 2025). For instance, crop mobile phones in Morocco and Niger, traders' mobile phones in Ghana, Kenya's wholesale traders' mobile phones, and the Zambia National Farmers Union's SMS (World Bank 2012). Alternative mobile extension delivery platforms such as e-Soko in Ghana, Kenya, Burkina Faso, Nigeria, Malawi, Zimbabwe, Benin, Madagascar, and Mozambique, Rural e-Market in Madagascar, M-Shamba, Kilimo Salama, and i-Cow in Kenya, and Nigerian-based Hello Tractor, among others (Emeana et al. 2020; Mavhunga 2017; Mugabi 2019). Thus, digital technologies present an opportunity to improve farmers' access to information on soil and crop management, biodiversity, pest and disease detection, weather forecasting, and improved financial services (Ndhlovu 2025). However, Ndhlovu (2025) highlighted the digitalization challenges in SSA, including poor technological infrastructure, high costs of technology, low levels of education and digital literacy, and limited access to facilities. Thus, in this chapter, we attempt to answer two core questions: i) Which mobile technology platforms are used by men and women smallholder farmers? ii) Which regenerative agricultural information is disseminated through mobile technology platforms? The chapter is organized into different chronological sections. The first section presents the introduction to the study. Section two describes the methodology, sections three and four are the discussion of study findings, conclusions, and recommendations, respectively.

5.2 Methodology

We adopted a convergent mixed-methods research design using both qualitative and quantitative methods. A survey questionnaire was administered to 390 (52% and 48% were women and men, respectively. Also, focus group discussion and key informant interviews data were collected concurrently at the same point in time. On one hand, the survey design is anchored in the positivist philosophy to measure relationships between variables and generate quantifiable, measurable data. On the other hand, the qualitative approach is embedded in social constructionism with a focus on understanding experiences and meanings, opinions, and attitudes of individuals about social phenomena (Creswell 2014; Flick 2018; Leech/Onwuegbuzie 2009). We conducted nine key informant interviews with district technical staff (six were men and three were women). We had seven focus group discussions with women and women farmers. In terms of gender, out of the seven focus group discussions, three were mixed, with both men and women[1] two were conducted with only women, and two with only men.

As far as data analysis is concerned, survey questionnaire data were collected using the Open Data Kit (ODK) platform. The uploaded data entries were then exported into Microsoft Excel and transformed into Statistical Package for the Social Sciences (IBM SPSS) Windows 25.0 computer data analysis software. Survey questionnaire data analysis relied on descriptive statistical analytics. While qualitative (KIIs and FGDs) data were analyzed using NVivo12 Computer-Assisted Qualitative Data Analysis Software (CAQDAS). NVivo software provides a measure of convenience and efficiency, increasing the overall level of organization of a qualitative project (Lu/Shulman 2008). All KIIs and FGDs recordings were transcribed and recorded in a Microsoft Word document. We comprehensively read through the transcripts and uploaded them to NVivo version 12. Using NVivo12, we coded and categorized qualitative data under different parent nodes, attributes, and classifications corresponding to thematic areas of our research objectives and questions.

This study was conducted in central Uganda in the greater Masaka sub-region in the districts of Bukomansimbi, Kalungu, Lyantonde, Lwengo, Masaka, Rakai, and Kyotera. The central region was chosen because it is an agricultural hub and epicenter

[1] While we initially planned to conduct separate FGDs for men and women, some participants requested mixed FGDs because they had previously been sensitized and trained on gender relations and gender mainstreaming. As a result, they felt comfortable expressing their opinions without our gender bias.

of diffusion of innovations, including mobile technologies. In addition, the districts share common geographical, socio-cultural, and economic characteristics, which helped to generate comparable results. In terms of ethical considerations, the study was reviewed and approved by an Institutional Review Board (IRB), specifically, Makerere University School of Social Sciences Research Ethics Committee (MAKSS REC). The study was further reviewed and cleared by the Uganda National Council of Science and Technology (UNCST), a research regulator for all studies involving human subjects in Uganda.

5.3 Findings and Discussion

The chapter is anchored on two central research questions, namely: first, which mobile technology platforms are used by men and women smallholder farmers? Second, what regenerative agricultural information is disseminated through mobile technology platforms?

Mobile technology tools and platforms used by men and women

Findings show that of the 390 survey respondents, 52% were women and 48% were men. The majority had primary levels of education at 60%, followed by 24% with secondary, 8% had no formal education, and 3% had finished advanced secondary level. While 2.6% had tertiary certificates, 0.8% and 1.3% had with diploma and an undergraduate degree, respectively. In terms of age, most participants were between 45-55 years and 35-44 years. Very few men and women smallholder farmers were between 18-24 years and above 65 years, as shown in Table 5.2 below.

We also noted that almost all farmers engaged in growing bananas, maize, beans, and coffee. While vegetables, fruits, and tuber crops were the least grown. In terms of

Table 5.1 Conceptualization of regenerative agriculture. *Source* Adopted from Dabalen et al. (2024:5)

Regenerative agriculture	Climate-Smart agriculture	Conservation agriculture
A farming approach that focuses on restoring and improving the health of soil, biodiversity, and ecosystems	An approach that seeks to address three main objectives simultaneously: increasing agricultural productivity; adapting and building resilience to climate change; and reducing and/or removing greenhouse gas emissions	An approach to maintain the soil from erosion and degradation
Minimum tillage, no-till, mulching with crop residue, crop rotation, legume-based cover crops, livestock rotations, use of compost and organic waste	Drought-resistant seeds, solar-powered pump systems	Minimum tillage, no-till, mulching with crop residue, crop rotation, legume-based cover crops

Table 5.2 Showing age brackets of smallholder farmers by Gender

Gender	Age bracket of respondents					
	18-24	25-34	35-44	45-54	55-64	+65
Men	3.8%	18.3%	27.4%	25.3%	16.1%	9%
Women	2.9%	15.2%	25%	31.9%	15.2%	8.8%
Overall	3.8%	16.7%	26.2%	28.7%	15.6%	9%

access to, use, and ownership of digital tools and platforms such as radio, television, mobile phone, and computer, all participants had access to or owned a mobile phone and radio, with digital skills to use them. However, access to television (17.7%) and computers (1%) was the very least, especially among female smallholder farmers. The gender equality in access to digital tools is a function of the project's emphasis on gender mainstreaming in the training modules for farmers. Mobile phones and radio are not only communication tools, but also facilitate access to agricultural information and extension services.

As far as access to regenerative agricultural information was concerned, it was noted that at least all men and women smallholder farmers had access to multiple sources. Overall, as shown in Fig. 5.1 below, the main sources of regenerative agricultural information and extension services were the Village Enterprise Agents (VEA) using *Kulima* mobile extension application (100%), mobile phone (100%), radio (17.9%), fellow farmers (98%), television (3.2%), and NGO/CSOs (10.7%). However, women had more access to television at 5.9% and NGOs (5.6%) compared to their male counterparts. Other sources of agricultural information were social media, though used minimally by both women and men. Fig. 5.1 summarizes the results.

To supplement the survey findings above, qualitative data also demonstrated that the most mentioned source of extension services was the VEA-Kulima mobile application, as explained by a male FGD Participant:

> … With the Kulima mobile application, we got information on good agricultural practices. This Kulima application is concerned with three items: maize, beans, and coffee. If your plantation has a problem, you touch the Kulima app, and it gives all the necessary information. Even if you have a small thing, you can ask and they give you the information which eventually helps you (A Man Farmer, Tukolere Wamu Farmer Group, Bukomansimbi district).

Radio was also a prominent source of agricultural information in the qualitative interviews, as narrated below.

> We also tune radio stations such as Central Broadcasting Station (CBS), which has a program of "Mwanyi Terimba", "Enkumbi Bugaga", and other programs (A woman SHF, Lwanda sub-county, Rakai District).

In addition, there were various CSO-supported technology platforms such as bulk SMS, Plant Clinic Platform, mobile money, Beans Innovation Platform, WhatsApp, and 2KUZE that enabled men and women farmers to access extension services. The

Table 5.3 Gendered agricultural information received. *Source* Author created

Gender	Access to digital tools							
	Proper site selection	Selec-tion of good plan-ting materials	Crop Management	Weather information	Manure Application	Soil Conserva-tion	Pest & di-sease con-trol	Market information
Men	84.4%	90.8%	64%%	47.8%	75.5%	87.5%	52%	64%
Women	78.5%	89.8%	66.7%	41.7%	70%	86.8%	44%	66.7%
Overall	81.7%	90.3%	65.4%	44.6%	72.9%	87.3%	48%	65.4%

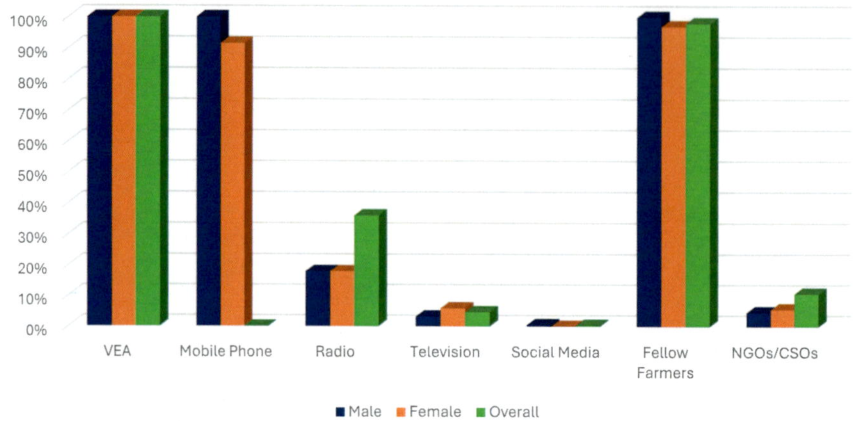

Fig. 5.1 Sources of regenerative agricultural information and extension services. *Source* Permission

Short Message Service (SMS) implemented by CIDI, where agricultural information is directly conveyed to the personal phones of SHFs in Kyotera and Rakai districts. The SMS program is built on the Unstructured Supplementary Service Data (USSD) system for agricultural information dissemination. SMS are sent to SHFs every week or fortnight, depending on the crop activity calendar. We found out that SHFs received SMS about pests and diseases, and market price information, especially for coffee. The system is swift to disseminate information directly to the handsets of SHFs without the influence of trained or lay extension workers. However, it was a unidirectional communication channel, where SHFs received SMS without giving their feedback or making any consultation.

There are also social media platforms such as WhatsApp, mostly used by professionals, the Uganda Agriculture Extension platform managed by UFAAS, the district and sub-county extension workers platform, and the farmer call center, among others. For example, the technical staff at the Masaka district local government highlighted the various ICT-based extension delivery tools available to SHFs.

> … there are also other platforms such as M-Mulimisa and beans platform by Makerere University. M-Mulimisa is an ICT program; using USSD technology farmers ask questions and are answered there and then. I see farmers using phones to calculate prices because when they take their coffee buyers, they are able to know how much they are going to get, depending on the market price. In addition, as an agricultural department at the district; we have a WhatsApp platform and also for the ministry of agriculture, where we share information (Male Staff, Masaka District Production Department).

In addition, Community Enterprises Development Network (CEDO), a local NGO, in partnership with Mastercard Foundation and the International Centre for Tropical Agriculture (CIAT) developed and piloted the 2KUZE payment system, a mimic of a mobile money platform. Using *2KUZE,* money is directly transferred to SHFs' mobile phone wallets. In this regard, one KI explained how *2KUZE* systems work.

> Previously, we were using mobile money for payments to farmers after the supply of bean produce. Money was paid to the lead farmer's phone, who was expected to send it to individual farmers' phones. But now we are using the 2KUZE payment system as a pilot for the last six months (Manager, CEDO Kyotera district).

The extension services and mobile finance services were tailored to the existing social systems and resources of producer organizations and social networks as channels of communication, awareness, trial, and diffusion. The existence of multiple mobile extension platforms created several substructures of diffusion of regenerative agricultural information, which gave alternative options from which women and men can choose based on their socio-economic conditions, so as to enhance financial inclusion.

Although the study focused on mobile technology platforms for extension service delivery, various sources of agriculture extension delivery exist at the community and national levels. The community extension delivery structures, such as VEA, radio, fellow farmers, and personal mobile phones, were more prominent among men and women. However, the structures are not harmonized as far as dissemination approaches, *modus operandi*, content package, and geographical coverage are concerned. Similar challenges were reported in Zambia, where there were no clear guidelines to guide the research direction of extension applications, always developed in response to ad-hoc problems and poor coordination among stakeholders (Phiri, Phiri, & Kunda, 2018). Thus, poor coordination of orthodox and mobile extension structures creates confusion among smallholder farmers. There are several pilot extension innovations and platforms being designed and implemented without a clear sustainability and integration strategy in the mainstream Minister of Agriculture, Animal Industry and Fisheries (MAAIF) single spine extension system.

We observed that mobile extension services are flexible and ever-changing based on the end users' extension needs and development agenda of CSOs' extension providers. The paradigm shift from public extension systems to pluralistic CSOs is attributed to challenges of the former. We are also cognizant of the fact that the proportion of households with access to agricultural extension services has increased to 49% in 2021/2022 from 11.7% in 2017/2018 (NPA 2025). On one hand, the prominence of VEAs-*Kulima* mobile extension services is because VEAs are local intermediaries, intentionally recruited, trained, and deployed within their community to support SHFs who hitherto hardly had access to extension workers.[2] VEA structures, just like CKWs and NAADs farmer groups, are premised on SHFs' participation, with strong social connections or belonging to pre-existing community groups (Hugh/White 2014; Isubikalu 2007; Ninsiima 2015). On the other hand, VEA-Kulima mobile extension delivery, fellow farmers, radio, and mobile phone SMS services are affordable and easy to access, especially to women, because of the personal 'touch' with SHFs' involvement. The in-person interactions of VEAs-SHFs strengthen the

[2] The Village Enterprise Agents (VEAs) as lay community extension agents (volunteers) use Smart mobiles to convey quality and reliable agricultural information, while farmers also enquire about new agricultural practices, crop varieties and disease among others (McCole et al. 2014, Kimbowa 2015).

confidence of extension users; trust in localized extension structures through timely delivery of information to guide strategic decision-making about the most appropriate practices to use. Complementary evidence revealed that many men and women, 62% have access to information through radio, and 49% mobile phones usage however, there is still low adoption of ICT-enabled agricultural extension applications, partly due to low internet coverage at 12% (NPA 2025).

In addition, most agricultural extension providers, such as the VEA-*Kulima* model, mobile phone SMS, and CSOs, are mainly project-based, donor-dependent, and have limited geographical scale to support the majority of SHFs in Uganda. The sources of extension services are often designed and developed using a top-down approach where agriculture scientists and experts, who are mostly men, design platforms with less participation of end users. The implication is that women's digital needs can easily be left out. No wonder most of these platforms are built on formal education systems and structures; they use English, yet in Uganda, many women are illiterate or have low levels of education. This alienates women end users in terms of language, nomenclature, and content package. This was expressed by most key informants, for example, a senior staff member at Masaka district local government noted that:

> More challenges always come as a result of our innovations. If they were coming from farmers, they would be more sustainable, but when they come from us, from the top, they come with some costs, which are owned by the projects, and once the project is over it becomes a very big problem. The platform itself may have technical challenges and breakdowns. In addition, farmers may not meet regularly, they cannot sustain activities, and you may find that few people, like three or four, are very active. So those are some of the challenges to its sustainability, because many of the jobs are done by the project (**Agriculture Officer, Masaka district).**

While the existence of multiple mobile platforms for disseminating regenerative agricultural information presents an opportunity, there are also challenges of unequal power relations of expertise, and men and women as end users within the social system. This breeds power and knowledge contestation between agricultural experts and local end users. Thus, access to *Kulima* mobile extension services, and indeed other extension services, is embedded in formal social systems and rules of CSOs and private sector players to which men and women smallholder farmers have less or no influence at all as far as *modus operandi* is concerned. Previous studies demonstrate that boundaries of formal extension social systems are, in most cases, insufficient to adequately reach all last-mile SHFs (McCole et al. 2014; Walakira et al. 2016). Hence, due to the deficiencies of formal extension structures, SHFs leverage multiple mobile platforms to reconstruct practices of orthodox extension by engaging peers (social learning) and knowledge exchange of good agricultural practices.

5.4 Access to and Utilization of Regenerative Agricultural Information

Men and women were asked to indicate the kind of regenerative agricultural information received using mobile technologies. Among the regenerative agricultural information disseminated were: proper site selection (81.7%), selection of good planting materials (90.3%), proper crop management (65.4%), market information (65.4%), weather information (44.6%), fertilizer and manure application (72.9%), soil conservation (87/3%), pest and disease control (48%), among others. There are differences between men and women as far as access to regenerative information and extension services. ore men accessed information on site selection, manure application, pest and disease control, as seen in the table 5.4 below.

Further, study findings revealed that for all regenerative information received, men applied the information more than women. The highest percentage difference (15%) was in the application of proper site selection information, where men were at 76.7% and women were at 61.8%. This was followed by a selection of good planting materials, where 78.5% of the men had applied the regenerative information compared to 66.2% of the women. For weather information, soil conservation, there were nearly similar application levels, with men stating at 76.7% while women were 65%. Seventy-seven percent of the men applied soil conservation knowledge compared to 66.5% of the women. More men (84.3%) applied pest and disease control information compared to 74.8% of the women. There was, however, a less than 10% difference in the application of information between men and women for both crop management and market information.

We also underscore previous studies, which classified extension information and needs into three categories: know-how information, market information, and contextual information (Mittal et al. 2010). Additional categories include seed information, land suitability, suitability of varieties, cultivation information, agrotechnology, and agricultural news (Arian et al. 2019). In this study, we used the survey questionnaire data to create a hybrid classification of five broader strategic extension information and needs, namely: agronomic information, climate change mitigation, market information, financial services, and other generalized extension need. Agronomic extension services involve agricultural knowledge and practices related to site selection, planting materials, weed control, pruning, crop management, pest and disease control, among others. Climate change mitigation services and practices include, but are not limited to, weather information, soil conservation, quality planting materials, and agroforestry. In addition, market information is concerned with post-harvest handling, storage, collective bulking, and collective marketing for better prices through primary cooperatives and SHF groups. Financial services are associated with access to credit for agriculture investment, mainly through Village Saving and Loan Associations (VSLA). Lastly, the generalized extension services were mainly poultry and livestock management.

In this regard, our study findings indicate that the overriding information and extension needs of SHFs were 95% and 56% agronomic practices and climate change mitigation. This is followed by market information (38%), financial services (32%), and 12% generalized extension services. We further found out that among the most critical agronomic information and extension needs are crop management, fertilizers/ manure application, and pest and disease control. The predominant climate change mitigation information was the use of good planting materials and soil conservation, while market information was mainly postharvest handling and prices, and access to financial services.

5.5 Conclusion and Recommendations

The adverse effects of climate change have trickled down to men and women smallholder farmers, worsening their precarious living and working conditions. To moderate the climate crisis, regenerative agriculture is vital to improve soil health, protect the climate and biodiversity, and increase farm productivity in harmony with nature. In this paper, we underscore the *Laudato Si* manuscript that nuances integral ecology efforts for a sustainable relationship between nature and the society that lives in it. The chapter presents findings on two research questions: Which mobile technology platforms are used by men and women smallholder farmers? Which regenerative agricultural information is disseminated through mobile technology platforms?

The fourth industrial revolution has transformed multiple means through which men and women SHFs receive and share information. As a result, men and women leverage diverse mobile platforms, especially radio and mobile phones, to have access to vital regenerative agricultural information and extension services. However, computers and television platforms were least diffused among both men and women farmers. The key enablers for the diffusion of mobile platforms are the VEA's lay extension workers, farmer groups, and producer organizations, and *Kulima* mobile application systems. At a macro level, the integration of information, communication technologies in the operations of all sectors facilitates the adoption of e-extension delivery platforms. For example, the Uganda National Development Plans IV (2024/2025-2029/2030) underscore the use of mobile technologies to enhance efficient service delivery, including in extension service delivery (NPA 2015). The main communication channels for mobile platforms are at the individual-end-user level, village enterprise agents, and SHF group structures. In addition, mobile platforms are enabled by the agency of men and women SHFs to access vital agricultural regenerative information and extension services to enhance soil health, protect the climate and biodiversity, and increase farm productivity.

Overall, the multiple agricultural information dissemination platforms present a timely opportunity for men and women SHFs to choose the most appropriate, effective, affordable, and easily accessible information source. However, mobile platforms are limited, project-based, and bogged down by information symmetry. Yet, in reality, robust mobile platforms and tools require a long period for ideation, trial, observation, and adoption by end users (Rogers 2010). Over time, the diffusion process is iterative for learning and improvement.

Our findings revealed that the current mobile extension platforms hardly address the critical issues of regenerative agriculture is hinged on six principles of (1) Minimal Soil Disturbance, (2) Diversity, (3) Protect Soil Surface, (4) Maintain Living Roots, (5) Livestock Integration, and (6) Understanding the Context of the Farm Operation (Moodley et al. 2024). The only information that was closely related to regenerative agricultural information was climate change information, which is also limited in scope. Yet, regenerative agricultural information is vital for long-term benefits to millions of resource-poor small-scale, subsistence, and commercial farmers worldwide (Moodley et al. 2024). Thus, there is limited regenerative agricultural information accessed by men and women SHFs in central Uganda. The available platforms were driven by CSOs donor projects, which were not intentional and deliberate on the core regenerative agricultural information; on soil health, protecting the climate and biodiversity, and increasing farm productivity in harmony with nature.

There were several uncoordinated mobile technology platforms used to disseminate regenerative agricultural information. The majority of platforms were mainly project-based, driven by start-ups, donor-funded through civil society organizations, and at different levels of ideation, trials, and not mainstreamed within the government agricultural extension systems. Therefore, we recommend that an effective coordination and harmonization of extension delivery platforms and structures is vital to create more synergies, co-learning, and knowledge co-creation among extension delivery players and the community. Furthermore, to nuance the call by *Laudato Si*, integral ecology with regenerative agricultural information dissemination initiatives, state and non-state actors should be intentional about regenerative agricultural content packages, should be mainstreamed in government extension systems, translate content into indigenous languages, and ensure quality assurance of content packages.

Reference

Brazal, A.M. (2021). Ethics of care in Laudato Si: a post-colonial ecofeminist critique. *Feminist Theology*. Vol. 29(3), 220-233.

Arian, N., Widowati, S., & Imrona, M. (2019). Agricultural Information Application Design using User-Centered Requirements Engineering. *Journal of Physics: Conference Series, 1367-012017*, 1-8.

Codur, A.-M., & Watson, J. (2018). Climate-smart or regenerative agriculture. *Global Development and Environment Institute, Climate Policy Brief, 9.*

Creswell, W. J. (2014). *Research design: qualitative, quantitative, and mixed methods approaches* (4th ed.). California SAGE Publications, Inc.

Dabalen, A., Goyal, A., & Song, R. (2024). *Regenerative Agriculture in Practice* (Vol. 10919): World Bank.

Emeana, E. M., Trenchard, L., & Dehnen, S. K. (2020). The Revolution of Mobile Phone-Enabled Services for Agricultural Development (m-Agri Services) in Africa: The Challenges for Sustainability. *Sustainability, 12*(2), 485.

Flick, U. (2018). *Doing Triangulation and Mixed Methods.* London: SAGE Publications, Inc.

Francis, P. S. (2015). Encyclical Letter Laudato Si' of The Holy Father Francis on Care for Our Common Home. *London: Catholic Truth Society.*

Giller, K. E., Jens A. Andersson, Marc Corbeels, John Kirkegaard, David Mortensen, Olaf Erenstein, & Vanlauwe., B. (2015). Beyond conservation agriculture. *Frontiers in Plant Science 6*(870).

Hugh, W., & White, H. (2014). *Farmer field schools: From agricultural extension to adult education.* Retrieved from London:

Isubikalu, P. (2007). *Stepping-stones to improve upon the functioning of participatory agricultural extension programmes*: Wageningen Academic Publishers, The Netherlands.

Leech, N. L., & Onwuegbuzie, A. J. (2009). A typology of mixed methods research designs. *Quality & quantity, 43*(2), 265-275.

Lu, C.-J., & Shulman, S. W. (2008). Rigor and flexibility in computer-based qualitative research: Introducing the Coding Analysis Toolkit. *International Journal of Multiple Research Approaches, 2*(1), 105-117.

Mavhunga, C. (2017). *What Do Science, Technology, and Innovation Mean from Africa?* : The MIT Press.

McCole, D., Culbertson, M. J., Suvedi, M., & McNamara, P. E. (2014). Addressing the Challenges of Extension and Advisory Services in Uganda: The Grameen Foundation's Community Knowledge Worker Program. *Journal of International Agricultural and Extension Education, 21*(1).

Mirembe, D. P., & Lubega, J. T. (2019). *State of Information Communications Technology (ICT) for Agricultural Innovations in Uganda 2019.* (2518-8623). Retrieved from Kampala:

Mittal, S., Gandhi, S., & Tripathi, G. (2010). *Socio-economic impact of mobile phones on Indian agriculture*: Indian Council for Research on International Economic Relations, New Delhi.

Moodley, V., Phophi, M. M., & Mafongoya, P. L. (2024). Regenerative agriculture: A paradigm shift for food production in South Africa. *Outlook on agriculture, 53*(4), 336-350.

Mugabi, N. (2019). Digitization of Agricultural Extension Services: A Case of Mobile Phone-based Extension Delivery in Central Uganda. *International Journal of Technoscience and Development (IJTD), 4*(2019), 24-40.

Mugabi, N., State, A. E., Omona, J., & Jansson, B. (2018). Revolutionalizing Agriculture Extension Delivery Through Mobile Telephony: the Experience of Village Enterprise Agent Model in Greater Masaka Area, Uganda. *Transactions on Ecology and the Environment, 217*(2018), 963-974.

Mpiima, D. M., & Mugabi, N. (2025). Gender Relations in the Utilisation of Digital Technologies in the Changing Food Production Systems in Northern Uganda. In *African Food Systems: Rethinking Prospects for Continental Sustainable Transformation* (pp. 151-178). Cham: Springer Nature Switzerland.

Ndhlovu, E. (2025). *Agriculture, Autonomous Development, and Prospects for Industrialization in Africa.* Switzerland AG: Springer Nature.

Ninsiima, D. (2015). *Factors affecting the adoption of information and communications technology systems for agriculture in Uganda.* Michigan State University,

NPA. (2025). *Fourth National Development Plan (NDPIV) 2025/26–2029/30.* Kampala: National Planning Authority.

NPA. (2015). *Second National Development Plan (NDPII) 2015/16-2019/20.* Kampala: National Planning Authority.

Phiri, H., Phiri, J., & Kunda, D. (2018). A Review of E-agriculture Application in Developing countries for the Youth: Challenges, Opportunities and Open Issues *Official Journal of the Zambia Information and Communications Technology Authority, 1*(1), 1-59.

Rodale, R. (1983). *Breaking new ground: The search for a sustainable agriculture* (Vol. 17): Futurist.

Rogers, M. E. (2010). *Diffusion of innovations*: New York, New York: Simon and Schuster.

Walakira, J. E., Denis Muhangi, Sam Munyuwiny, Fred Matovu, Eric Awich, Ismael Ddumba Nyanzi,... Ruiz-Rodriguez, M. (2016). *The State of the Ugandan Child: An Analytical Overview.* From Kampala/Washington DC.

Chapter 6
Integral Ecology as Praxis of Hope: Dreams from Bethany Land Institute, Uganda

Emmanuel Ojeifo

Abstract *Global* ecological challenges are having a negative impact on young people's sense of happiness and hope for a better future. Recent research has shown that many young people around the world feel that the adult generation has destroyed the planet and is leaving behind a debris of serious ecological catastrophe for them to deal with. Faced with this problem, what does hope for a better ecological future look like and where can one find it? Responding to this question requires an empirically grounded practice that cultivates a vision of ecological hope. Drawing on surveys conducted on current students ("interns") and graduates ("caretakers") of Bethany Land Institute (BLI), a faith-based integral ecology formation school in Uganda, this essay demonstrates that hope is cultivated, nurtured, and grows through rootedness in a place one calls "home," through spiritual practices, forms of practical engagement in creation care, finding purpose in meaning work, and encountering the space to imagine and dream of an alternate and better future. By showing how the vision, spirituality, lifestyle, and practice of Integral Ecology is implemented at BLI, this article reinforces the conviction that integral ecology is not a set of ideas but a praxis and a way of life, which yields specific results and outcomes.

Keywords Hope · Spiritual conversion · Caretakers · Ecological anxiety

6.1 Introduction

Amidst our present global environmental crisis, the question of hope has emerged as one of the defining issues of our time. Faced with widespread environmental degradation and the anxieties generated by predictions of ecological doomsday, the tendency for people to be paralyzed by fear has become so common and so real. The German theologian Jürgen Moltmann (2012, pp. 51-53) noted that in the face

E. Ojeifo (✉)
University of Notre Dame, Notre Dame, US
e-mail: eojeifo@nd.edu

R. Marcantonio et al. (eds.), *Integral Ecology in the Anthropocene: Africa and Beyond*, The Anthropocene: Politik—Economics—Society—Science 49,
https://doi.org/10.1007/978-3-032-32125-1_6

of the global dangers, including ecological catastrophe, which threaten us all today, human reaction runs a spectrum from panic attack, alarmism, and fatalism to social rigidity, psychic numbing, and escapism. To therefore speak of hope in these times or to ask people to be hopeful may seem attractive and comforting, but it can also be frustrating and problematic. The eco-ethicist Norman Wirbza (2024, p. 15) captures this problematic situation when he writes:

> Admonitions to "Be hopeful," can like a soporific, lull people into accepting the status quo. Suitably pacified, hopeful individuals simply wait for the miracle that will in some hypothetical, perpetually deferred future make everything all right. The Indigenous philosopher Kyle Whyte instructs us to be cautious around these (often comfortable and satisfied) purveyors of hope. Their calls to hopefulness can generate what he calls the "ultimate bystander effect," by giving people an excuse not to do the hard work of correcting the injustices that create hopelessness in the first place. Think, for instance, of the generations of religious leaders who have told their followers to accept suffering and endure every hardship because God will, in the end, make everything right. Pains will eventually be healed and injustices corrected, provided that people are patient and put their hope in God. Think, too, of the business tycoons and technology gurus who tell their followers that a technological fix will always, inevitably, get people (at least those who can afford it) out of whatever troubles they find themselves in.

Wirzba's point is not to deny God's providence or to thrash technology, but to demonstrate that "people have not been well served by these expressions of 'faith' that keep them passive or reduced to being little more than spectators of their own lives, mostly deaf to the world's cries for healing and help, and thus exempt from the work of creating a more hopeful future" (Wirzba 2024, p. 16). In light of this, how possible is it to speak about hope in a way that makes people attentive and responsive to the cry of the earth while also affirming the possibilities of a better future? Or, to paraphrase Wirzba, how can we speak of hope more as a *verb* and less as a *noun*?

6.2 Ecological Anxiety and the Crisis of Hope

I begin this article with a brief reflection on hope in order to underscore a matter of fact: our world is presently experiencing a crisis of hope. When Pope Francis declared a Jubilee of Hope for the Catholic Church from 2024 to 2025, he desired to draw attention to this global crisis and to propose a path of renewal that extends beyond the Catholic Church. In the inaugural document of the jubilee year, *Spes Non Confudit*, ("Hope does not disappoint"—Rom 5:5), Pope Francis (2024, p. 1) remarks: "Often we come across people who are discouraged, pessimistic and cynical about the future, as if nothing could possibly bring them happiness." In his autobiography titled *Hope,* Pope Francis (2025, p. 290) notes: "Many people today, for various reasons, seem not to believe that a happy future is possible." In these two brief statements, Francis identifies a serious problem facing contemporary humanity: the lack of hope for the future. When he referred to young people, he affirmed that they are "the very embodiments of hope," but he also expressed worry that "they often see their dreams and aspirations frustrated," adding that "it is sad to see young people who are without hope, who face an uncertain and unpromising future" (Pope Francis 2024, p. 12).

A survey conducted in 2021 among 10,000 children and young people (aged 16-25) in ten countries of the world, including one country in Africa, revealed widespread evidence of climate anxiety:

> Respondents across all countries were worried about climate change (59% were very or extremely worried and 84% were at least moderately worried). More than 50% reported each of the following emotions: sad, anxious, angry, powerless, helpless, and guilty. More than 45% of respondents said their feelings about climate change negatively affected their daily life and functioning, and many reported a high number of negative thoughts about climate change (eg, 75% said that they think the future is frightening and 83% said that they think people have failed to take care of the planet)… Climate anxiety and distress were correlated with perceived inadequate government response and associated feelings of betrayal (Hickman 2021, p. 863).

In the same year 2021, UNICEF declared that the climate crisis is a child rights crisis. The global body provided evidence of how many children were exposed to a variety of environmental/climate hazards globally: 820 million children were highly exposed to heatwaves, 400 million to cyclones, 570 million to riverine and coastal flooding, 920 million to water scarcity, 600 million to vector-borne diseases, 2 billion to air pollution, and 815 million to lead pollution (UNICEF 2021, p. 10). Mental health challenges globally have also been linked to the growth of eco-anxiety and eco-grief (Jones and Lucas 2023; Kluger 2025).

Faced with our present ecological crisis and the crisis of hope that it generates, especially among young people, I propose the praxis of Integral Ecology, a theological framework articulated by Pope Francis in *Laudato Si'*, as an antidote. To provide a concrete example of this praxis, I discuss the work of Bethany Land Institute (BLI), a faith-based integral ecology formation school in Uganda, where young people from the local communities are trained in the practice of sustainable creation care. By showing how the BLI formation model is impacting on the holistic development of the young people, this article argues that hope is cultivated and grows through a life of prayer, practical engagement in creation care, finding purpose in meaningful work, and encountering the space to imagine and dream of a better ecological future with others.

6.3 Integral Ecology and Hope: The Story of Bethany Land Institute

The notion of Integral Ecology expresses the interconnectedness of varied aspects of reality. In his encyclical letter *Laudato Si'*, Pope Francis (2015, para. 137) emphasizes that "everything is closely interrelated, and that today's problems call for a vision capable of taking into account every aspect of the global crisis." In other words, responding to the twin cry of the earth and the poor demands a broader vision and a holistic approach that integrates various dimensions: environmental, social, political, economic, religious, educational, cultural, scientific, and technological. Based on this multidimensionality, Elizabeth Gandolfo (2023, p. 128) notes that we should speak about Integral Ecology in the plural ("integral ecologies are pluriversal ways of being human") rather than in the singular.

From a Christian theological point of view, the care for creation that is central to Integral Ecology is borne out of love for all that God has created and which God found "very good" (Gen 1:31).[1] The biblical Book of Wisdom says in relation to God: "You love all that exists, you hold nothing of what you have made in abhorrence, for had you hated anything, you would not have formed it" (Wis 11:24-26). Love of something gives rise to a praxis of care and care itself is a sign of hope in the possibilities of being and becoming. This means that hope sits on the foundation of love. This explains why Wirzba (2024, p. 17) argues that, amidst our ecological crisis, instead of asking "What gives you hope?" we should rather ask, "What do you love?" Love is what grounds and concretizes hope, because "people who live in hope do not seek to shield themselves from the pains and problems of this life. Their love compels them to engage the trouble and work for a better future. Hope grows, and its meaning is more fully discovered, when love goes to work in the world" (Wirzba 2024, p. 17). Love saves hope from sinking into meaningless abstraction.

When Pope Francis proposed the framework of Integral Ecology, he did so as a way of offering hope in the possibility of change rooted in love. As he states, "social love moves us to devise larger strategies to halt environmental degradation and to encourage a 'culture of care' which permeates all of society" (Pope Francis 2015, para. 231). These strategies are rooted in ecological conversion which gives rise to changes in personal and communal habits and lifestyles from a culture of waste, excessive consumption, and lack of concern for others to a culture of shared life, respect for the environment, less consumption, moderation in the use of resources, and simplicity of life. Pope Francis (2015, para. 212) stresses: "We must not think that these efforts are not going to change the world. They benefit society, often unbeknown to us, for they call forth a goodness which, albeit unseen, inevitably tends to spread." If human beings experience profound interior conversion and change from environmentally destructive habits and lifestyles practices to sustainable habits and practices, then there is reason to hope for a new and better ecological future for humanity. Indeed, "all is not lost. Human beings, while capable of the worst, are also capable of rising above themselves, choosing again what is good, and making a new start, despite their mental and social conditioning" (Pope Francis 2015, para. 205).

When the Ugandan Roman Catholic priest, Father Emmanuel Katongole, and his two colleagues, Father Cornelius Ssempala and Father Tony Rweza, began to think and reflect together about how to respond to the environmental and social problems facing their local communities, they had no idea where the discussions would lead them. Katongole (2022a, pp. 142-143) recounts:

> At a retreat for our twenty-fifth priestly anniversary celebration in June 2012, I asked Fr. Cornelius, my classmate, if he knew a piece of land I could buy. I told him I wanted to plant a forest to replace the forest which had all been destroyed near the home where I grew up. Every time I would drive to my home village, I was shocked at the rate of deforestation and,

[1] All biblical citations are from *The Jerusalem Bible*. New York: Doubleday and Company, Inc., 1966.

> overall, at the ecological devastation underway. On his part, Cornelius (an educationist), was concerned about the kind of education that brainwashes young people into looking at the village as a place of backwardness and losers and so lets them pin their hope on finding employment in cities and towns—which has only contributed to the growing number of slums. Tony, our other classmate who later joined the conversation, drew our attention to the levels of poverty not only in the villages, but even more so, in the slums. The three of us together bought some land in Luweero (central Uganda) and started thinking about an initiative that would bring together these three *Es*: environment, education, and economics.

Insights from *Laudato Si'* helped these three friends to gain some clarity about their joint efforts. Katongole (2022a, p. 143) recalls: "In *Laudato Si'* , Pope Francis had not only put words to some of the restlessness that I and my classmates were already feeling; he had also provided both a vision and urgency for the Bethany Land Institute." Located on over 300 acres of land in the village of Nandere, Uganda, near the capital city of Kampala, BLI brings together many young people from rural communities across Uganda and forms them in regenerative agriculture, agroecology, animal husbandry, conservation and reforestation practices, and economic entrepreneurship. BLI's mission is to transform rural ecologies, economies, and livelihoods by attending to issues of poverty, ecological degradation, and food insecurity. After undergoing a two-year residential formation program, the interns are commissioned as integral ecology ambassadors with a mandate to return to their local communities and set up their regenerative agriculture farms, while also helping to transform environmental practices in eco-sustainable ways.

BLI's theological vision is shaped by the biblical Bethany, a village outside Jerusalem in the time of Jesus where the poor and lepers lived. Jesus frequented this village to spend time with a family of three siblings: Mary, Martha, and Lazarus (Luke 10:38-42; John 11:1-25). In an earlier work, Katongole (2012, p. 9) notes that the story of Bethany offers an innovative way of doing theology and thinking about the mission of the church in Africa amidst "social, historical, contextual, and practical exigencies." For him, Bethany is a motif of hope in the possibility of a new beginning and a new future, particularly in the story of the raising of Lazarus. Using the gospel story of the raising of Lazarus as a theological model for an emerging African ecclesiology, Katongole (2015, p. 194) argues that "like Lazarus, Africa is ill and in distress, and the Church must therefore draw on its prophetic, pastoral, and teaching traditions to name, describe, and denounce the many forms of Africa's sickness." This strategy is a way of lamenting Africa's condition of social, political, economic, and ecological distress that opens up fresh and liberating possibilities for a new African future of wholesome and abundant life. Stan Chu Ilo (2024, p. 10) has observed that:

> The question in the hearts of all Africans today in their strong faith and commitment to God and the Church is: where can we find hope for the future? Faced with many challenges, but also surrounded with boundless opportunities and possibilities, many Africans are longing for a better future. No activity of the Church in Africa, no synodal process will make much sense to the men and women of our continent if it is not translated into hopeful praxis for a more abundant life.

In founding BLI, Katongole and his friends were driven by the conviction that things can change and that a new African future of abundant life is possible. In seeking to raise an army of young African ecological practitioners who will go on to transform environmental practices in their local communities, the BLI founders were providing a model for what hope might look like in Africa amidst social and ecological distress. At the very least, the insertion of the story of Lazarus in the BLI vision energizes this model in the sense that it outlines a path of African ecological and social resurrection. "A church committed to the resurrection of Africa," Katongole (2015, p. 198) notes, "must be in the business of community mobilisation." This mobilisation takes its cue from Jesus' instruction after Lazarus came out of the grave: "Unbind him" (John 11:44). The church in Africa has the task of continuing to help unbind and free Africa from all that threaten to snuff life out of the continent. Focusing on the task of unbinding Africa's young people who disproportionately bear the burden of the continent's social predicament, Katongole (2015, p. 199) writes: "A ministry of unbinding and setting free the youth in Africa will need fresh experiments in educational initiatives for empowerment as well as pastoral programmes that engage the youth beyond drama and entertainment." BLI is the ministry of hope that the BLI founders are offering to unbind and set Uganda's, and by extension Africa's, young people free.

6.4 Bethany Land Institute: A Laboratory of Hope through Integral Ecology

Between March 2024 and July 2025, I undertook four field trips to BLI Uganda, each lasting between a week and three weeks. These trips are connected with a research project, *Sowing Hope*, which was conceived and initiated by BLI co-founder and president, Father Emmanuel Katongole. The project seeks to assess the effectiveness and impact of BLI formation program, through a longitudinal study of the BLI caretakers while they are in the training program and after graduation when they return to their local communities. During each of the trips to BLI, I spent between 4 and 8 hours each day interacting with and interviewing the interns and caretakers. I administered a questionnaire on them, focusing on their personal, family, and farming histories, their experience of the BLI formation program, learning outcomes, challenges, opportunities, and future goals. Within the period of the research, 54 persons from five different cohorts of students were surveyed, including ten graduate caretakers in various local communities. In terms of gender and religion distribution, there were 30 males and 24 females, 28 Catholics, 16 Protestants (mostly Anglicans), 6 Pentecostals, and 4 Muslims. The media age is 21.2.

While it would be desirable to analyze the result of the interview with each intern or caretaker, it is impossible within the limited scope and space of this essay. What I rather focus on here is some key insights that emerged from the interviews, using the interview responses from specific persons to buttress the point about Integral

Ecology as a praxis of hope. The identity of the persons mentioned in this discussion is not pseudonymized.[2] Paul Perrin (2020, p. 177) has argued that one of the ways of recognizing and enacting human dignity in project monitoring and evaluation is through narrative approaches rather than quantitative methods "due to their human touch and their far more grounded approach to discussing impacts." Unlike quantitative methods, narrative, story-telling methods refuse to turn people's lives, struggles, and stories into statistical data points that "rob local populations of their voice and the opportunity to share their experiences in their own ways" (Ibid, p. 176). In light of this, a key contribution of this essay is its elevation of the voices of young people at the grassroots who are caring for creation and investing their lives, time, energy, and resources in working for a more ecologically sustainable future. My task as an ethnographic theologian is simply to serve as a midwife, "to re-present what we have seen and heard, smelled and touched" (Whitmore 2019, p. 5).

I discuss the emerging insights from the surveys under five senses:

1. Sense of place
2. Sense of growth
3. Sense of purpose
4. Sense of mission
5. Sense of spirituality

6.4.1 *Sense of Place*

It is practically impossible to talk about Integral Ecology without a sense of place. Place is where we live, and move, and have our being. It is the setting where people build their lives. At the heart of the sense of place are these questions: Where do you call "home?" What ethical commitments do you have to this place that you call "home"? I raise these questions because we are living in a hypermobile world. Globalization and improvements in transportation and communications have made it possible for people to move from one place to another with ease and at breathtaking speed. This has made many people dislocated and uprooted, without a place they can really call "home." To use the words of Pope Francis (2013, para. 29), many people today are "nomads without roots." If Integral Ecology is not merely a speculative endeavor but a praxis and a form of engagement, then each person needs a stable place where they can connect with the natural environment more intimately. Integral Ecology is "an invitation to enter into a specific location, to be rooted in a place, to touch and feel the energies of life present in the place, and to make that place the center of one's way of being in the world" (Ojeifo 2025a, p. 12).

In conversations with BLI interns, I came to terms with the significance of place and how this shapes their approach to the care for creation. When I asked Isa, 20,

[2] They were informed and verbally consented to the use of the data for research purposes. The research protocol is covered by IRB approval from the University of Notre Dame, in 2024. The IRB protocol number is 24-10-8888.

what impact the formation at BLI is having on him, he told me: "In our place, they use a lot of pesticides. You cannot grow mangoes without using pesticides to control the fruit flies. However, here at BLI, we are taught how to make organic pesticides. They also teach us how to make organic manure from chicken wastes. In our village, these wastes are just thrown away, but here we are taught how to make wastes useful." Jovia, 19, explained what she liked about BLI in these words: "What I like the most at BLI is the forest. I came here and saw that all the trees are green. There at our village, they cut all the trees. There are no green trees there." Reflecting on what she is learning, Racheal, 19, said her plan is "to go back home and change my mom. She is rearing chickens for three weeks, and takes them to the market, but they are not good for people. The time is too short, and they affect the body of people."

Whether they refer to *our place*, *our village*, *back home*, *my community*, *here*, or *there*, what is clear is how they are articulating a sense of place with respect to the predominant ecological practices in their local communities in relation to and in contrast with what they are learning *here* (meaning BLI campus). This growing connection to place is quite significant because it reflects the difference that the BLI formation is making in their lives as they try to make sense of and respond to the socio-ecological challenges in their local communities. This suggests that the place where they are grounded (*roots*) informs the future paths and journeys (*routes*) that they may take.

6.4.2 Sense of Growth

Integral Ecology involves the moral development of persons. It is not merely a change in patterns and systems of development. While these external changes in politics, economics, and policy are significant, Pope Francis (2015, para. 202) reminds us that "it is we human beings above all who need to change." In calling for ecological conversion, Pope Francis places interior renewal of the human mind, heart, and spirit at the center of the call for "the development of new convictions, attitudes, and forms of life" (Ibid). Talking about growth, Tracy, 20, reveals that one of the greatest lessons she has learnt and internalized at BLI is self-confidence: "Before now, I was a person who was very shy. If I needed something, I could never ask for it. Also, when someone does something wrong to me, I could hardly speak out. But right now, I can talk for myself, and I defend what is right."

Nantumbwe, 18, is quite conscious of the ways that she has changed in her care and concern for God's creation as a result of the BLI formation: "Before coming here, I could torture any other living organisms, like caterpillars. I never liked them, but with my knowledge of integral ecology, I now even call mosquitoes 'Sister Mosquito.' So, I have changed. I expect that when I go out, I will be capable of making others love those creatures. And I hope that I will teach many people also to do organic farming. Because all the people know the repercussions of using synthetic chemicals, only that they just pretend and think it is the only way to go. But good enough I came to BLI which is teaching that."

Others like Mike 24, and Frank, 26, speak about the changes that has occurred in their mindset in respect of their belief in the necessity and possibility of regenerative agriculture. Coming to BLI from conventional agriculture backgrounds where chemical fertilizers are being used to grow food, Mike reflects on the ambition that his 'conversion' to the gospel of regenerative agriculture is driving him towards: "Integral ecology has the capacity to cater to needs of small families, to provide them with food and income. So, to me, integral ecology is my future and the future of my country and the world at large. Conventional agriculture will cater for the rich only. Only the rich are able to cultivate huge plantations, which they pump with chemicals and sell the toxic food to the poor, while they [the rich] go and buy organic food. If the poor can produce their own food, then there will be no way they will go and buy poisoned food."

6.4.3 Sense of Purpose

If our present socio-ecological crisis is the result of our reckless and irresponsible use and abuse of (the goods of) God's creation, to think about the consequences of our present lifestyles, choices, and actions on future generations is to enter into the realm of ethics where we wrestle with deeper questions about who we are as persons and what sort of persons we are called to be. In *Laudato Si'*, Pope Francis (2015, para. 160) states that our lives here on earth has a goal and that each one of us must think about that goal by engaging with some fundamental questions: "What is the purpose of our life in this world? Why are we here? What is the goal of our work and all our efforts? What need does the earth have of us?" Although human beings have engaged these questions for countless generations, in this age of ecological catastrophe, they have become more pressing because the ecological crisis we are facing today is one that results from anthropological amnesia. We have forgotten who we are and where we have come from. We have forgotten that we are "earth-lings" (Myers 2016, p. 120), "dusty earthlings" (Mustol 2017, p. 1), formed from the dust of the earth and destined to return to the earth from which we came.

Without reflecting on the purpose of our lives here on earth and the goal of all our work, we risk not only missing our way but also leaving behind an uninhabitable planet for future generations (see Pope Francis 2015, para. 161). This is an issue "which dramatically affects us, for it has to do with the ultimate meaning of our earthly sojourn" (Ibid, para. 160). BLI formation focuses on inculcating values that will help caretakers reflect on this sense of purpose in their lives. Many of them acknowledged that they were taught to see all creatures as their brothers and sisters and that they themselves as part of the community of creation and not lords over it. In terms of practical engagement, the interns engage regularly in community outreach programs where they share the knowledge and practices of Integral Ecology that they are acquiring in their BLI formation with local community members. This same pattern was reported by caretakers in the local communities who stated that they actively promote the dissemination of regenerative agriculture practices with local farmers in their communities whenever they have the opportunity to do so.

When I visited Vincent, 22, one of the BLI caretakers in his organic farm, he shared how the BLI formation has impacted him: "I think the greatest impact is knowing who I am and knowing my responsibility to the world. I used to think about myself only, but BLI helped me to discover myself and my responsibility towards making the planet a better place." What Vincent says about self-discovery is important for the practice of Integral Ecology. Without self-discovery it is difficult to discern the meaning and purpose of our lives. Joylin, 20, understands this well. Having come to a better understanding of the importance of Integral Ecology practices, she strives to promote tree planting. She noted, however, that due to the urge to make quick money from ecological resources, earth care does not seem to make much appeal to many people. "Once I tried to talk to the priest in my church," she said, "but he told me he didn't understand what I was saying. 'How can I plant a tree that will take 20 or 30 years to mature? When will I harvest it?' I told him it is not for us but for future generations. But you see, he was only interested in hearing that aspect of money."

Despite this obstacle, Joylin has no plan of backing down: "If I keep talking to people one by one, I might be able to change some people's minds, and when I do that in a good way, I am pleasing God." Joylin's conviction is quite striking: to preach the gospel of care for creation and be able to positively change people's minds is a way of "pleasing God." At the heart of this conviction is a deep sense of the purpose of life and how this is connected to care for creation. One cannot practice Integral Ecology if one does not entertain the hope that the world can be better than it presently is. This requires time, energy, and commitment. It requires the patience of hope. This means that some of the actions and changes that one canvasses will probably only bear fruit in the future. There is an African proverb that states: "A society grows great when leaders plant trees under whose shade they may never sit."

In discussing the integral development of society, Pope Francis (2013,) articulated four specific principles for interpreting and evaluating social phenomena. One of these is *time is greater than space*. To build something that is lasting and that benefits the common good, the principle of time is greater than space "enables us to work slowly but surely, without being obsessed with immediate results… Giving priority to time means being concerned about initiating processes rather than possessing spaces… What we need, then, is to give priority to actions which generate new processes in society and engage other persons and groups who can develop them to the point where they bear fruit in significant historical events" (Pope Francis 2013, para. 223). Hope for a better planetary future is not like turning the switch and the light comes on. It is starting a process whose lifespan might well extend beyond one's lifetime.

6.4.4 Sense of Mission

If BLI interns are concerned only about acquiring the formation in Integral Ecology so that they can go out and make a living for themselves, that would be too narrow a vision, however plausible that aspiration is. What makes the formation program rich is that it embodies a broader vision of social transformation that connects ecology with faith and mission. Most of the interns that I spoke with who are Christians believe that the church has a key role to play in Africa's socio-ecological transformation. They see the biblical mandate to "till the earth and care for it" (Gen 2:15) as "central to the church's mission in this age of ecological destruction, and they are bringing that message to the churches in their localities" (Ojeifo 2025b). While it may take some time for the full impact of BLI mission-based approach to Integral Ecology to be felt, its praxis already testifies to the fact that doing Integral Ecology challenges Christians to rethink Christian themes such as church and mission (Katongole 2022b, p. 166).

Gladys, 20, understands this very well. She told me that, among her many aspirations, she looks forward to transmitting the knowledge she is gaining at BLI to the church: "I want to use the knowledge I have gotten here to help people in my church. When I go back, I know I will be called to tell them what I have learnt. I want to put into practice what I have learned when I go back." For Tracy, 20, this sense of mission is not an aspiration but something that is already being realized. "When I joined BLI, Father Ssempala [one of the co-founders] taught us how to be confident to speak to other people. So, when I went home, I went to the church and told the catechist that I wanted to say something. He gave me the microphone, and I told the people, 'You have to plant trees.' It was the first time I was speaking in my church. All my friends were surprised to see me, and the catechist affirmed that what I said was good."

I consider this mission-based approach to Integral Ecology significant because of the way it centers the church in the task of ecological renewal facing Africa and the world (see Katongole 2022b). Andrew J. Spencer (2023, p. 160) has argued that practicing creation care is not a distraction from the mission of the church but rather part and parcel of the mission. He envisions care for creation as an essential dimension of "holistic discipleship." He insists that the world will listen to what the church has to say when the church become a "pilot plant," a concrete embodiment of the healing of the human person and the healing of nature that our time so direly needs (Spencer 2023, p. 157).

6.4.5 Sense of Spirituality

At BLI, spirituality is an integral part of the formation that interns and caretakers receive. The life of the institute follows the rhythm of prayer, study, work, and recreation. Interns gather together to pray at the start of each day and end the day with Bible study, reflection, and prayer. Communal Mass is celebrated weekly on campus and in the nearby parish church on Sunday. On July 22, 2024, I participated in the evening prayer session, which was attended by all the interns and instructors. The session leader, one of the interns, took the scripture reading from Genesis 1:1-3, focusing on the gift of light at creation. Before offering some brief reflections on light, he invited participants to freely share what stood out for them in the passage. When it came to his turn, he said: "Light enables us to see, but for light to be produced, many factors have to come together." He used the example of a charcoal stove: "For it to give you light to cook, you need many factors. One piece of charcoal cannot do it. You need a bunch. We need to come together, to work together, so that our light may shine out." Referring to the founding of BLI, he noted that it started with three priests: "One person couldn't do it alone. They had to come together. It was a joint force. That is how it should be with us. We need to have the love that attracts one another so that our dreams and visions may shine out."

The Bible reading and reflection was followed by a moment of examination of conscience and intercessory prayers. During the intercessory prayers, one intern thanked God for the gift of rain that has fallen that day, "this shows that nature is happy," she concluded. From the Bible reflection on the gift of light at creation and the divine cooperation that brings forth creation into existence, to the need for human cooperation in advancing the work of creation care at BLI, and then the intercessory prayer about rainfall and the happiness of nature, it is evident that the formation at BLI is permeated with spiritual sense. This further became clear when I spoke with Francis, 24: "I have seen an impact on the spiritual level both personally and collectively. Personally, I have learnt a culture of praying every day and I have also learnt to pray with my fellow interns. We are journeying together. You know before someone comes here, they have their own lifestyle, but when you come here, you see that you are slowly absorbed into the culture of daily prayer. It is easy for someone to convert while here. But if there is no spiritual aspect, then integral ecology becomes hard."

The conversion that Francis speaks about is at the heart of the spirituality of Integral Ecology that Pope Francis wrote about in *Laudato Si'*. For Pope Francis, Integral Ecology has a deep spiritual dimension grounded in the convictions of our faith, which "can motivate us to a more passionate concern for the protection of our world" (Pope Francis para. 216). What drives Francis' notion of an ecological spirituality is the conviction that creation care cannot be sustained by doctrine alone "without an interior spirituality capable of inspiring us, without an 'interior impulse which encourages, motivates, nourishes, and gives meaning to our individual and communal activity'" (Ibid). Although the Pope did not work out in detail what this spirituality entails, he underscores the value of a spirituality that leads us to acknowledge our faults and failures in regard to care for creation, heartfelt repentance, and a desire to change (Ibid, para. 218). He speaks also about the importance of the sacraments as "a privileged way in which nature is taken up by God to become a means of mediating supernatural life" (Ibid, para. 235), inviting us to enter into a deeper relationship and "live in communion with God, with others, and with all creatures" (Ibid, para. 240).

6.5 Conclusion

While proposing the model of Integral Ecology as a framework for addressing our ecological crisis, Pope Francis was keenly aware that there are several radiant examples of people and communities that are already pointing us in the direction of sound ecology. This is precisely why he says that all hope is not lost. Pope Francis is not saying that the earth is doomed and that there is nothing that can be done. If that were the case, the only option would be to wriggle our hands in fatalism and wait for doomsday to come upon us. On the contrary, Pope Francis is saying that change is possible, but that for this to happen we need a greater mindset shift. BLI people are the sort of people who believe that change is possible and are willing, ready, and committed to doing the sort of thing that *Laudato Si'* is calling for, namely: roll up your sleeves and get to work! This commitment to the care for creation that BLI people exemplify has potential for shifting the needle of research in the field of environmental ethics from theory to praxis.

From the perspective of World Christianity studies, I find the BLI model of Integral Ecology quite illuminating because it speaks to the quiet revolution taking place in Christianity in the global south, particularly in Africa, where the key agents of social change are not religious power elites such as cardinals, bishops, senior apostles, or even general overseers of megachurches, but faith-filled, energetic, and creative young believers at the grassroots who are putting their faith into action in order to better respond to the cry of the earth and of the poor. What is more interesting is that this quiet transformation is not taking place in the big cities and urban centers of Africa, but at the frontier of an African village. As Katongole (2011, p. 144) has argued, Africa's social revival will come about through the church's embrace of an "ecclesiology of relocation," from heaven to earth, and from the city to the village, in ways that embody the incarnational realism of the God who enters into the messiness of our human condition in order to redeem his whole creation. A church that is grounded in the Incarnation takes such places—villages and slums in Africa inclusive—seriously. These are the places where hope tends to be lacking, where hope is solely needed, where hope is surprisingly found. It is hard to see how hope has a future unless we love and cherish such places. At the end of *Laudato Si'*, Pope Francis (2015, para. 244) left us with these immortal words of encouragement: "Let us sing as we go. May our struggles and our concern for this planet never take away the joy of our hope."

References

Francis, Pope. 2013. *Apostolic Exhortation Evangelii Gaudium on the Proclamation of the Gospel in Today's World*. Washington DC: USCCB.

Francis, Pope. 2015. *Encyclical Letter Laudato Si' on the Care for Our Common Home*. New Jersey: Paulist Press.

Francis, Pope. 2024. *Bull of Indiction of the Ordinary Jubilee of the Year 2025*; at: https://www.vatican.va/content/francesco/en/bulls/documents/20240509_spes-non-confundit_bolla-giubileo2025.html (Accessed: 12 July 2025).

Francis, Pope. 2025. *Hope: The Autobiography* (with Carlo Musso). New York: Random House.

Gandolfo, E.O. 2023. *Ecomartyrdom in the Americas: Living and Dying for Our Common Home.* Maryknoll, NY: Orbis Books.

Hickman, C. et al. 2021. 'Climate anxiety in children and young people and their beliefs about government responses to climate change: a global survey', *Lancet Planet Health* 5(12), pp. 863-873.

Ilo, S.C. 2024. "Church-Family of God in Africa on a Pilgrimage of Hope', in Ilo S.C. and Katunge L. (eds.) *Journeying Together in Hope: A Theological and Pastoral Reflection for a Synodal Church in Africa.* Nairobi: Paulines Publications Africa, pp. 1-12.

Jones, A.C. and Lucas, C. 2023. "Listen to me!': Young people's experiences of talking about emotional impacts of climate change', *Global Environmental Change* 83, pp. 1-15.

Katongole, E. 2011. *The Sacrifice of Africa: A Political Theology for Africa.* Grand Rapids: Eerdmans.

Katongole, E. 2012. *Stories from Bethany: On the Faces of the Church in Africa.* Nairobi: Paulines Publications Africa.

Katongole, E. 2015. 'Of Coffins and Churches: Seven Marks of an Emerging African Ecclesiology', in Orobator A.E. (ed.) *The Church We Want: Foundations, Theology and Mission of the Church in Africa.* Nairobi: Paulines Publications, Africa, pp. 190-202.

Katongole, E. 2018. 'Rooting the Church in African Soil and the Bethany Land Institute: A Theological Experiment', in Cavanaugh W.T. (ed.) *Fragile World: Ecology and the Church.* Eugene, OR: Cascade Books, pp. 219-232.

Katongole, E. 2022a. 'Mission as Integral Ecology: Doing Theology at Bethany', *Mission Studies* 39, pp. 163-185

Katongole, E. 2022b. 'Enacting *Laudato Si'* in Africa', in Ilo S.C. (ed.) *African Ecological Ethics and Spirituality for Cosmic Flourishing: An African Commentary on Laudato Si'*. Eugene, OR: Cascade Books, pp. 142-147

Kluger, J. 2025. "Climate Anxiety Is Taking Its Toll on Young People." *Time,* 29 April [online]. Available at: https://time.com/7280989/climate-anxiety-mental-health-young-people/ (2 September 2025).

Moltmann, J. 2012. *Ethics of Hope.* London: SCM Press.

Mustol, J. 2017. *Dusty Earthlings: Living as Eco-Physical Beings in God's Eco-Physical World.* Cambridge: The Lutterworth Press.

Myers, C. 2016. *Watershed Discipleship: Reinhabiting Bioregional Faith and Practice.* Eugene: Cascade Books.

Ojeifo, E. 2025a. 'Integral Ecology as Theology and Ethics of Place', *Religions* 16(659), pp. 1–14

Ojeifo, E. 2025b. 'It's A New Day!': Faith-based ecological revival in an African village', *The Occasional*, 23 June [online]. At: https://omsc.ptsem.edu/its-a-new-day-faith-based-ecological-revival-in-an-african-village/ (Accessed: 5 September 2025).

Perrin, P. 2020. 'Contextualized Participatory Monitoring and Evaluation', in Carozza, P.G. and Sedmak, C. (eds.) *The Practice of Human Development and Dignity.* Notre Dame: University of Notre Dame Press, pp. 163-188.

Spencer, A.J. 2023. *Hope for God's Creation.* Brentwood, TN: BH Academic.

UNICEF. 2021. *The Climate Crisis is a Child Rights Crisis: Introducing the Children's Climate Risk Index*. New York: United Nations Children's Fund.

Whitmore, T.D. 2019. *Imitating Christ in Magwi: An Anthropological Theology.* London: T&T Clark.

Wirzba, N. 2024. *Love's Braided Dance: Hope in a Time of Crisis.* New Haven: Yale University Press.

Chapter 7
Becoming Integral Ecologists: Amazonian Encounters, Ecological Conversions, and Recovering the Commons

Ryan Juskus

Abstract Calls for an integral approach to human ecology are premised on the recognition that such a framework is currently lacking in ecological thought and practice. According to the proponents of integral human ecology, current frameworks tend to be anthropocentric and technocratic in nature, assuming ethical and ontological distinctions between humanity and the natural world and privileging technoscientific forms of knowledge and management of environmental challenges rather than transformations in the social and cultural systems that produce ecological crises. Shifting from anthropocentric and technocratic frameworks to integral human ecology therefore requires a fundamental change that Pope Francis has called "ecological conversion." Ecological conversion requires changes in how individuals and communities relate to the world around them. This chapter examines the case of Peruvian peace and justice organization Paz y Esperanza's change from an exclusive focus on human rights to an integral human ecology framework to try to ascertain the conditions that foster ecological conversion. Analyzing fieldwork with Paz y Esperanza, it gives an account of how they evolved from a focus on addressing political and interpersonal violence to defending the Amazonian commons and its Indigenous protectors through a sustained practice of ethical encounter. This practice of encounter and accompaniment over time with defenders of the commons created the conditions in which mestizo human rights activists could rediscover and refashion a theological commitment to commoning and co-defending the commons from enclosure. Based on analysis of this case, the paper promotes cross-border, neighbor-loving encounters on the front lines of the ecological crisis that can propel the rediscovery and refashioning of an integral human ecology capable of transforming the ecological crisis.

Ryan Juskus, University of Notre Dame, rjuskus@nd.edu.

R. Juskus (✉)
University of Notre Dame, Notre Dame, USA
e-mail: rjuskus@nd.edu

R. Marcantonio et al. (eds.), *Integral Ecology in the Anthropocene: Africa and Beyond*, The Anthropocene: Politik—Economics—Society—Science 49,
https://doi.org/10.1007/978-3-032-32125-1_7

Keywords Ecological conversion · Peru · Peace · Commons · Environmental defenders

7.1 Introduction

In 1999, just a few years after their founding in Lima, Peru, the peace and justice organization Paz y Esperanza opened a provincial office in the upper jungle of the Peruvian Amazon. Paz y Esperanza ("Paz"), an ecumenical Christian organization with largely mestizo leadership, had spun off from the National Evangelical Council of Peru in 1996 as an independent human rights NGO. When they started operations in the Amazonian city of Moyobamba three years later, it was the tail end of two decades of political violence that had rocked the country's interior since 1980 (Comisión 2004). Paz Moyobamba's initial goal in the Amazon was to address the effects of this political violence as well as sexual and domestic violence. Yet, within a few years, they were engaged in environmental justice, equipping citizen-led environmental monitoring, resolving land conflicts, obtaining legal recognition and land title for Indigenous communities, facilitating alternative models of development, strengthening Indigenous peoples' consent rights, developing mechanisms for intercultural governance, opposing "green landgrabbing," advocating for climate action, suing industrial agribusinesses over deforestation laws, and promoting Indigenous-led conservation and sustainable development. In short, this Christian peacebuilding organization that was initially focused on political and intrafamilial violence became key contributors to Indigenous-led protection of Amazonia's lifeways and ecosystems. The story and import of Paz Moyobamba's evolution into defenders of the Amazonian commons is my focus in this essay. As a story of ecological conversion on the front lines of an Anthropocene resource frontier, it offers wisdom about the conditions and possibilities for making integral ecologists out of proprietary and technocratic ecologists.

Calls from religious leaders and grassroots communities for an integral approach to human ecology are premised on the recognition that such a framework is currently lacking in ecological thought and practice. According to the proponents of integral human ecology, current frameworks tend to be anthropocentric and technocratic in nature, reproducing dualisms between humanity and the natural world and privileging proprietary and technoscientific forms of knowledge and management of environmental challenges rather than transformations in the social and cultural systems that reproduce ecological crises. Making the shift from anthropocentric and technocratic frameworks to integral human ecology is what Pope Francis (2015, para. 217) has called "ecological conversion," meaning changes in how individuals, communities, and peoples relate to the world around them.

This chapter examines the case of Paz Moyobamba's conversion from an exclusive focus on human rights to an integral human ecology framework to try to ascertain the conditions that foster ecological conversion. Analyzing fieldwork with Paz y Esperanza, it gives an account of how they evolved from a focus on addressing political and interpersonal violence to defending the Amazonian commons and its Indigenous protectors through a sustained practice of ethical encounter. This practice of encounter and accompaniment over time with defenders of the commons created the conditions in which mestizo human rights activists could recover and refashion a theological commitment to commoning and co-defending the commons from enclosure. Based on analysis of this case, the paper promotes cross-border, neighbor-loving encounters on the front lines of the ecological crisis that can propel the rediscovery and refashioning of an integral human ecology capable of transforming, and not simply managing, the ecological crisis.

This case study from the Amazon belongs in a volume focused on Africa because in the Anthropocene age, they are mere inches apart. What happens in the Peruvian Amazon affects the earth system. From a climate perspective, the Amazon sequesters more carbon than it emits, though this may soon tip in the other direction (Carrington 2021). Amazonia is rightly called "the lungs of the earth," and its fate has planetary significance. And yet, those who have learned over generations to breathe together with the earth's lungs and defend them from enclosure and degradation are under assault from many quarters. These "land and environmental defenders" carry intergenerational, place-based wisdom about what it means to be integral ecologists who breathe with Amazonia. Yet, Amazonia is the world's most dangerous place to defend the environment. According to the NGO Global Witness, "As guardians of the forest, land and environmental defenders are on the frontline of the Amazon's devastating exploitation. They face dangerous companies acting with impunity, ruthless state security forces and contracted killers. As they stand up against powerful agribusiness, mining and logging interests, defenders are systematically intimidated, criminalised, attacked and murdered. Every year, defenders of this priceless biome pay with their lives for protecting their homes, livelihoods and the health of our planet" (Global Witness 2023, p. 15). In 2022, one in every five individuals murdered globally for defending their environment was in Amazonia, with Peru being among the top ten deadliest countries (Global Witness 2023, p. 15, 24). What is more, strategies developed in Latin America to criminalize and suppress environmental defenders are on the rise in places like Australia, Europe, India, Uganda, and the U.S. (Canineu 2024; Lakhani/Beaumont 2024). In short, what happens in Latin America impacts Africa and beyond.

In this context, the story of Paz Moyobamba's evolution into an integral node in the defense of the Amazon and its Indigenous peoples is a story of repair and conversion on the conflictual frontlines of climate change. Paz has been led by mestizo Peruvian inheritors of a missionary-derived evangelical Christianity that had stripped down its ecological elements to, at best, anemic and proprietary teachings about environmental stewardship. However, through their purposive encounters with Indigenous defenders of the commons in the Amazon, some of whom are also evangelical, they discovered a pathway of repair. By focusing on Paz Moyobamba's journey, my study is simultaneously about a way of repair for those beyond Amazonia who may care to address the climate crisis, but who do so with an enclosed imagination that has been braided around violent historical processes of enclosing the commons for private property or public management.

As an ethicist, my aim is to contribute to reparative ethics. As Kristen Heyer argues, "ethics scholarship often remains in the mode of *critique*—where we unmask, problematize, and subvert—rather than undertaking the work of *repair*." Whereas critique is often immediately rewarded, the work of repairing a broken moral imagination tends to be slow and far less glamorous and nonlinear. Yet, it is necessary for wider transformation. As Heyer adds, "renewing moral imagination is not disconnected from advancing structural change" (Heyer 2024, pp. 6–7).

And today, climate politics needs a renewed vision of structural change. As it becomes more apparent every year that the international mechanisms created to address climate change are being captured by the very actors driving the crisis, meaningful structural changes will need to be driven by an imagination that is liberated from the shackles of violent enclosure and its closely associated state-market binary.[1] Together these render far too many climate concerned individuals passive denialists. Although they may not deny climate science, their denial of agency is perhaps more pernicious—denying that there is anything meaningful common people can do other than pray for an awakening in the Silicon Valleys, Wall Streets, and Beltways of the world, where tech, business, and governmental elites fashion themselves as the primary agents of resolving big problems. A renewed imagination is needed. The wager of my study is that a repaired imagination will come about through sustained, ethical encounters with those who carry traditional wisdom about how to steward the commons for the common good.

In what follows, I first describe the case of Paz y Esperanza in Moyobamba, Peru, and tell the story of how they developed an integral ecology. Then, shifting toward analysis, I argue that Paz's commitment to Samaritan encounters is what propelled their learning about the cultural, economic, and environmental causes of violence and eventually their contribution to an Indigenous-led movement to resist enclosure and defend the commons. In conversation with Indigenous studies, I theorize this practice of encounter as conditioning the retrieval and refashioning of a largely suppressed Christian commons framework that interacts with the Indigenous commons in politically generative ways. These distinct approaches to commoning suggest that a shared response to socio-environmental conflicts in the Amazon is focused on defending and constructing an Amazonian commons. Finally, I consider what my study of this case entails for integral ecological conversions beyond Amazonia.

7.1.1 *The Case of Paz y Esperanza in the Peruvian Amazon*

The following account of Paz Moyobamba's evolution from addressing violence to defending the Amazonian commons is based on document analysis, interviews, focus groups, and participant observation conducted on an intermittent basis since 2004.

[1] In DiGirolamo and Rachel Donald (2025), Christiana Figueres, the executive secretary of the UN Framework Convention on Climate Change, the world's leading body for addressing climate change, recently urged the UN to overhaul the COP process, calling it "no longer fit for purpose."

Paz y Esperanza is a Christian peace and justice organization with *evangélico* roots in Protestant missionary activity.[2] Founded in Lima during a period of civil conflict, their mission is to accompany victims of violence and vulnerable communities so that they may live lives that are free of violence and injustice. They realize this mission through a three-pronged approach: preventing violence and injustice through building cultures of peace; intervening in cases of violence and injustice through restorative justice practices; and reforming policies and structures through advocacy. While many of Paz's leaders, practices, and frameworks are *evangélico* in nature, the organization defines itself as 'ecumenical,' meaning it collaborates with anyone who resonates with its biblically engaged vision of peace and justice.

Theirs is a whole-person, whole-society theological approach to peace and justice. Their vision of the human person and of the conditions for human flourishing is based on the Hebrew concept of *shalom* and the Johannine concept of *vida plena*, or "full life." They therefore approach human rights in social terms, contrasting somewhat with a liberal, proprietary account of rights vis-a-vis states. Rights also attach to human persons as members of associations like families, churches, and cultural groups. Differentiating their mission from that of the church, Paz sees their role as akin to that of the prophet Nehemiah rebuilding the city's walls—its laws, infrastructure, and citizenship—vis-a-vis Ezra's role to rebuild Israel's religious life and foundations. Jorge Arboccó, Paz's director, specifies that their "role is to reconstruct the walls of society destroyed by corruption, exclusion, intolerance, discrimination, and violence" (Focus Group 2018). In short, they aim to create conditions in which individuals and communities can realize justpeace and abundant life.

Paz's pathway into integral ecology has followed a typically Latin American trajectory. Scholars have noted that Latin American environmentalism is more focused on "environmental justice" or "social ecology" than on the "deep ecology" or "land ethic" frameworks that have predominated among middle-class North American and European environmentalists (Lorentzen 2020; Peterson 2016). Centering on the intersection of human rights and environmental degradation, Latin American environmentalisms have been shaped by leftist politics, liberation theology, the Indigenous movement, and the pursuit of sustainable alternatives to exploitative development. Paz y Esperanza is no exception to this general story. Their environmentalism emerged from their prior theological commitment to the human person as one to whom individuals and the whole society owe justice. It grew from a concern for individual persons, eventually incorporating their ecological and cultural embeddedness in ecosystems and groups.

Paz's environmental work evolved through their programs in intercultural development and intercultural governance in three phases. In their first phase of environmental activity, *Paz y Esperanza* opened a provincial office in 1999 in the upper Amazon department of San Martín, where they first learned of intercultural conflicts between Indigenous groups and migrant mestizo farmers (Paz y Esperanza

---

[2] While I use the Spanish *evangélico* and the English *evangelical* interchangeably, it is important to note that "evangelical" does not carry the same meaning in Latin America as it does in North America; it broadly refers to all Protestant Christians.

2009). They chose this region because although Indigenous peoples had been greatly affected by political violence, they were invisibilized in Peruvian society. The region also lacked an NGO presence.

Initially focused on political and interpersonal violence, Paz Moyobamba was soon invited to address conflicts between Aguaruna Indigenous communities and mestizo farmers migrating from the highlands. According to Jorge, "During these years many Indigenous women, above all Awajún women, came to us seeking help, denouncing cases of sexual abuse and domestic violence. But then the tribal leaders began coming, seeking our assistance with getting recognition of their territories" both to resolve conflicts with invading mestizo 'settlers' and to strengthen their advocacy with the state. In this way, "*poco a poco*," Jorge says, Paz "began with a very basic agenda and then widened it in response to demand from those around us" (Interview 2023).[3]

More than just conflicts over land, Paz realized these were also conflicts over visions of the Amazon and its development. The mestizo newcomers, traveling on highways newly paved by the government and lured by agricultural subsidies, were largely unwitting foot soldiers in a strategy to integrate the upper jungle into national and international markets.[4] Aguaruna leaders invited Paz Moyobamba to collaborate with them to address these political-ecological conflicts. In response, Paz Moyobamba started the Program for Mestizo-Aguaruna Intercultural Development ('PRODIMA'). PRODIMA supported Aguaruna groups to pursue a development model that was unique to their cultures and their agricultural and forest management practices (Paz y Esperanza 2009). The PRODIMA team also helped identify far more Indigenous communities in the region than had previously been thought. There were 28 Indigenous federations recognized at the time. Eventually PRODIMA helped identify and obtain recognition of nearly 200 more, roughly 60 of whom have since gained collective title to their land (Interview 2013; Arboccó 2021).

In a second phase, encompassing about a decade beginning in 2003, Paz Moyobamba's analysis of the drivers of environmental conflicts expanded to include wider global forces, and they responded through strengthening Indigenous and agrarian communities' ability to reject exploitative development models and pursue alternative trajectories.

The story of Paz Moyobamba's campaign to address pollution from a cement factory in the mestizo evangelical town of Segunda Jerusalén exemplifies both their growing commitment to faith-driven environmental justice and its limits. In 1997, a cement factory opened on the outskirts of Segunda Jerusalén, and residents soon grew concerned about the effects of pollution on child health and crop yields. Paz organized citizen participation to hold the factory accountable to the town's citizens. Support by the town's large Pentecostal church was integral in successfully pressuring

[3] Here, and in what follows, all translations are mine.

[4] More precisely, according to Orihuela (2019, pp. 140–41, 147), this was a second wave of global market integration. The first wave was during the late nineteenth- and early twentieth-century rubber boom occasioned by high demand for tires during the exponential growth of the auto industry. However, that wave crashed and subsided decades before this second wave.

the factory to install filters and invest in the town. But the cement factory then began exchanging financial gifts to the community, including building a large new church building, for opposition to Paz Moyobamba. The Pentecostal church severed ties with Paz by 2006. While the campaign was partially successful in improving environmental conditions, it also taught Paz the power of distant economic interests to shape the region's theology and political ecology (Paz y Esperanza 2006).

Burned by this experience with mestizo evangelicals, Paz strengthened its support for Indigenous groups protecting the Amazon from extractive development pressures and "fortress conservation" concessions (Brockington 2002). These tensions, exacerbated by a new free trade agreement with the U.S., reached a boiling point in 2009. Peru's president pushed a liberalizing policy package through Congress meant to remake Amazonia as a resource base for global markets. Indigenous groups blocked the highway and therefore the flow of resources from the jungle to the coast. The president sent the military to unblock the flow through violence. Thirty-three people were killed. This 'Baguazo' massacre was a wake up call regarding the need for greater protections. Paz increased their work for Indigenous recognition, citizenship, and land titling as a bulwark against exploitative development. They helped unite Indigenous federations into confederations in response to decades of state and engineer-led efforts to divide them into individual administrative units to more easily facilitate land sales and leasing. Paz lobbied for Indigenous peoples' rights to free, prior, and informed consent. They also supported the creation of Indigenous journalism and communications networks to visibilize Indigenous peoples' goals, experiences, and efforts in a country whose journalism, like its politics, was centered in coastal, mestizo-led cities.

During this period, Paz Moyobamba developed a theological interpretation of the political-ecological conflict over Amazonia akin to the "two ways" framework of Deuteronomy 30:20: "I have set before you life and death, blessings and curses. Now choose life." For example, a 2010 guide to mining and hydrocarbons published by Paz in collaboration with Indigenous organizations, rejected extractivism as a curse. One section asks, "Is mining the salvation of our economy?" It answers No, stating further, "Having a resource-based mineral economy has been more of a curse than a blessing." It ruins Amazonia to feed a larger global-economic machine (Aguilar et al. 2010, pp. 7–8). While the guide does not specify a positive account of what *is* salvific, one Paz leader points in this direction when he says, "The historical role of Paz is to break the chains of death and motivate life, life in abundance; not that of the god of money, but rather that of the Lord of life and justice." For Paz, at stake in conflicting development models was the meaning of 'life' and what might be sacrificed to achieve it. This leader continues, "[Ours] is a spirituality based on the right and duty of all human beings to respect life and to foster it under norms of life where we all have the same rights" (Focus Group 2018).

Jorge, the Paz director, sums up Paz's evolution during these years. He says, "Honestly, in every place where we have given attention to one issue, another issue would inevitably come into view, and clearly, the environmental issue has been a constant: reductions in water levels, soil loss, or some other specific topic. However, these specific issues have each been localized. What has most demanded our attention over the years is the issue of Indigenous territory. Why? Because it is the majority of the territory [in this region], and because it is the most affected, and because it is also the least seen" (Interview 2023). This was a journey from seeing discrete environmental problems to strengthening Indigenous territorial sovereignty.

Paz Moyobamba's third phase has continued this trajectory in an increasingly dangerous environment, responding to the consolidation of a violent extractivist mafia in the Amazon and fortress conservation with more sophisticated tools, support for "integral territory," and engagement with climate change. There is now little daylight between industrial agribusinesses, narcotraffickers, illegal gold miners, police forces, and government officials, making environmental defense more dangerous than ever. During my brief visit in 2023, Paz personnel traveling with Indigenous leaders had a riverboat stolen, presumably by mafia agents, and then an Aguaruna pastor and his family in a remote community were forced at gunpoint to flee because of their vocal opposition to illegal gold mining in the river that serves as the community's livelihood. Nevertheless, Paz Moyobamba has continued its support for environmental defenders by publishing annual alternative development reports, using aerial and geospatial imaging to build legal cases and equip sympathetic government agents with maps of forest cover, and promoting sustainable alternatives to the politics of enclosure's binary of extractivism, on one side, and fortress conservation, on the other. During this phase, they have also followed the Indigenous movement's lead by expanding their focus from land titling to "integral territory," which I will explicate shortly. At the same time, Paz began to engage directly with climate change, portraying environmental defenders as defending not only their territory and culture but also the planet's 'lungs' from those who would sacrifice them for profit.[5]

Jorge, a trained anthropologist, published a report in 2021 that reflects on Paz's learning and champions an Amazonian political ecology grounded in ecojustice spirituality. Drawing again on the Deuteronomistic "two ways" framework, he paints the Peruvian Amazon as caught up in a struggle between the forces of life and the forces of death, with extractivist development, global market integration, and fortress conservation on one side and a constellation of spiritually rooted ecojustice paradigms in dialogue with one another on the other. These include the Indigenous concept of *buen vivir*, Pope Francis's "integral ecology," particularly as it was developed at the 2019 Amazon Synod, and Paz y Esperanza's *vida plena*, or "abundant life" theology as it developed in an ecological direction through Amazonian encounters. He writes, "We are not just talking about a clash of opinions; we're talking about sustaining the defense of the natural and socio-economic environment that is key to humanity's subsistence as well as that of the planet's" (Arboccó 2021, p. 9). Jorge's report, full primarily with empirical data, history, and analysis, ultimately promotes a dialogical spirituality of ecojustice capable of effectively countering Amazonia's sacrifice.

[5] During this period, Paz y Esperanza embraced Pope Francis's concept of "integral ecology" and the Catholic Church's affirmation of the unique cultures and spiritualities of Amazonia in documents related to the 2019 Amazon Synod.

7.1.2 Encounter and the Commons

What should we make of Paz Moyobamba's evolution? One could interpret it as an example of "The Greening of Religion Hypothesis," which posits that religious ideas, like anthropocentrism, drive environmental behaviors; the world's religions have ideas that drive pro-environmental behaviors; therefore, these should be promoted to make societies more environmentally sustainable (Taylor 2016, p. 296). However, this theory foregrounds ideas in both its problem description and its proposed solution, presuming to know what sustainability is and which religious ideas support it. Specifically, it fails to account for how Paz Moyobamba centers the human person and the practice of encounter that catalyzed their defense of the commons against efforts to enclose them, including enclosure for conservation. I submit that Paz's evolution is better understood in terms of the commons and commoning: it is an evolving Christian political ecology of the commons that is driven by a border-crossing, neighbor-making practice of encounter. Paz Moyobamba's mission to accompany vulnerable persons to live free of violence and injustice led them to sustain encounters with Indigenous communities on the frontiers of a globalizing enclosure movement. Efforts to integrate the Amazon into global markets as a resource base for state-making and money-making projects and efforts to use Amazonian nature as a fix for sustainability problems clash with Indigenous peoples' unique relationships to their ancestral territory. Over the past quarter century, and propelled by their commitment to Samaritan encounters with these communities, Paz has discovered a commitment to defending and managing the commons in collaboration with Indigenous stewards. This has become essential to their enactment of *shalom* at a time of violent enclosure-caused environmental changes driven by agents of the market and state. Paz Moyobamba therefore demonstrates the cultural and political transformation that can happen through encounters on the front lines of systematic, ideologically driven efforts to enclose what has been—and can be—realistically and collectively stewarded as a common good, or, following Pope Francis, a "common home."

This argument to subordinate the concept of sustainability to the practices of commoning and encounter is indebted to a growing interest in studying and promoting the commons. Having originated in the environmental movement in the 1960s, the field of environmental policy initially treated the commons as a straw man to establish a public-private binary. Garrett Hardin (1968), in his influential 1968 essay "The Tragedy of the Commons," put economistic theorizing ahead of empirical study when he argued that commons would naturally be overused and would thus degrade over time, ending in tragedy; the only strategies to prevent tragedy were privatization and government regulation. Elinor Ostrom's (1990) empirical approach to the matter a few decades later showed that Hardin both confused commons with open-access resources, thus distorting his conclusions, and neglected to study how communities around the world successfully managed common-pool resources with collectively-established rules. Ostrom's research revealed the public-private binary to be an ideological construct. It also inaugurated a wave of research on the commons that has not subsided, some of it going beyond Ostrom's own theoretical limits by examining strategies of commoning beyond the management of common-pool resources (Silber-Byrne 2024).

The ideological public-private binary is currently being dismantled on the front lines of enclosure—the *avant-garde* of the state-market binary. Key to understanding Paz Moyobamba's evolution is recognizing that they opposed not only Amazonia's global market integration but also its conscription into fortress conservation through green landgrabbing, whether by private or public entities. Several international programs, such as debt-for-nature swaps and the UN's REDD program, have converted Amazonian natures into solutions to economic and planetary problems through the mechanism of enclosure. Paz Moyobamba witnessed the use of programs like these to dispossess Indigenous and agrarian communities, enclosing their former territories as conservation concessions to solve the problems of distant wealthy elites and petrocultures in the Global North. Furthermore, enclosure for conservation does not conserve forested landscapes. By removing the land's protectors, conserved areas become more vulnerable to exploitation. No longer in the hands of local communities, forest protection becomes dependent on perennially fickle national budget procedures; it also becomes ripe for corruption.

For these reasons, Paz has more recently followed the lead of Indigenous groups by supporting the pursuit of "integral territory," in addition to land titling. Researcher Roger Merino (2021, p. 136) describes the differences between these, writing "integral territory" is a "new model for territorial governance [that] is aimed at transforming indigenous peoples from ethnic communities with property entitlements to nations with territorial rights." By following this evolution from titling to territory, Paz has positioned itself as a defender of the commons against enclosure through promoting intercultural territorial governance. According to commons scholars, "In Latin America, recognition of indigenous commons is the result of social movements in which indigenous peoples struggle for land and resources and pressure governments to recognize their rights, a process that has had significant influence on environmental governance" (Monterroso et al. 2019, p. 377). In this sense, the state has been forced by social movements to recognize Indigenous commons as an alternative to enclosure, though this recognition is still far from established. What is not yet understood is how the Indigenous commons interact with other commons frameworks on these front lines.

The work of Susan Hegeman and Dina Gilio-Whitaker on Indigenous-settler encounters in North America helps fill in this gap, allowing us to understand both what is happening through these encounters on the front lines as well as their politically generative fruits. In her research on the Standing Rock encampment to oppose the Dakota Access Pipeline, Hegeman argues that while the concept of the commons is Western in origin, Indigenous groups have found common cause with settler groups by using language that connects multiple traditions of commoning. In the Standing Rock case, these are an "Indigenous commons" and "a wide variety of customary ideas about land use that derived from medieval Europe" (Hegeman 2019,

p. 137). Hegeman points out how the Standing Rock Sioux activists projected themselves as water protectors, not protesters—protecting water against private expropriation and publicly sanctioned pollution (Hegeman 2019, p. 134). Many non-Indigenous people who joined the encampment in solidarity were driven by concerns about climate change, which is a concern with the commons at a planetary scale. Sioux activists incorporated these concerns into their encampment and messaging. Hegeman concludes that when the water protectors at Standing Rock "link[ed] their local protests to multiple ideas of commons, in their rhetoric and in the practices of the encampment, they simultaneously invoked ways of thinking long suppressed in the settler-colonial imagination and asserted a fully indigenous political principle" (Hegeman 2019, p. 139). And re-engaging long suppressed thought is what Indigenous scholar Dina Gilio-Whitaker identifies as one of the goals of decolonization when she writes, "colonization has steered the colonizers away from their own ancestral wisdom," which is why "[d]ecolonizing the colonizers is necessary so that they can once again learn how to respect themselves and others" (Gilio-Whitaker 2019, p. 26). In short, these scholars observe that encounters between colonizers and colonized, within the context of efforts to protect environmental commons, creates favorable conditions for retrieving concepts that predate the enclosure movement, and they think this holds great promise for climate politics. It provides a context in which a moral imagination braided with enclosure can be repaired, not by appropriating Indigenous cultures, but by revisiting long suppressed traditions that had nurtured a commons wisdom long before the violent enclosure movement that began in early modern Europe and now threatens to collapse "the lungs of the earth."[6]

I think something like this phenomenon of encounter between different traditions of commoning leading to reconnection with ancestral wisdom is operative in Jorge's evolving understanding of human rights. It is worth quoting him at length. Jorge says, "The spirituality of land defenders grows out of their sense that they are the land, that it is their mother, their father, their family." In contrast, those of us who have separated ourselves from nature need to re-learn how to be "human beings, not productive machines. We are not producers of things. Rather, we are gregarious more than anything, and as gregarious beings we try to achieve a kind of balance. And when we make ourselves into machines we feel isolated, self-absorbed, we feel that we are nothing, that we don't matter." But Jorge sees transformative potential in encounters with Indigenous cultures. "So when we look back and see the spirituality of these communities and see how they treat their environment … I think it gives us back a few things. I think it even gives us back more rights." He says there is a backwardness in 'our' [Western] thinking about rights, putting economic, social, and cultural rights before environmental rights. But Indigenous peoples help correct this by reversing it.

We should probably reverse this list and begin with environmental rights. Ultimately, we live in an environment, and our first right is when we are born, to breathe, because when we're given a tap and made to cry the first thing we do is take a breath. And maybe it is for this reason that in the Bible it talks about how God breathed and

[6] For the beginnings of the enclosure movement, see Polanyi (1944).

gave us that first breath. I believe that that breath ought to call our attention also to understanding that this breath of life is our first right. And this breath of life is our first relationship with our family, our relationship with our environment, which is nature. Nature is also our family.

Jorge thinks mestizo Christians have lost this sense because they have treated religion as a commercial thing, something to be bought and sold as-is, failing to grapple with it in dialogue with others. Instead, he argues that it needs to be personalized. The evangelical tradition he inherited did not teach these things, but it has become part of his spirituality through encounters with Indigenous communities in the upper Amazon. "What the missionary, the pastor, the church leader gives is a path, but how you go walking along that path is another thing. … We have to go about learning in community with others how to give shape to that path." He continues, "In order to be salt and light in reality we need to be able to learn or re-learn the Bible with our brothers and sisters, understanding it not only from an enclosed perspective but understanding it in a completely open way, that is with a vision that can help us understand things about creation" (Interview 2023). In Jorge's case, this led him to revisit the priority of the breath of life in the creation story and therefore the priority of environmental rights over economic, social, even cultural rights. Before the creation of the human being and the creation mandate, there is the breath of life. Without a common space to breathe, there can be no life. In this sense, rights to property should be embedded within a more primary right to the ecological conditions that support life.

Jorge's reflection is evidence that sustained encounters on the front lines of enclosure led Paz to revise their understanding of rights while they were becoming co-defenders of the Amazonian commons. Jorge's encounter with the Indigenous commons led him to invert his prior understanding of human rights: environmental rights are primary; cultures, economies, and social systems ought to fit within them.[7]

In addition to this revised understanding of rights, Jorge says that his encounters in the upper Amazon have reshaped his notion of Christian stewardship. When I asked Jorge what guides his thinking about environmental topics, I was initially surprised by his conventional response. He replied, "stewardship of creation." But then he clarified that it is an understanding of stewardship that has to be formed through Samaritan encounters. For him, Samaritan encounters help us first see the plank in our own eye before we can see the splinter in our brother's eye. "I have to be like the Samaritan, and I have to see that when my brother suffers we all suffer, and I have to see also that when it talks about good news it means good news for all. When Jesus tells his disciples go and spread good news to the whole creation, it is for everything created." Through Samaritan encounters he has learned that stewardship does not mean 'domination,' 'subjugation,' or 'annihilation.' He says that by entering into the problems of Amazonian communities he has learned that creation rejects human efforts to dominate it. Now he believes stewardship of creation means aiming for balance, protecting the breath of life that circulates through the ecosystem. The

[7] It is worth noting that this is remarkably similar to the medieval teachings of Thomas Aquinas and the tradition of Catholic Social Teaching on the common gift and destination of created goods.

plank in his eye was a malformed theological practice of stewardship. The only way to see and repair it was through Samaritan encounters.

From one perspective, my positive appraisal of an ethical practice of encounter is quite rudimentary, perhaps even banal. Environmental ethicist Willis Jenkins has argued that one of the most ecologically significant Christian practices that should be cultivated today is that of cross-border, neighbor-making encounters oriented by a notion of love that is shaped by the cross. Moving "across boundaries to the far and forgotten places of alienation to seek God's solidarity with those suffering powers of evil, violence, and exclusion" helps build "moral creoles" and "middle axioms," like sustainability from below, that have enough shared moral content across moral worlds to enable collective action at multiple scales. However, as Jenkins recognizes, border-crossing, "neighbor-making love" that "cultivate[s] shared commitments is the most difficult and most important task for global ethics" (Jenkins 2013, p. 128). As the Paz case suggests, it forces people committed to the politics of enclosure to confront the shortcomings of their preexisting concepts and categories, like when Jorge repaired the concept of stewardship by detaching it from enclosure and reattaching it to the commons.

If it is accurate to argue, as I have, that Paz Moyobamba's evolution was driven by a Christian practice of encounter, then my analysis of the Paz case may also add something to Jenkins's conclusion. Jenkins wants to promote sustainability as a middle axiom—a bridging concept across moral worlds capable of addressing planetary challenges at the scales at which they form (Jenkins 2013, pp. 140–42). My study largely affirms this. However, the Paz case also suggests the need to anchor sustainability to what may be a more emergent middle axiom worth promoting: the commons. Paz has seen the concept of sustainability used to justify Indigenous peoples' dispossession and the making of conservation areas that follow the logic of enclosure, creating the conditions in which lands that were previously managed as commons have become available for exploitation and extraction. In other words, the concept of sustainability may not yet sufficiently challenge the binary of privatization and government regulation—of markets and states—that undergirds climate violence. What is more, unlike the concept of sustainability, the concept of the commons has a much deeper history of meaning and practice in Christian tradition. There is ancestral wisdom about it to be recovered in places where people are joining together to defend the commons.

7.2 Conclusion

In closing, I draw three conclusions from this study. The first is that a politics of integral ecology prevents climate 'solutions' that merely reproduce the causes of climate violence, such as the violent politics of enclosure. As exemplified by Paz Moyobamba's evolution into a co-defender of the Amazonian commons, a politics of integral human ecology puts people before program.[8] It's opposite—putting

[8] On putting people before program in politics, see Bretherton (2019, pp. 16–48).

program before people—is what fuels climate violence. For example, the ideological framers of the Paris Climate Agreement decided to sacrifice people to sustain an economic program when they agreed to overshoot the target of 1.5°C warming over pre-industrial averages, blindly hoping that untested and not-yet-existing technologies would bring the temperature back down toward the end of the century.[9] They did this in full knowledge of its antihumanistic implications, which were articulated in the *Economist*: "Overshooting 1.5°C does not doom the planet. But it is a death sentence for some people, ways of life, ecosystems, even countries" (Malm and Wim 2024, p. 97). Perhaps to the framers this was an acceptable price. By contrast, the practice, exemplified by Paz Moyobamba, of putting people before program opens up a space for more transformative politics. Paz is far from the only example of this, and encounters with the commons can happen almost anywhere. There are creative experiments in commoning emerging in places as distant from one another as tech-savvy urban communities in the U.S. and agrarian communities in rural Amazonia. I have tarried with the Paz Moyobamba case to demonstrate the power of ethical encounters on the front lines of the climate crisis. My study provides evidence to conclude that it is not necessary to acquiesce to a logic of overshoot that saves an economic program only by sacrificing human and other kinds of life. Other options are available. Commoning is one of them, and it may become a more viable solution the more those with an enclosed imagination encounter its knowledges, workings, and benefits.

Second, climate politics should support and build on communities that sustain the types of encounters that create the conditions and possibilities for reparative ethics, ecological conversions, and transformations in the drivers of ecological crisis. I have argued that ethical encounters can condition the discovery of politically powerful common norms and goods. The Standing Rock case examined earlier is one example. However, it was short lived. The Paz Moyobamba example shows what is to be gained by sustaining encounters over time, beyond the moment of crisis. This sustained commitment provided Paz leaders an opportunity to repair a broken model of stewardship that had been braided around a logic of enclosure by returning it to more primary understandings of creation as a common good to be stewarded so that it sustains life for all members of a community. This is stewardship as commoning. The pursuit of environmental politics, as opposed to the depoliticizing tendencies of established climate management, is facilitated by sustaining cross-border, neighbor-making encounters that can catalyze the repair of a broken stewardship ethic.

Finally, a political vision for integral ecology should replace the misbegotten politics of enclosure, and its ideological state-market binary, with a realistic politics of the commons and commoning. Renewing a commitment to the commons is both radical

---

[9] Andreas Malm and Wim Carton (2024) document how the state-based, expert-led system created to address climate change has championed a set of analytics and strategies that put the earth system at grave risk, recklessly overshooting the 1.5°C target established by representatives of the world's nation-states in Paris in 2015, in order to save the very economic system that is driving the crisis. Overshoot is the plan to continue with business as usual while hoping to bring temperatures back down to 1.5°C in the distant future.

and yet not as radical as it seems. Commoning is not utopian. Successful commoning entails agreed-upon rules, institutions, and enforcement. Commoning takes a realistic view of matters like scarcity while spreading risks, benefits, and access more equitably than market- and state-based models of environmental governance. But it also depends on trust, wisdom, and other social goods formed by communities over time. As they are practiced in real world contexts, commons are not opposed to markets and states. Empirical studies of the commons show that they often intersect with markets and states in productive ways. In this sense, promoting the commons is neither anti-state nor anti-property. Jessica Silber-Byrne (2024) summarizes their advantage over the binary of markets and states well when writing that "successful commons governance emancipates people from passive roles towards their surroundings, and endows them with the responsibility of stewardship for the common good." Commoning ultimately increases participation and communal agency. At a time when climate politics often has us look up at global elites both for our understanding of that problem and for solutions, the most liberating approach may instead be to look down, around, and into our own hearts in search of more capacious ways of being alive together and holding things in common.

References

Aguilar, A.R.; Morales, P.A.; Habich, M. S.; and Contreras, R.M. (2010) *Minería e Hidrocarburos en el Peru*. Lima: Asociación Paz y Esperanza.

Arboccó, J. (2021) *Volviendo a la Maloca: Eco Justicia y Espiritualidad Indígena desde la Amazonía Andina del Nor Oriente Peruano*. Lima: Paz y Esperanza.

Bretherton, L. (2019) *Christ and the Common Life: Political Theology and the Case for Democracy*. Grand Rapids: W.B. Eerdmans.

Brockington, D. (2002) *Fortress Conservation: The Preservation of the Mkomazi Game Reserve, Tanzania*. Bloomington, IN: Indiana University Press.

Canineu, M.L. (2024) 'Environment Defenders Face New Wave of Oppression', *Newsweek*, 29 April. 29 [online]. At: https://www.newsweek.com/environment-defenders-face-new-wave-oppression-opinion-1894433 (9 July 2025)

Carrington, D. (2021) 'Amazon rainforest now emitting more CO2 than it absorbs', *The Guardian*, 14 July [online]. At: https://www.theguardian.com/environment/2021/jul/14/amazon-rainforest-now-emitting-more-co2-than-it-absorbs (31 December 2024)

Comisión de Entrega de la CVR (2004) *Hatun Willakuy: Versión Abreviada del Informe Final de la Comisión de la Verdad y Reconciliación*. Lima, Peru.

DiGirolamo, M. and Donald, R. (2025) 'Former UN climate chief Christiana Figueres remains optimistic despite disappointing COP process', *Mongabay*, 7 January [Online]. At: https://news.mongabay.com/podcast/2025/01/former-un-climate-chief-christiana-figueres-remains-optimistic-despite-disappointing-cop-process/ (8 January 2025)

Francis (2015) Laudato si': On Care for Our Common Home [Encyclical Letter]. At: https://www.vatican.va/content/francesco/en/encyclicals/documents/papa-francesco_20150524_enciclica-laudato-si.html (10 November 2025)

Gilio-Whitaker, D. (2019) As Long As Grass Grows: The Indigenous Fight for Environmental Justice, from Colonization to Standing Rock. Boston: Beacon Press.

Hardin, G. (1968) 'The Tragedy of the Commons', *Science* 162(3859), p1243–1248.

Hegeman, S. (2019) 'The Indigenous Commons', *Minnesota Review* 93.

Heyer, K.E. (2024) 'Moral Imagination and the Future of Ethics', *Journal of the Society of Christian Ethics* 44(1), p1–9.
Jenkins, W. (2013) *The Future of Ethics: Sustainability, Social Justice and Religious Creativity.* Washington, DC: Georgetown University Press.
Lakhani, N. and Beaumont, H. (2024) '"Fear and intimidation": how peaceful anti-pipeline protesters were hit with criminal and civil charges', *The Guardian*, 27 September [online]. At: https://www.theguardian.com/us-news/2024/sep/27/mountain-valley-pipeline-protest (: 9 July 2025)
Lorentzen, L.A. (2020) 'Christianity and Ecology in Latin America' in Orique, D.T. (ed.) *The Oxford Handbook of Latin American Christianity.* Oxford: Oxford University Press, pp. 458–472.
Malm, A. and Carton, W. (2024) *Overshoot: How the World Surrendered to Climate Breakdown.* New York: Verso.
Merino, R. (2021) 'Buen vivir and the Making of Indigenous Territories in the Peruvian Amazon', *Latin American Perspectives* 48(3), p136–151.
Monterroso, I.; Cronkleton, P.; and Larson, A.M. (2019) 'Commons, Indigenous Rights, and Governance' in Hudson, B.; Rosenbloom, J.; and Cole, D. (eds) *Routledge Handbook of the Study of the Commons.* London: Routledge, pp. 376–391.
Orihuela, J.C. (2019) 'Embedded Countermovements: The Forging of Protected Areas and Native Communities in the Peruvian Amazon', *New Political Economy* 25(1), p140–155.
Ostrom, E. (1990) *Governing the Commons: The Evolution of Institutions for Collective Action.* Cambridge: Cambridge University Press.
Asociación Paz y Esperanza (2006) *Detras de las nubes.* Lima: Paz y Esperanza.
Asociación Paz y Esperanza (2009) *Experiencias de desarrollo desde una visión intercultural y de derechos: Programa de Desarrollo Intercultural Mestizo Aguaruna (PRODIMA).* Lima: Paz y Esperanza.
Peterson, A. (2016) 'Latin America' in Jenkins, W.; Tucker, M.E.; and Grim, J. (eds) *Routledge Handbook of Religion and Ecology.* London: Routledge, pp. 190–198.
Polanyi, K. (1944) *The Great Transformation: The Political and Economic Origins of Our Time.* Boston: Beacon Press.
Silber-Byrne, J. (2024) *The Commons: More Than Just Resources* [Online]. Available at: https://seas.umich.edu/news/commons-more-just-resources (Accessed: 6 January 2024)
Taylor, B. (2016) 'The Greening of Religion Hypothesis (Part One): From Lynn White, Jr and Claims That Religions Can Promote Environmentally Destructive Attitudes and Behaviors to Assertions They Are Becoming Environmentally Friendly', *Journal for the Study of Religion, Nature, and Culture* 10(3), p268–305.
Global Witness (2023) *Standing Firm: The Land and Environmental Defenders on the Frontlines of the Climate Crisis.* Global Witness.

Chapter 8
Integral Ecology as a Framework for Soil Restoration: Insights from Regenerative Agriculture at the Bethany Land Institute, Uganda

Mitchell J. Liddick, Jennifer L. Tank, and Richard Marcantonio

Abstract Accelerating global soil degradation exemplifies the central crisis of the Anthropocene: the fracture between human and ecological systems. This chapter explores how integral ecology (IE), a framework rooted in the unity of environmental, social, and spiritual dimensions, can be operationalized to restore soil health. At the Bethany Land Institute (BLI) in Luwero District, Uganda, IE is expressed through regenerative agriculture (RA), which emulates ecological processes to restore soil health. Using a baseline dataset from the end of the dry season (March 2024), we evaluated how RA influenced soil chemistry relative to conventional agriculture at BLI. We sampled soils across three agroecosystems representing a gradient of ecological and moral renewal: conventional farmland (control), regenerative monocropping, and regenerative intercropping, reflecting a gradient of IE-informed implementation of RA. To provide a reference for how these managed systems compared to an undisturbed ecological baseline, we also sampled soils from a restored native forest at BLI. We analyzed soil chemistry to compare how IE-informed practices shift managed lands toward greater ecological integrity while sustaining food production. Analyses of total organic carbon (TOC), total nitrogen (TN), soil test phosphorus (STP), exchangeable potassium (K^+) and calcium (Ca^{2+}), and pH suggest improved soil health under IE-informed regenerative practices. Notably, regenerative systems exhibited significantly higher exchangeable K^+, elevated TOC, TN, and STP, and

Mitchell J. Liddick, University of Notre Dame, mliddick@nd.edu; Jennifer L. Tank, University of Notre Dame, jtank@nd.edu , and Richard Marcantonio, University of Notre Dame, rmarcant@nd.edu.

M. J. Liddick (✉) · J. L. Tank · R. Marcantonio
University of Notre Dame, Notre Dame, USA
e-mail: mliddick@nd.edu

J. L. Tank
e-mail: jtank@nd.edu

R. Marcantonio
e-mail: rmarcant@nd.edu

R. Marcantonio et al. (eds.), *Integral Ecology in the Anthropocene: Africa and Beyond*, The Anthropocene: Politik—Economics—Society—Science 49,
https://doi.org/10.1007/978-3-032-32125-1_8

more neutral soil pH conditions, closer to those of the forest. Our results indicate that the operationalization of IE has fostered ecological regeneration that mirrors moral regeneration at BLI. The recovery of soil chemical indicators reflects not only a biophysical process but also a renewed alignment between human management and ecological function. In this way, BLI demonstrates the potential of IE to serve as a restorative practice in the Anthropocene, suggesting that shifts in human values (interior) and land stewardship of soils (exterior) are closely connected.

Keywords Soil analysis · Restoration · Environmental impacts · Regenerative farming

8.1 Introduction

8.1.1 Soil, Spirit, and the Anthropocene

The concept of the Anthropocene highlights the close connection between human flourishing and ecological integrity (Steffen et al. 2007). We inhabit a moment when environmental degradation and climate change signal not merely a global decline but a deeper disordering of the relationship between humans and the Earth (Francis 2015, §2; Eisenhauer et al. 2024). In his writings, Pope Francis emphasized that the present social and environmental issues are not separate crises, but instead, they reflect "one complex crisis which is both social and environmental" (Francis 2015, §139). Within this context, the search for integrative frameworks that can reconcile ecological knowledge with ethical and spiritual renewal for sustainable development has become urgent. Rooted in catholic social teaching, integral ecology (IE) emerges as one such framework which seeks to understand the world as a network of interdependent relationships among humans, the land, and the transcendent (Christie et al. 2019).

8.1.2 Soils as a Mirror

In their current state, soils reflect both the brokenness and the potential for healing in the relationship between humans and Earth. Alterations in land use, including agriculture, deforestation, and urbanization, have led to significant soil erosion, depleted soil organic matter, biodiversity loss, and reliance on synthetic fertilizer inputs to increase crop productivity (Lu/Tian 2017; Barros-Rodríguez et al. 2021; Beillouin et al. 2022; Biru et al. 2025). Although often perceived as inert, soils are a living interface rich with microbial life that controls carbon cycling, nutrient turnover, and stabilization of organic matter, which are processes relevant to soil fertility and climate resilience (Bahram et al. 2018; Wu et al. 2024). Yet modern agricultural systems have treated soil as substrate for extraction to maximize crop yields for human use. Viewed in this way, degraded soils become both a symptom and symbol of disconnection. Across religious and philosophical traditions, humans are described as emerging from the Earth. In Hebrew scriptures, the human is formed from the dust of the ground (NRSVue 2021, Gen. 2:7). Therefore, the degradation of soil not only threatens ecological function but also mirrors the erosion of our interior selves and our sense of belonging to the land.

Regenerative agriculture (RA) offers a practical embodiment of IE by treating soil not as a resource to be exploited but as a living and integral part of an interconnected system. Rather than maximizing yields through chemical inputs and mechanical disturbance, RA seeks to restore soil structure and function through ecological processes, aiming to rebuild and even increase soil capacity rather than depleting it. While RA is difficult to reduce to any single method or approach, its core practices, including minimal soil disturbance, continuous ground cover, use of organic amendments, and diversified cropping, arise from the principle that the health of land and people are inseparable (Khangura et al. 2023). Through these practices, RA results in increased soil organic matter, which improves soil structure, water retention capacity, and fuels microbial activity (Khangura et al. 2023; Akhil et al. 2025). Practices such as intercropping further increase plant and microbial biodiversity, supporting increased soil moisture and fertility while maintaining or improving crop yields (Stefan et al. 2020). The benefits of RA often extend beyond the soil itself, reinforcing the interconnected vision of IE. For example, cover cropping during fallow periods can improve soil function while reducing nitrogen and phosphorus losses to waterways, thereby protecting local water resources (Christopher et al. 2021). The commitment to farm using RA becomes a tangible expression of the ethical and spiritual commitments of IE by farming in ways that honor the relationships among soil, water, plants, and people. Yet despite this conceptual alignment, empirical data linking IE to measurable improvements in soil health remains limited, which is a gap that our research seeks to address.

8.1.3 Bethany Land Institute: A Living Laboratory of Integral Ecology

The Bethany Land Institute (BLI), located in central Uganda, embodies an experiment in living out IE as both pedagogy and practice in restoring soil health while maintaining agricultural productivity. Uganda currently faces intersecting crises of land degradation, soil infertility, and food security that are driving rural youth to out-migrate in search of a living wage (Nabyonga et al. 2022). However, this environmental displacement frequently results in resettlement under less favorable living conditions, including poor sanitation and overcrowding (Ssemugabo et al. 2020). In response, BLI has implemented regenerative farming practices grounded in IE principles to answer both "the cry of the Earth and the cry of the poor." (Francis 2015, §49). The BLI was founded to cultivate leaders in IE for rural transformation in Uganda, training a new generation of farmers, called caretakers, through a two-year residential program designed to integrate RA, agribusiness, agroforestry, and personal formation.

Across Uganda, agricultural land use predominates and is projected to continue expanding through 2040, a trajectory that carries significant implications for the persistence of tropical high forests across the country (Mwanjalolo et al. 2018). The components of the BLI caretaker program are implemented on working lands in the Nandere Parish, Luwero District, Uganda, where much of the landscape is dominated by conventional farmland characterized by heavy tillage, synthetic fertilizers, and monocropping. This intensive land-use pattern, combined with unsustainable extractive methods, has contributed to soil degradation, underscoring the need for smallholder farmers to adopt regenerative practices to restore soil health and function (Nabyonga et al. 2022).

In response to these linked ecological and social challenges, BLI grounds its training in three core pillars that support the residential caretaker program. Mary's School emphasizes classroom and hands-on learning that shapes future caretakers in RA, working alongside ecological processes to rebuild and sustain soil function. Martha's Market provides training in agribusiness and upscaled regenerative food production that supports livelihoods and strengthens community food systems. Lazarus' Forest integrates principles of agroforestry while protecting the only 250 acres of remaining natural forest in Luwero District. Together, these three components offer a holistic framework for restoring both the communities and their land through a natural laboratory for testing whether IE can yield tangible improvements in soil health.

8.2 Strategy and Methodology

8.2.1 Question and hypothesis

Through this study, we evaluated whether the moral and spiritual vision of IE, as it is being implemented at BLI, has had a measurable biophysical effect by analyzing soil chemical properties across agricultural management types that represented a gradient of RA intensity (Fig. 8.1). To guide our research, we asked the question: "Can the operationalization of IE on working agricultural lands improve soil health?"

Specifically, we examined whether RA management grounded in IE principles improved soil chemical metrics relative to conventional farming practices, collecting soils from four management types within and adjacent to BLI that represented a gradient of ecological and moral renewal. The three agricultural systems, in order of increasing RA intensity, included conventional agriculture (as a control), regenerative organic monocropping, and regenerative organic intercropping. To provide a comparative reference for how these agricultural systems differed from an ecological baseline, we also sampled soils from the Lazarus' Forest. While we did not expect

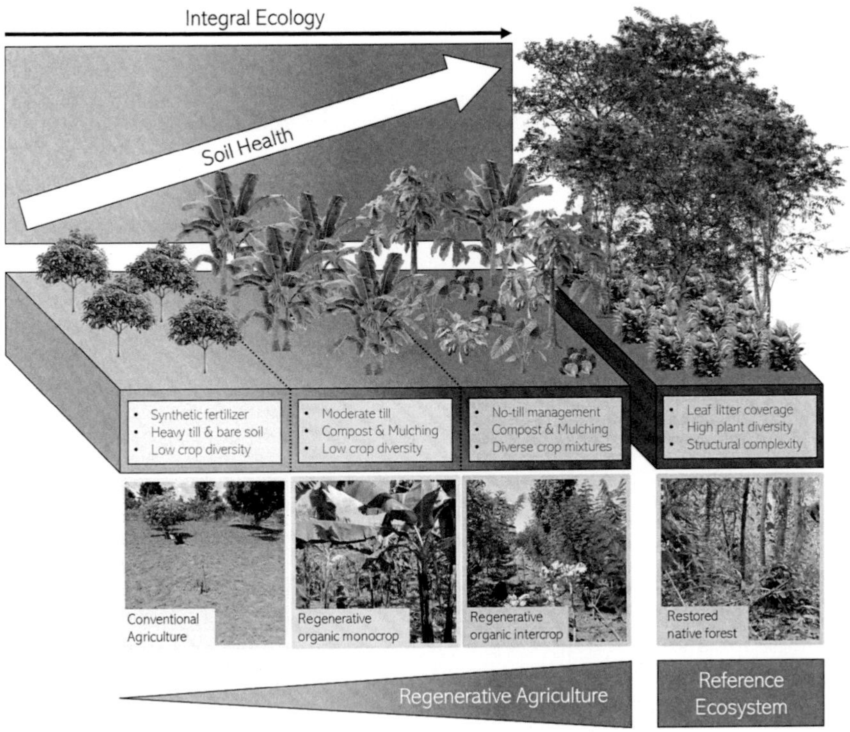

Fig. 8.1 Conceptual diagram illustrating agricultural management gradients, from conventional agriculture (parish farmland adjacent to BLI; far left) to RA (Martha's Market left center; Mary's School right center) compared to the restored forest reference (Lazarus' Forest; far right). The diagram depicts our predicted improvement in soil quality metrics with increases in RA and alignment with IE principles

regenerative systems to replicate the structure or function of the restored forest, since farming must balance ecological integrity with sustained food production, this comparison allowed us to illustrate the direction and magnitude of change associated with practices grounded in IE. The four management types are characterized as follows:

1. Conventional Agriculture (n=5 sites)—This regional control consisted of parish farmland adjacent to BLI that was managed using heavy tillage, synthetic fertilizer inputs, with no crop residue ground cover. These fields typically produced single crops such as robusta coffee, cassava, and eggplant.
2. Regenerative Organic Monocropping (n=5 sites)—Managed according to the BLI agribusiness pillar (Martha's Market), using regenerative practices to cultivate single crops such as banana, cassava, and groundnut with compost and mulch replacing chemical inputs.
3. Regenerative Organic Intercropping (n=10 sites)—Managed according to the BLI educational pillar (Mary's School), serving as a living classroom for RA,

to cultivate diverse mixtures of herbs, vegetables, fruits, and trees designed to emulate aspects of natural ecosystem structure, function, and resource complementarity.

4. Restored Native Forest (n=12 sites)—Managed according to the BLI agroforestry pillar (Lazarus' Forest), which has been actively restored through native species planting and natural regeneration and serves as a comparative reference for assessing the extent to which regenerative systems can move towards greater ecological integrity.

We hypothesized that soils under RA management at BLI would show increased organic matter and nutrient content, higher cation exchange capacity, and soil pH values closer to the restored forest reference compared with conventionally managed soils, reflecting the cumulative influence of IE-inspired practices.

8.2.2 *Soil Sampling and Laboratory Analyses*

We collected soil cores from the top 0–5 cm of the soil profile at each site using a compact one-piece soil probe (Oakfield Apparatus, USA). Depending on plot size, we sampled one to three transects per site, collecting 15 cores per transect, which were then composited for analysis. Following standard methods outlined in Okalebo, Gathua and Woomer (2002), sample processing and analyses were conducted by the Soil, Plant, and Water Analytical Laboratory at the College of Agricultural and Environmental Sciences, Makerere University, Uganda. We chose a number of soil health metrics commonly used to assess agricultural soils; description and utility included in Table 8.1. Briefly, the samples were air-dried, sieved (2 mm), and analyzed for total organic carbon (TOC, %), total nitrogen (TN, %), soluble test phosphorus (STP, mg kg^{-1}), exchangeable potassium (K^+, $cmol_c$ kg^{-1}) and calcium (Ca^{2+}, $cmol_c$ kg^{-1}), and pH (Table 8.2).

8.2.3 *Data Analyses*

We evaluated treatment effects for each soil metric across the four management types by fitting generalized linear mixed models (GLMM) using the *glmmTMB* package in R (Brooks et al. 2017). We included management type as a fixed effect and site as a random effect to account for variation among sites. We used Akaike information criterion to select the appropriate model family (e.g., Gaussian or Gamma) and link function (e.g., log or square root), and evaluated whether including a dispersion formula improved model fit (based on management type when necessary). We checked model assumptions with the *DHARMa* package in R to confirm that residuals did not deviate from uniformity, overdispersion was absent, and outliers had minimal influence (Hartig 2024). We conducted post-hoc comparisons among management

Table 8.1 Description of quantified soil metrics, analytical methods, and relevance to soil health. *Source* Author created

Soil metric	Method	Relevance to soil health
TOC	Measured using Walkley–Black wet oxidation method.	Central to soil structure, water retention, nutrient cycling, microbial activity, and resilience to stress.
TN	Determined calorimetrically after digestion with concentrated sulfuric acid, selenium powder, and salicylic acid.	Key for plant growth and microbial processes. Strongly linked to TOC. Higher TN supports soil fertility and microbial diversity.
STP	Extracted with Bray 1 solution and measured spectrophotometrically at 882 nm using ammonium molybdate–ascorbic acid reaction.	Indicates plant-available phosphorus. Essential for plant growth and microbial processes. Higher STP supports soil fertility and microbial diversity.
Exchangeable K^+	Extracted with ammonium acetate and quantified by flame photometry.	Essential for plant metabolism, water regulation, and stress tolerance. Indicative of soil fertility.
Exchangeable Ca^{2+}	Extracted with ammonium acetate and quantified by flame photometry.	Vital for soil structure through TOC accumulation and stabilization. Influences K^+ availability and pH buffering capacity.
pH	Measured in 1:2.5 soil-water suspension using a pH meter.	Controls nutrient availability, microbial activity, and cation exchange. Optimal range (i.e., 6-7) maximizes nutrient uptake by plants.

types using estimated marginal means (EMM) derived from the fitted GLMMs. We performed pairwise contrasts using the *emmeans* package in R, with p values adjusted for multiple comparisons (Piaskowski 2025). For models with a link function, we back-transformed EMMs to maintain interpretable units for visualization.

Additionally, we visualized relationships among all soil health metrics using principal component analysis (PCA) and tested for differences among management types using permutational analysis of variance (PERMANOVA, vegan package in R; Oksanen et al. 2025). We also used the PCA to guide analyses using Pearson correlation to test the significance of relationships between TOC and TN, and between exchangeable Ca^{2+} and pH. For the bivariate relationships between TOC and TN, and exchangeable Ca^{2+} and pH, we calculated group centroids as the mean values of each variable within a given management type. We were then able to calculate the Euclidean distance from each agricultural management centroid to the restored native forest centroid to quantify how closely regenerative and conventional sites resembled forest soils in bivariate space. We created all figures using the *ggplot2* package in R (Wickham 2016), and all analyses were performed using R version 4.3.0 (R Core Team 2023).

Table 8.2 Summary table of EMMs (and 95% confidence intervals) for all soil metrics across conventional agriculture, regenerative monocropping, regenerative intercropping, and restored forest reference soils. Statistical significance of management effects are bolded as determined by GLMMs (significance, $p<0.05$). Additional columns indicate whether regenerative systems exceeded conventional soils and the observed directional trend across management types

Soil metric	Management effect p-value	Conventional mean (95% CI)	Reg. monocrop mean (95% CI)	Reg. intercrop mean (95% CI)	Rest. forest mean (95% CI)	Regenerative > conventional?	Trends
TOC (%)	**<0.001**	**3.0 (2.4–3.9)**	**3.7 (3.3–4.0)**	**3.3 (2.8–3.7)**	**4.7 (4.3–5.2)**	Yes	**Regenerative monocrop > conventional; intercropping slightly higher than conventional but lower than monocrop; forest highest**
TN (%)	**<0.001**	**0.21 (0.16–0.28)**	**0.27 (0.24–0.31)**	**0.25 (0.22–0.29)**	**0.39 (0.35–0.43)**	Yes	**Regenerative systems higher than conventional; forest highest**
STP (mg kg^{-1})	0.577	10 (3.4–21)	15 (11–20)	14 (9.1–20)	12 (8.7–15)	Yes	Trends show regenerative > conventional, but non-significant

(continued)

Table 8.2 (continued)

Soil metric	Management effect p-value	Conventional mean (95% CI)	Reg. monocrop mean (95% CI)	Reg. intercrop mean (95% CI)	Rest. forest mean (95% CI)	Regenerative > conventional?	Trends
Exchangeable K^+ (cmol kg^{-1})	**<0.001**	**0.75 (0.51–1.1)**	**1.2 (0.99–1.5)**	**1.6 (1.3–2.1)**	**0.69 (0.62–0.78)**	Yes	**Both regenerative systems higher than conventional; intercropping highest; forest similar to conventional**
Exchangeable Ca^{2+} (cmo kg^{-1})	0.597	5.7 (5.3–6.1)	5.8 (5.7–6.0)	6.0 (5.8–6.2)	5.9 (5.7–6.0)	Yes	Slight trend of regenerative > conventional; all close to forest; not significant
pH	0.063	5.9 (5.5–6.2)	6.1 (6.0–6.3)	6.3 (6.2–6.5)	6.2 (6.1–6.3)	Yes	Regenerative soils higher than conventional; intercropping highest; forest in between

8.2.4 *Linking Soil Health Metrics to Integral Ecology*

8.2.5 *Individual Soil Chemistry Results*

We observed a gradient in organic matter and key soil nutrients, with restored forest and regenerative systems generally exhibiting higher TOC, TN, and STP than conventional agriculture, although statistically significant differences were infrequent. We found that mean TOC was highest in the restored native forest, followed by regenerative monocropping and regenerative intercropping, which both exceeded conventional agriculture (Fig. 8.2 left). Post-hoc comparisons confirmed that TOC values in forest soils were significantly higher than all agricultural systems (EMM, $p<0.01$), while regenerative monocropping and intercropping exhibited intermediate TOC values that were not statistically different from conventional agriculture (EMM, $p>0.05$). We found that mean TN displayed a similar pattern to TOC with the highest values in forest soils, followed by regenerative monocropping and intercropping, which were higher than conventional agriculture (Fig. 8.2 middle). Similar to TOC, only the forest TN was significantly higher than conventional agriculture (EMM, $p<0.001$), while regenerative systems were intermediate but not significantly different from conventional agriculture (EMM, $p>0.05$). In contrast, we found that STP was not statistically different among systems (GLMM, $p=0.58$), however, both regenerative systems had mean STP that trended above conventional agriculture (Fig. 8.2 right). In general, regenerative practices improved mean TOC, TN, and STP availability.

We also found that regenerative systems had higher mean exchangeable cations and pH values that were closer to neutral compared to conventional agriculture, although results were not statistically significant in all cases. We found that mean exchangeable K^+ increased across the agricultural management gradient, with the

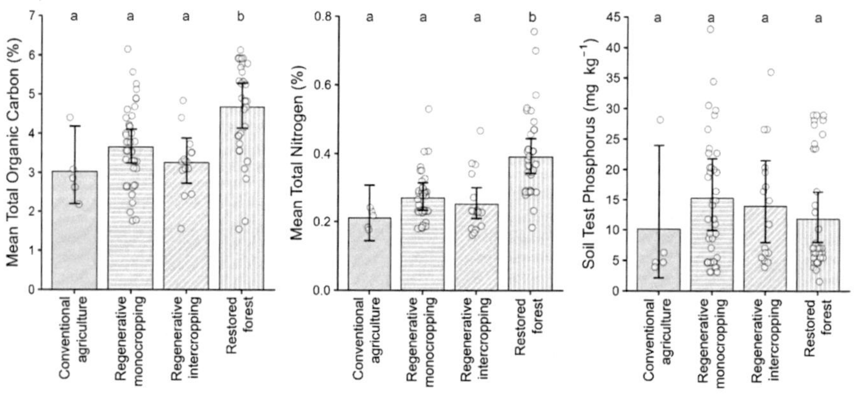

Fig. 8.2 Mean soil organic matter and nutrient content across management types at BLI. TOC (left), TN (middle), and STP (right). Bars show EMMs with 95% confidence intervals, letters above bars indicate results of post-hoc pairwise comparisons, and jittered points represent individual samples collected across sites

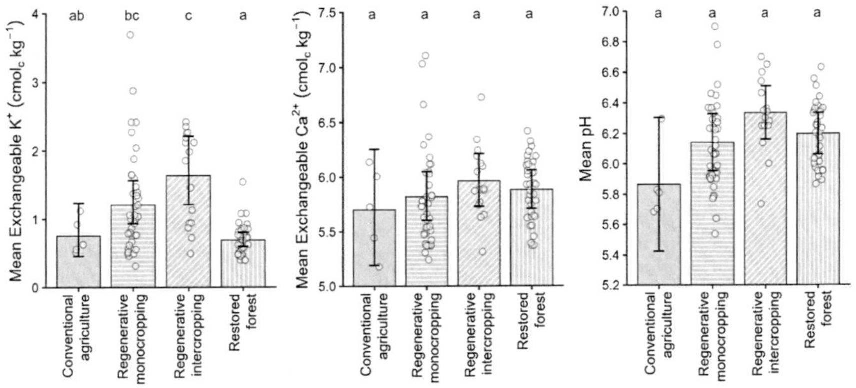

Fig. 8.3 Mean soil cation exchange capacities and pH across management types at BLI. Mean exchangeable K^+ (left), mean exchangeable Ca^{2+} (middle), and mean soil pH (right). Bars show EMMs with 95% confidence intervals, letters above bars indicate results of post-hoc pairwise comparisons, and jittered points represent individual samples collected across sites

highest values observed in regenerative intercropping and the lowest in conventional agriculture and forest soils (Fig. 8.3 left). Regenerative intercropping soils were significantly higher in exchangeable K^+ than conventional agriculture and forest soils (EMM, $p<0.01$), which were not significantly different from one another (EMM, $p>0.05$). Regenerative monocropping exhibited intermediate exchangeable K^+, having statistical overlap with conventional agriculture and regenerative intercropping (EMM, $p<0.05$). While not statistically significant, we found that both forms of regenerative management and forest soils had mean exchangeable Ca^{2+} values that trended slightly above conventional (Fig. 8.3 middle, GLMM, $p>0.05$). Lastly, we found that soil pH was higher in both regenerative monocropping and intercropping compared to conventional agriculture, with values falling in the more neutral range (6–7) similar to the restored forest (Fig. 8.3 right). In contrast, soils managed under conventional agriculture remained below pH 6. While we did not find soil pH to differ significantly among management types overall (GLMM, $p>0.05$), regenerative intercropping soils exhibited marginally higher pH than conventional agriculture (EMM, $p=0.06$).

8.2.6 *Results Integrating Multiple Soil Metrics*

Multivariate analyses revealed that soil chemistry differs among management types, with regenerative systems generally clustering closer to forest soils than conventional systems do. Visualizing these data using a PCA (Fig. 8.4 top left), we found that regenerative monocropping and intercropping soils clustered closer to forest soils than did the conventional soils. These clusters were driven by exchangeable Ca^{2+} and pH on the first principal component (PC1, explaining 31.7% variation), and TOC and

TN on the second principle component (PC2 explaining 26.3% variation). Moreover, we found that management type significantly altered soil chemistry (Fig. 8.4 top left, PERMANOVA, $F_{3,94}$=6.26, R^2=0.17, p=0.001), indicating that regenerative practices can influence combined soil metrics. Despite these trends, all subsequent pairwise comparisons among management types were not significantly different (unadjusted p>0.05), even prior to adjustments for multiple comparisons.

When examining the relationship between individual soil metrics, we found that TOC and TN were positively correlated across all sites (Fig. 8.4 bottom left, Pearson Correlation, r=0.40, p<0.0001). Based on centroid positions for each management type when correlating TOC and TN, both regenerative systems fell closer to the

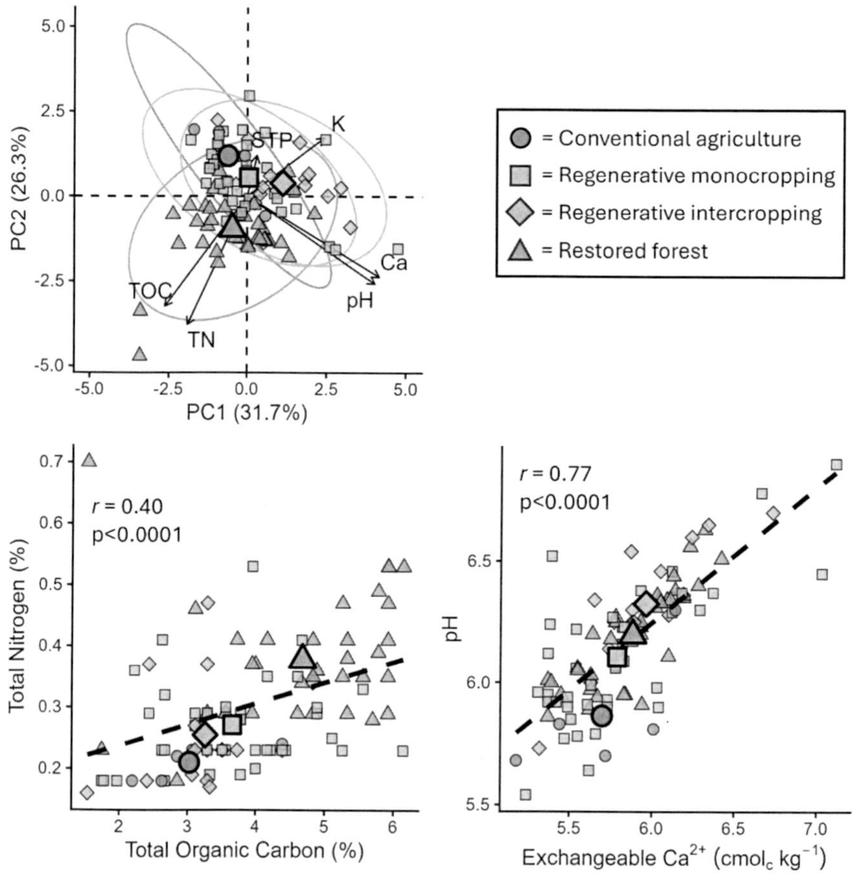

Fig. 8.4 Multivariate and bivariate analyses of soil chemistry across management types. PCA of all soil metrics showing individual sample points colored/shaped by management type, 95% confidence ellipses for each group, and group centroids indicated by enlarged symbols (top left). Relationship between TOC and TN (bottom left) and between exchangeable Ca^{2+} and pH (bottom right), with individual samples colored/shaped by management type and group centroids indicated by enlarged symbols

restored forest, whereas conventional soils were more distant (Fig. 8.4 bottom left). Specifically, the Euclidean distances from the forest centroid were 1.03 for regenerative monocropping, 1.43 for regenerative intercropping, and 1.66 for conventional agriculture, suggesting that regenerative management shifted TOC and TN toward levels observed in forest soils. We found that exchangeable Ca^{2+} and pH were also positively correlated across sites (Fig. 8.4 bottom right, Pearson Correlation, r=0.77, p<0.0001). Again, based on centroid positions for each management type when correlating exchangeable Ca^{2+} and pH, regenerative systems fell closer to the restored forest, while conventional soils were more distant (Fig. 8.4 bottom right). Euclidean distances from the forest centroid were 0.13 for regenerative monocropping, 0.15 for regenerative intercropping, and 0.38 for conventional agriculture. These patterns suggest that regenerative management shifted Ca^{2+} and pH toward levels observed in forest soils.

8.3 Discussion

Soil sampling across sites with contrasting management practices showed shifts in carbon, nitrogen, phosphorus, exchangeable cations, and pH towards values indicating soil health improvement. At BLI, RA operates through intercropping, composting, mulching, conservation tillage, green manures, and seeking alternatives to avoid use of synthetic fertilizers, but these practices function as more than a checklist of preferred techniques. Together, these practices form an integrated system that emphasizes continuous organic matter inputs, diverse rooting structures, minimal soil disturbance, and the steady recycling of nutrients (Khangura et al. 2023). Although our study did not isolate the effect of each individual practice, the soils managed using RA approaches, grounded in an IE-inspired framework, showed consistently improved soil health metrics compared to conventionally managed soils. This improvement aligns with ecological expectations for regenerative systems (Khangura et al. 2023; Buckton et al. 2023), where increased plant diversity, organic matter returns, and reduced reliance on synthetic inputs foster improved nutrient cycling, greater microbial activity, and stronger soil structural development. In this sense, the shifts in soil chemistry that we have documented, including general trends that regenerative systems more closely resemble restored forest soils compared to conventional systems, represent the cumulative expression of a management philosophy that treats soils as a living system rather than solely as a substrate for production (Sofo et al. 2022).

The RA practices used at BLI since 2020, based on principles of IE, are beginning to yield observable improvements in soil carbon and nutrients; increased TOC and TN likely reflect steady inputs of organic matter from crop residues and compost amendments, which stimulate microbial activity and strengthen nutrient retention. Other regenerative systems have shown similar results during the first decade after conversion, where organic amendments accelerated soil organic matter accumulation, increased microbial biomass, and improved soil structure (Lekberg et al. 2024).

While we did not measure soil microbial activity in this study, increases in TOC and TN may support microbial processing, which enhances nutrient mineralization, making nitrogen and other nutrients more available for plant uptake (Feng et al. 2022). At BLI, continuous organic amendments to both regenerative monocropping and intercropping fields, including mulch, crop residues, and plant litter, promote the recycling of carbon and nitrogen back into the soil, enhancing microbial communities in ways that create larger pools of accessible nutrients (Liu et al. 2023). The organic matter inputs also help increase the availability of exchangeable K^+ and STP, in part because organic amendments promote mineralization and reduce nutrient fixation in soils (Bader et al. 2021). We found a similar pattern in the regenerative plots at BLI, with higher exchangeable K^+ and STP relative to conventional systems, consistent with nutrient release from decomposing organic materials and reduced leaching under continuous ground cover and mulching.

We also observed a shift toward a more neutral pH (6–7) in regenerative plots, likely driven by the buffering capacity of organic amendments and the absence of acidifying inputs (Hao et al. 2020; Shabtai et al. 2023). This shift approximates the chemistry of the reference soils found in the restored forest, where higher exchangeable Ca^{2+} indicated stronger buffering capacity, which stabilizes pH within an optimal range. The outcomes of maintaining soils in this pH window include enhanced nutrient availability for plants (Penn/Camberato 2019), and reduced risk of aluminum toxicity, which becomes increasingly soluble under acidic conditions (Wang et al. 2025). Collectively, these changes reflect the reciprocal flows of energy and matter among soil, microbes, and plants that RA, grounded in IE principles, seeks to cultivate, where soils are part of a living system in which each component contributes to the resilience and nourishment of the whole (Rhodes 2017).

Core principles of IE are evident in the regenerative framework at BLI. Interdependence is expressed through the circular flows of matter and energy, such that composted garden waste nourishes crops, crop residues are maintained to protect soil, and organic amendments cycle nutrients back into the system, tightening connections between decaying organic matter, microbes, and plant growth. These recursive flows echo what Pope Francis describes as the inseparable connection between nature and society (Francis 2015, §139). Humans are part of this larger whole, and our engagement with the land extends beyond simple material feedbacks, and instead includes the application of local knowledge, collective labor, and spiritual formation, illustrating how human relationships and social participation are deeply intertwined with ecological outcomes, a central tenet of IE (Messias 2024). Regeneration, as a guiding principle, is reinforced by the soil chemistry trends of the regenerative sites at BLI moving away from the chemistry of conventional plots and towards the chemistry of the forest reference. Our results confirm that intentional management can restore soils such that they approximate a high-functioning reference condition. Early improvements in the soil chemistry profile at BLI do not represent solely agronomic gains; they signify a system in which human practices are reintegrated into ecological cycles, and that ecological health, human flourishing, and cultural engagement are mutually reinforcing (Prescott et al. 2022). However, we recognize that IE calls for patience and consistency, even under pressing environmental pressures, as transformation unfolds gradually over time (Francis 2015, §201) and soil health metrics often require a decade or more to change significantly (Li et al. 2021; Jordon et al. 2022). We plan continued longitudinal monitoring to definitively document the divergence from conventional systems as organic matter and nutrient pools continue to accumulate. In this sense, we characterize the soil as a beacon of hope, recording the cumulative effects of sustained, mindful stewardship and exemplifying how ethical frameworks can translate into tangible ecological outcomes.

8.4 Conclusion and Implications

This case study from BLI in Luwero District, Uganda demonstrates that the operationalization of IE is not merely a philosophical aspiration; it yields tangible ecological outcomes. Regenerative practices inspired by IE enhanced soil carbon, nitrogen, exchangeable cations, and pH, moving the chemical profile of agricultural soils toward that of a restored forest system. Beyond biophysical metrics, our study shows how IE can serve as a framework for ecological restoration, combining science, ethics, and spirituality in a coherent praxis. The implications for this research extend beyond Uganda to other settings where faith-based ecological initiatives engage in material restoration, cultivating new forms of social and environmental resilience (Marcantonio 2024). Our future work will build on these findings through longitudinal soil monitoring and expansion to additional IE-informed regenerative sites. By continuing to bridge measurable soil transformations with cultural and moral renewal, such work can help realize the aspirations of IE for healing both land and life in the Anthropocene.

References

Akhil, K., Renukaswamy, N. S., Venugopala, K. M., Jakkampudi, S., Kishore, S. M., & Bharthisha, S. M. (2025) 'Revitalizing Soil Health through Regenerative Agriculture: History, Principles, Practices and Challenges', *International Journal of Environment and Climate Change, 15*(8), 249–262. https://doi.org/10.9734/ijecc/2025/v15i84969

Bader, B. R., Taban, S. K., Fahmi, A. H., Abood, M. A., & Hamdi, G. J. (2021) 'Potassium availability in soil amended with organic matter and phosphorous fertiliser under water stress during maize (Zea mays L) growth', *Journal of the Saudi Society of Agricultural Sciences, 20*(6), 390–394. https://doi.org/10.1016/j.jssas.2021.04.006

Bahram, M., Hildebrand, F., Forslund, S. K., Anderson, J. L., Soudzilovskaia, N. A., Bodegom, P. M., Bengtsson-Palme, J., Anslan, S., Coelho, L. P., Harend, H., Huerta-Cepas, J., Medema, M. H., Maltz, M. R., Mundra, S., Olsson, P. A., Pent, M., Põlme, S., Sunagawa, S., Ryberg, M., … Bork, P. (2018) 'Structure and function of the global topsoil microbiome', *Nature, 560*(7717), 233–237. https://doi.org/10.1038/s41586-018-0386-6

Barros-Rodríguez, A., Rangseekaew, P., Lasudee, K., Pathom-aree, W., & Manzanera, M. (2021) 'Impacts of Agriculture on the Environment and Soil Microbial Biodiversity', *Plants, 10*(11), 2325. https://doi.org/10.3390/plants10112325

Beillouin, D., Corbeels, M., Demenois, J., Berre, D., Boyer, A., Fallot, A., Feder, F., & Cardinael, R. (2023) 'A global meta-analysis of soil organic carbon in the Anthropocene', *Nature Communications, 14*(1), 3700. https://doi.org/10.1038/s41467-023-39338-z

Biru, M. K., Gadisa, C. W., Dirbaba, N. B., & Nunes, M. R. (2025) 'Land Use Changes and Their Impacts on Soil Erosion in a Fragile Ecosystem of the Ethiopian Highlands', *Land, 14*(7), 1473. https://doi.org/10.3390/land14071473

Brooks, M., E., Kristensen, K., Benthem, K., J. ,van Magnusson, A., Berg, C., W., Nielsen, A., Skaug, H., J., Mächler, M., & Bolker, B., M. (2017) 'glmmTMB Balances Speed and Flexibility Among Packages for Zero-inflated Generalized Linear Mixed Modeling,' *The R Journal, 9*(2), 378. https://doi.org/10.32614/rj-2017-066

Buckton, S. J., Fazey, I., Sharpe, B., Om, E. S., Doherty, B., Ball, P., Denby, K., Bryant, M., Lait, R., Bridle, S., Cain, M., Carmen, E., Collins, L., Nixon, N., Yap, C., Connolly, A., Fletcher, B., Frankowska, A., Gardner, G., … Sinclair, M. (2023) 'The Regenerative Lens: A conceptual framework for regenerative social-ecological systems', *One Earth, 6*(7), 824–842. https://doi.org/10.1016/j.oneear.2023.06.006

Christie, I., Gunton, R. M., & Hejnowicz, A. P. (2019) 'Sustainability and the common good: Catholic Social Teaching and 'Integral Ecology' as contributions to a framework of social values for sustainability transitions', *Sustainability Science, 14*(5), 1343–1354. https://doi.org/10.1007/s11625-019-00691-y

Christopher, S. F., Tank, J. L., Mahl, U. H., Hanrahan, B. R., & Royer, Todd. V. (2021) 'Effect of winter cover crops on soil nutrients in two row-cropped watersheds in Indiana', *Journal of Environmental Quality, 50*(3), 667–679. https://doi.org/10.1002/jeq2.20217

Eisenhauer, N., Frank, K., Weigelt, A., Bartkowski, B., Beugnon, R., Liebal, K., Mahecha, M., Quaas, M., Al-Halbouni, D., Bastos, A., Bohn, F. J., Brito, M. M. de, Denzler, J., Feilhauer, H., Fischer, R., Fritsche, I., Guimaraes-Steinicke, C., Hänsel, M., Haun, D. B. M., … Quaas, J. (2024) 'A belowground perspective on the nexus between biodiversity change, climate change, and human well-being', *Journal of Sustainable Agriculture and Environment, 3*(2). https://doi.org/10.1002/sae2.12108

Feng, X., Qin, S., Zhang, D., Chen, P., Hu, J., Wang, G., Liu, Y., Wei, B., Li, Q., Yang, Y., & Chen, L. (2022) 'Nitrogen input enhances microbial carbon use efficiency by altering plant–microbe–mineral interactions', *Global Change Biology, 28*(16), 4845–4860. https://doi.org/10.1111/gcb.16229

Francis. (2015). *Laudato si': on care for our common home* [Encyclical letter]. At: https://www.vatican.va/content/francesco/en/encyclicals/documents/papa-francesco_20150524_enciclica-laudato-si.html

Hao, T., Zhu, Q., Zeng, M., Shen, J., Shi, X., Liu, X., Zhang, F., & Vries, W. de. (2020) 'Impacts of nitrogen fertilizer type and application rate on soil acidification rate under a wheat-maize double cropping system', *Journal of Environmental Management, 270*, 110888. https://doi.org/10.1016/j.jenvman.2020.110888

Hartig, F. (2024) *DHARMa: residual diagnostics for hierarchical (multi-level/mixed) regression models.* At: https://cran.r-project.org/web/packages/DHARMa/vignettes/DHARMa.html (Accessed: 1 December 2025)

Jordon, M. W., Willis, K. J., Bürkner, P.-C., Haddaway, N. R., Smith, P., & Petrokofsky, G. (2022) 'Temperate Regenerative Agriculture practices increase soil carbon but not crop yield—a meta-analysis', *Environmental Research Letters, 17*(9), 093001. https://doi.org/10.1088/1748-9326/ac8609

Khangura, R., Ferris, D., Wagg, C., & Bowyer, J. (2023) 'Regenerative Agriculture—A Literature Review on the Practices and Mechanisms Used to Improve Soil Health', *Sustainability, 15*(3), 2338. https://doi.org/10.3390/su15032338

Lekberg, Y., McLeod, M., Bullington, L. S., DuPre, M. E., Roca, G. D. L., Greenbaum, S., Rousk, J., & Ramsey, P. W. (2024) 'Substantial and Rapid Increase in Soil Health across Crops with Conversion from Conventional to Regenerative Practices', *Sustainability, 16*(13), 5509. https://doi.org/10.3390/su16135509

Li, X.-F., Wang, Z.-G., Bao, X.-G., Sun, J.-H., Yang, S.-C., Wang, P., Wang, C.-B., Wu, J.-P., Liu, X.-R., Tian, X.-L., Wang, Y., Li, J.-P., Wang, Y., Xia, H.-Y., Mei, P.-P., Wang, X.-F., Zhao, J.-H., Yu, R.-P., Zhang, W.-P., … Li, L. (2021) 'Long-term increased grain yield and soil fertility from intercropping', *Nature Sustainability, 4*(11), 943–950. https://doi.org/10.1038/s41893-021-00767-7

Liu, B., Dai, Y., Cheng, X., He, X., Bei, Q., Wang, Y., Zhou, Y., Zhu, B., Zhang, K., Tian, X., Duan, M., Xie, X., & Wang, L. (2023) 'Straw mulch improves soil carbon and nitrogen cycle by mediating microbial community structure and function in the maize field', *Frontiers in Microbiology, 14*, 1217966. https://doi.org/10.3389/fmicb.2023.1217966

Lu, C., & Tian, H. (2017) 'Global nitrogen and phosphorus fertilizer use for agriculture production in the past half century: shifted hot spots and nutrient imbalance', *Earth System Science Data, 9*(1), 181–192. https://doi.org/10.5194/essd-9-181-2017

Marcantonio, R. D. (2024) 'How regenerative agriculture can foster peacebuilding in conflict areas' *Bulletin of the Atomic Scientists* [online]. At: https://thebulletin.org/2024/12/how-regenerative-agriculture-can-foster-peacebuilding-in-conflict-areas/

Messias, T. (2024) 'From Ecotheology to Ecospirituality in Laudato sí—Ecological Spirituality beyond Christian Religion', *Religions, 15*(1), 68. https://doi.org/10.3390/rel15010068

Mwanjalolo, M. G. J., Bernard, B., Paul, M. I., Joshua, W., Sophie, K., Cotilda, N., Bob, N., John, D., Edward, S., & Barbara, N. (2018) 'Assessing the Extent of Historical, Current, and Future Land Use Systems in Uganda', *Land, 7*(4), 132. https://doi.org/10.3390/land7040132

Nabyonga, L., Basamba, T. A., Nyakoojo, C., & Ssenku, J. E. (2022) 'Impediments to agricultural production in Uganda and measures to enhance soil fertility utilizing organic soil amendments: A review', *Cogent Food & Agriculture, 8*(1), 2113051. https://doi.org/10.1080/23311932.2022.2113051

Okalebo, J. R., Gathua, K. W., & Woomer, P. L. (2002) *Laboratory Methods for Soil and Plant Analysis: A Working Manual.* 2nd edn. Nairobi, Kenya: TSBF-CIAT and SACRED Africa.

Oksanen, J., Simpson, G. L., Blanchet, F. G., Kindt, R., Legendre, P., Minchin, P. R., O'Hara, R. B., Solymos, P., Stevens, M. H. H., Szoecs, E., Wagner, H., Barbour, M., Bedward, M., Bolker, B., Borcard, D., Borman, T., Carvalho, G., Chirico, M., De Caceres, M., Durand, S., Evangelista, H. B. A., FitzJohn, R., Friendly, M., Furneaux, B., Hannigan, G., Hill, M. O., Lahti, L., Martino, C., McGlinn, D., Ouellette, M.-H., Ribeiro Cunha, E., Smith, T., Stier, A., Ter Braak, C. J. F., & Weedon, J. (2025). *Package 'vegan'.* At: https://cran.r-project.org/web/packages/vegan/vegan.pdf (Accessed: 1 December 2025)

Penn, C. J., & Camberato, J. J. (2019) 'A Critical Review on Soil Chemical Processes that Control How Soil pH Affects Phosphorus Availability to Plants', *Agriculture, 9*(6), 120. https://doi.org/10.3390/agriculture9060120

Piaskowski, J. (2025) *Package 'emmeans'.* At: https://cran.r-project.org/web/packages/emmeans/emmeans.pdf (1 December 2025)

Prescott, S. L., Logan, A. C., Bristow, J., Rozzi, R., Moodie, R., Redvers, N., Haahtela, T., Warber, S., Poland, B., Hancock, T., & Berman, B. (2022) 'Exiting the Anthropocene: Achieving personal and planetary health in the 21st century', *Allergy, 77*(12), 3498–3512. https://doi.org/10.1111/all.15419

R Core Team. (2023). *R: A Language and Environment for Statistical Computing.* At: https://www.r-project.org/

Rhodes, C. J. (2017) 'The Imperative for Regenerative Agriculture', *Science Progress, 100*(1), 80–129. https://doi.org/10.3184/003685017x14876775256165

Shabtai, I. A., Wilhelm, R. C., Schweizer, S. A., Höschen, C., Buckley, D. H., & Lehmann, J. (2023) 'Calcium promotes persistent soil organic matter by altering microbial transformation of plant litter', *Nature Communications, 14*(1), 6609. https://doi.org/10.1038/s41467-023-42291-6

Sofo, A., Zanella, A., & Ponge, J. (2022) 'Soil quality and fertility in sustainable agriculture, with a contribution to the biological classification of agricultural soils', *Soil Use and Management, 38*(2), 1085–1112. https://doi.org/10.1111/sum.12702

Ssemugabo, C., Nalinya, S., Lubega, G. B., Ndejjo, R., & Musoke, D. (2020) 'Health Risks in Our Environment: Urban Slum Youth' Perspectives Using Photovoice in Kampala, Uganda', *Sustainability, 13*(1), 248. https://doi.org/10.3390/su13010248

Stefan, L., Hartmann, M., Engbersen, N., Six, J., & Schöb, C. (2021) 'Positive Effects of Crop Diversity on Productivity Driven by Changes in Soil Microbial Composition', *Frontiers in Microbiology, 12*, 660749. https://doi.org/10.3389/fmicb.2021.660749

Steffen, W., Crutzen, P. J., & McNeill, J. R. (2007) 'The Anthropocene: Are Humans Now Overwhelming the Great Forces of Nature', *AMBIO: A Journal of the Human Environment, 36*(8), 614–621. https://doi.org/10.1579/0044-7447(2007)36[614:taahno]2.0.co;2

The Bible: New Revised Standard Version Updated Edition (NRSVue). (2021). At: https://www.biblegateway.com/versions/New-Revised-Standard-Version-Updated-Edition-NRSVue-Bible/

Wang, T., Yu, L., Wang, Z., Yang, C., Dong, F., Yang, D., Xi, H., Sun, Z., Bol, R., Awais, M., Yang, L., & Fu, H. (2025) 'Effect of simulated acidification on soil properties and plant nutrient uptake of eggplant in greenhouse', *Frontiers in Plant Science, 16*, 1558458. https://doi.org/10.3389/fpls.2025.1558458

Wickham, H. (2016). *ggplot2: Elegant Graphics for data Analysis.* At: https://ggplot2.tidyverse.org (1 December 2025)

Wu, H., Cui, H., Fu, C., Li, R., Qi, F., Liu, Z., Yang, G., Xiao, K., & Qiao, M. (2024) 'Unveiling the crucial role of soil microorganisms in carbon cycling: A review', *Science of The Total Environment, 909*, 168627. https://doi.org/10.1016/j.scitotenv.2023.168627

Part II
Interconnected Ecological and Social Crises and Justice

Chapter 9
Laudato Si and Care for Our Natural Environment: The Catholic Church's Response to the Environmental and Health Crises Caused by Artisanal Small-Scale Gold Mining in Ghana

Ugochukwu Okoye

Abstract Pope Francis' audacious encyclical, Laudato Si, issued on May 24th, 2015, is now 10 years old in 2025. The concerns raised by the Holy Father in this encyclical, especially the need to care for our common home, still resonate in today's world. This paper focuses on the central theme of Integral Ecology in Laudato Si, highlighting the interconnectedness of all creation and how the exploitation of the environment inevitably leads to the exploitation of human beings. It offers a critical examination of Artisanal Small-scale Gold Mining (ASGM) in Tarkwa and the environmental dangers it poses. The paper demonstrates that the call to care for our common home is closely linked to the call to respond to the needs of the poor. It recognizes the efforts of the Catholic Church in Ghana to raise awareness about environmental destruction and highlights the initiatives, dialogue, and collaboration with political leaders. It reiterates the need for a renewed response to the call for 'ecological conversion' in Laudato Si. It concludes by emphasizing that our role as stewards of creation is one of service and solidarity.

Keywords Pastoralism · Mining · Ecological degradation · Livelihoods

9.1 Tarkwa: My Pastoral Journey in New Atuabo, Tarkwa

I arrived in Ghana on November 14, 2016, as a missionary from Nigeria to the Spiritan Province of Ghana. Ghana, formerly known as the Gold Coast, is renowned for its abundant gold deposits and ranks second in gold production in Africa, after South Africa (Basu et al. 2015, p. 11684). I was officially assigned to Ss. Peter & Paul

U. Okoye (✉)
University of Notre Dame, Notre Dame, United States
e-mail: uokoye@nd.edu

R. Marcantonio et al. (eds.), *Integral Ecology in the Anthropocene: Africa and Beyond*,
The Anthropocene: Politik—Economics—Society—Science 49,
https://doi.org/10.1007/978-3-032-32125-1_9

Parish, New Atuabo, Tarkwa, as the Parochial Vicar. Tarkwa is an ancient city in the Tarkwa-Nsuaem Municipal District in the Western Region of Ghana. The Tarkwa Nsuaem Municipal is situated in the Western Region and lies between latitude 4°5′ and longitude 5°5′, sharing borders with Prestea Huni-Valley district to the north, Nzema East municipality to the west, Ahanta West district to the south, and Mpohor Wassa East district to the east (Obiri-Yeboah et al. 2021, p. 2). It has a total land area of 905.2 square kilometers and a population of 218,664 (Ghana Statistical Service, retrieved August 2025). The region is mainly agricultural and mining-based, with cash crops like rubber and cocoa, and food crops such as cassava and plantain being commonly cultivated. Ghana's small-scale gold miners number roughly between 500,000 and 1 million (Clifford 2017, p. 498). The Tarkwa-Prestea area is especially famous for its large gold deposits. Gold mining in the area dates back to the 10th century, and other minerals mined include manganese, bauxite, diamonds, iron ore, asbestos, and chromite. It is documented that "the earliest European attempts to extract gold on a large scale were concentrated in Tarkwa and Prestea regions in the late 19th century" (Bhattacharya et al. 2012, p. 72). Upon entering Tarkwa and the surrounding towns, one immediately sees mining company sites, heavy-duty vehicles on the roads, numerous excavators, and sites for artisanal small-scale gold mining. The parish is located in a settlement called New Atuabo, and as the name suggests, there was an older settlement called Old Atuabo where residents lived before being resettled. That old settlement is now a mining site operated by an international company called Gold Fields Ghana. Additionally, several other mining companies operate in the area, such as AngloGold Ashanti at Bankyim and the Ghana Manganese Company (GMC) at Nsuta. These are public companies that employ many local people. However, many wealthy individuals also acquire mining sites, often with inadequate safety measures, and these are usually artisanal small-scale miners, also known as Galamsey locally. The St. Peter & Paul New Atuabo parish has three outstations: Huniano, Samahu, and Pepesa. These villages, especially Pepesa, are mainly populated by farmers who grow substantial amounts of cocoa. During visits and interactions with parishioners, I observed issues such as displacement from original settlements, air pollution from mining activities, land degradation, deforestation, destruction of farmlands, damage to water bodies, and lack of potable water due to the use of toxic chemicals like mercury, lead, and cyanide.[1]

[1] This project was made possible through the support of De Nicola Center for Ethics and Culture (Sorin Fellows) and the Zahm Professional Development Fund, through the Graduate School (GSPDA).

9.2 Land Degradation and Exploitation of the Poor

The artisanal and small-scale mining sector is expected to employ around 25 million people across about 50 countries worldwide, with gold deposits being the most sought-after resource (Clifford 2017, p. 497). In the Tarkwa Nsuaem District, small-scale mining is rife, both in the forest and in water bodies (Bhattacharya et al. 2012, pp. 72–73). There are about 20,000 small-scale mines in the Wassa West district, and 90% of them operate illegally (Bhattacharya et al. 2012, p. 73). The ASGMs employ various techniques to mine depending on the amount of gold deposit; however, they indiscriminately use special machines and excavators to achieve quick results, thereby destroying farmlands, causing "rapid deforestation, land degradation, and water pollution" (Blay 2022, p. 106). The excavators are so efficient that "the Chinese people take just a day or two to mine a piece of land that might take six months to a year for Ghanaians to mine" (Blay 2022, p. 106). Consequently, the "mechanization has intensified the degradation of both land and water in areas of alluvial mining; farmlands, in particular, have been extensively destroyed during the scramble to acquire land for small-scale mining from 2008 onwards, affecting both food crop production and cash crops (such as cocoa), and resulting in adverse effects for both food" (Botchwey et al. 2018, p. 317).

From the foregoing, we see the necessity of Pope Francis' concern that the earth cries out because of the harm and destruction our irresponsible activities have inflicted on her. He states that "we have come to see ourselves as her lords and masters, entitled to plunder her at will" (Francis 2015, § 2). Thus, he appeals that humanity should come together "to seek a sustainable and integral development" (Francis, 2015, § 13). Pope Francis believes that the ecological crises we face today are constantly engendered by human beings and human actions in the world. Whilst appreciating the fact that science and technology have benefited the universe in various ways by arresting certain evils that plague humanity (Francis 2015, §102), and improving "the quality of human life" (Francis 2015, § 103), he, however, maintains that they come with many risks. Humanity has devised various superficial mechanisms to subdue evil and improve lives, "but we cannot claim to have a sound ethics, a culture, and spirituality genuinely capable of setting limits and teaching clear-minded self-restraint" (Francis 2015, § 105).

The cocoa plantations, however, are being destroyed, as miners are caving through and purchasing farmlands with gold deposits for mining. The impoverished farmers, who struggle to make ends meet despite contributing significantly to the nation's economy, are often persuaded to sell to make quick money (Blay 2022, p. 118). At the root of this is economic hardship, which makes the people vulnerable, because "poverty is the arch pusher to *galamsey*" (Blay 2022, p. 119). This is a very clear example of the concerns of Pope Francis when he speaks of "Integral Ecology" that we must recognize that environmental or ecological crises are deeply dovetail with social, economic, and cultural issues.

Furthermore, in *Laudato Si,* one of the concerns of Pope Francis is the exploitation of Third World countries by the Western Superpowers, who, driven by capitalist principles, sacrifice the good of the environment and the lives of the poor on the altar of greed and avarice. The cost of engaging in Artisanal Small-Scale Gold Mining is high; thus, poor people lack the financial means to initiate illegal mining, and they are mainly exploited due to poverty. The poor enlisted in it and were taken advantage of because they tried to eke out a living. Despite the large gold deposit, the Ghana Statistical Service states that in Tarkwa-Nsuaem Municipal, 18.2 per cent of the population lives in multidimensional poverty, and the average intensity of poverty is 43.5 per cent. This means that in Tarkwa-Nsuaem Municipal, the Multidimensional Poverty Index is estimated to be 0.079 (GSS, retrieved August 2025). Pope Francis, thus, warns that there is a unique dignity to the human, a kind of good not reducible to the good of use for the satisfaction of desire (Francis 2015, §48). The negative effects of these unwholesome activities affect the poor and the vulnerable, especially women and children, and the miners, who, over time, experience adverse health challenges. Therefore, their integral development and self-actualization are put in jeopardy.

He avers that "when we speak of the 'environment', what we really mean is a relationship existing between nature and the society which lives in it" (Francis 2015, § 139). Humans are part of nature and interact with it constantly. Therefore, we should pay attention to the social issues, which in most cases are the root cause of environmental crises, rather than treating them as isolated cases. Therefore, we need a new paradigm that will prioritize both care for the environment and the poor.

9.3 Water Contamination and Health Crises in Tarkwa

Artisanal Small-Scale Gold Mining has undeniably contributed about 35% of Ghana's gold exports and has provided livelihoods for around 4.5 million people; however, it also causes land degradation, water pollution, and mercury poisoning (Blay 2022). Artisanal gold miners in Tarkwa and nearby areas almost exclusively use mercury (locally called 'med') to extract gold from ore (Clifford 2017, p. 501). These miners have little to no technology to recover mercury used in gold refining, so the majority of it is released into the environment (Clifford 2017, p. 49), and mercury, it has been documented, can cause many harmful effects on humans and the environment.

Additionally, the World Health Organization (2007) states that "mercury is a volatile element that diffuses easily into the air and, in most cases, makes its way into water bodies." Mercury is so poisonous that when humans inhale the vapor, it leads to damaging effects on the nervous, digestive, and immune systems and the lungs and kidneys, and may be fatal" (WHO 2007). Thus, seafood from a contaminated water body adversely affects the central and peripheral nervous system (Gibb and O'leary, 2014, p. 667). Accordingly, mercury is said to be classified among "the ten most dangerous chemicals and is a priority concern for the 128 countries of the Minamata Convention" (Celia Deane-Drummond & Hilda P. Koster, eds. 2023, p. 131).

Groundwater in the Tarkwa–Prestea area is known to be susceptible to pollution from mercury contamination that eventually causes major human health challenges

(Bhattacharya et al. 2012, p. 73). In gold mining areas, like the Tarkwa-Prestea area, it has been documented that "sulphide oxidation leads to low pH in the groundwater that encourages the mobility of trace metals, which are found in the groundwater in very high concentrations" (Bhattacharya et al. 2012, p. 73). In addition, studies have shown that "women and children are especially vulnerable and may be exposed directly by eating MeHg-contaminated fish, because the methylmercury (MeHg) bioaccumulates in fish, when consumed by pregnant women, may lead to neurodevelopmental problems in the developing fetus" (Gibb/O'leary 2014, p. 667). From the studies carried out on water samples in the Tarkwa-Prestea area, it was concluded that although results generally indicate some amount of good drinking water quality, "contamination of groundwater from mining activities has been found at some locations and further contamination is possible" (Bhattacharya et al. 2012, p. 83).

This contamination exposes the local communities in the Tarkwa-Prestea area to waterborne diseases. One major river in the Tarkwa-Prestea area that has been contaminated is the Bonsa River, which "serves major domestic and industrial needs while supporting aquatic life for the entire Tarkwa Nsuaem Municipality" (Obiri-Yeboah et al. 2021, p. 2). After sampling water from the Bonsa River, Obiri-Yeboah et al. concluded that "the physicochemical characteristics revealed that the Bonsa River had turbidity and total suspended solids above the recommended WHO values" (Obiri-Yeboah et al. 2021, p. 7). The lead and mercury levels were significantly higher than the permissible limits set by the Ghana EPA (Environmental Protection Agency) and the WHO (Obiri-Yeboah et al. 2021, p. 7). Therefore, "the likely potential contamination of the Bonsa River could be due to the use of mercury by illegal miners, erosion of natural deposits, discharge from refineries and factories, and runoff from landfills and croplands" (Obiri-Yeboah et al. 2021, p. 7).

From the foregoing, it is no surprise that Pope Francis decries the fact that the contemporary industrial system has not developed the capacity to reuse waste and byproducts. Accordingly, he asserts that "some forms of pollution are part of people's daily experience, and exposure to atmospheric pollutants produces a broad spectrum of health hazards, especially for the poor, and causes millions of premature deaths" (Francis 2015, § 20). Again, one observes how the interaction between the people and the environment produces negative resultant effects on the most vulnerable. It is no doubt then that Pope Francis submits that to understand "the reason why a given area is polluted requires a study of the workings of society, its economy, its behavioral patterns, and the ways it grasps reality" (Francis 2015, § 139). In this instance, the unbridled quest for quick profit-making without care for the consequences of one's actions on others led the miners to use harmful chemicals, like mercury, to mine.

I was able to access good water because the parish house had a steady supply of treated water. There are, however, many who cannot afford treated water daily due to the high cost of living. Pope Francis insists that

> Fresh drinking water is an issue of primary importance since it is indispensable for human life and for supporting terrestrial and aquatic ecosystems. Sources of fresh water are necessary for health care, agriculture, and industry. Water supplies used to be relatively constant, but now in many places, demand exceeds the sustainable supply... Water poverty especially

affects Africa, where large sectors of the population have no access to safe drinking water (Francis 2015, § 28).

Therefore, it is evident that environmental degradation and pollution, the social, economic, and physical well-being of the people are inevitably interconnected. These are not isolated issues and, as such, cannot be decisively dealt with without acknowledging that all combine to form the complex crisis. Hence, whatever plan or strategy we employ in solution to the crisis must reflect an integral approach to combating poverty, restoring dignity to the excluded and vulnerable, and at the same time protect nature from destruction (Francis, 2015, § 139).

9.4 Environmental Degradation and Forcing People to Move

In places where environmental degradation occurs, such as in Tarkwa and its surrounding areas, people are often displaced from their original settlements to accommodate full-scale mining activities in the area. Again, this instance is a veritable example of how exploitation of the environment cannot be isolated from exploitation of the human person. As earlier cited, the parish is located at a place called 'New-Atuabo', as one of the elders in my parish took time to describe, there used to be an 'Old-Atuabo'. Today, he affirms, it has become a major gold mining site. When one reflects on the psychological effects of displacing people from their homeland, we discover the feeling of alienation that displacement brings. This carries the same weight as migration, that is movement of people. The homeland of a people means so much to them, and their relationship to the place is closely tied by the inherited memory, history, tradition, and veneration that make them want to stay and prosper in their homeland (Berry 1977, p. 3).

Consequently, when people lose the connection with their homeland because their traditional communities have been demolished, they face the erasure of the "beginnings of domestic cultures" (Berry 1977, p. 3). The exploiters often find reasons to justify their unethical activities. There is that impression created that what they destroyed was "outdated, provincial, and contemptible," just to garner support from fellow citizens of the exploiting country (Berry, 1977, p. 3). This is how deeply rooted the mentality of exploitation is, where the exploiter thinks only in terms of numbers and quantity" (Berry 1977, pp. 10–11).

According to Berry, "the first principle of the exploitive mind is to divide and conquer" (Berry 1977, p. 17). This is evident in the way many foreign nationals were able to financially induce the chiefs and other public officials, and lands were yielded to them, in most cases against the wishes of the people. These migrants have devised ways to integrate and establish close relationships with the locals in Ghana by first enticing the chiefs. According to Botchwey et al. there were instances where disputes occurred between local artisanal miners and Chinese miners over access to mining sites (Botchwey et al. 2018, p. 311). Nonetheless, "with the connivance of chiefs and other public officials, in particular, the Chinese miners have been accused of displacing and outcompeting local alluvial gold miners" (Botchwey et al. 2018, p. 311).

Similarly, Pope Francis maintains that although we are at risk of losing our natural environment, however, "there is also an historic, artistic, and cultural patrimony which is likewise under threat" (Francis, 2015, p. 143). Thus, when people are displaced from their homelands to make way for various mining activities, the environmental destruction is not the only resultant effect. There is also the loss of history, memories, culture, and architecture, which erases the original identity of the displaced people. The solution is not just about building new settlements after tearing down old settlements, but "preserving the history, culture, and architecture of each place, thus preserving its original identity" (Francis 2015, §143). Additionally, he opines that varied forms of environmental exploitation and destruction not only exhaust the resources of the people, but they also "undo social structures which, for a long time, shaped cultural identity and their sense of meaning of life and community" (Francis 2015, §145). Since exploitation thrives wherever there is greed, a lack of respect for human dignity, and a total disregard for moral values, Pope Francis calls for dialogue and transparency in decision-making, which must be grounded in the principle of the common good, recognizing the views of everyone (Francis, 2015, §182).

9.5 Integral Ecology and Ecological Conversion: The Need for a Holistic Solution

Pope Francis espoused the concepts of 'Integral Ecology' and 'Ecological Conversion' to enable us to understand the complexity of environmental degradation and the need for a change in attitude. Integral Ecology recognizes the interconnectedness of all creation, both living and nonliving things, are "interrelated in a network which we will never fully explore and understand" (Francis 2015, §138). Accordingly, to address environmental crises, we must take cognizance of the social dimension, because there is a relationship between the environment and the society that lives in it (Francis 2015, §139). Care should be taken to avoid superficial solutions and rather "adopt an integrated approach to combating poverty, restoring dignity to the excluded, and at the same time protecting nature" (Francis 2015, § 139). Integral Ecology, Pope Francis avers, "is inseparable from the notion of the common good, a central and unifying principle of social ethics...it is the sum of those conditions of social life which allow social groups and their individual members relatively thorough and ready access to their own fulfillment" (Francis 2015, §156).

The encyclical not only addresses environmental issues but also clarifies that our lack of care for the environment affects the poor and marginalized. Thus, our care and compassion for the environment inadvertently imply care and compassion for our fellow human beings. If the earth cries, then the poor cry too, and if the poor cry, the earth cries as well. In the world today, injustices and various forms of exploitation abound, and the vast majority of people are conspicuously deprived of basic human rights and regarded as disposable (Francis 2015, § 158). Accordingly, the principle of the common good immediately becomes, logically and inevitably, a summons to solidarity and a preferential option for the poorest of our brothers and sisters" (Francis 2015, § 158). Therefore, decisions or resolutions should be made having the poor countries in mind, and "the priorities must be to eliminate extreme poverty and to promote the social development of the people" (Francis 2015, § 172). Pope Benedict XVI raised similar ecological concerns in *Africae Munus,* on how environmental degradation is driven by greed. Accordingly, Paulinus Odozor, in *Africae Munus as Christian Policy Statement for an African Renaissance*, articulates Pope Benedict XVI's concerns on ecology stating that "environmental degradation in Africa has roots in greed whose 'opulence…shocks human conscience' and the poverty and impoverishment of many misguided poverty alleviation programs which end up aggravating the situation" (Odozor 2022, p. 14).

Furthermore, he calls for ecological conversion, dialogue, and participation for all persons and sectors of society, including scientists, religious leaders, political leaders and policymakers, and local communities. By 'Ecological Conversion', the Pope enjoins us to develop a spirituality to generously care for our environment, with gratitude to God for the gift of creation (Francis 2015, §220). By shaping and reshaping our attitude toward the environment without neglecting any aspect, we are inspired to understand that our superiority is not "a reason for personal glory or irresponsible dominion, but rather as a different capacity which in its turn, entails a serious responsibility stemming from our faith" (Francis 2015, §220).

Consequently, there is a need to deemphasize the human-centered interpretation of God's creation, so that we can appreciate the significance of God's creation, since the Bible describes "all creation as good, as all creation is loved by God and gains its worth through this love" (Deane-Drummond 2017, p. 25). We are stewards, not because humans are of a higher good with moral sensibilities than other creatures. God grants humanity dominion over the earth (cf. Gen 1:28); however, the command does not imply that we should exploit and destroy the earth, but we should be stewards, taking care of our common homeland. God tells us to "till and keep", which implies we should cultivate, as well as protect and preserve the earth (cf. Gen 2:15). Thus, as stewards, we respond to God's divine call on humanity to play a special role in God's redemptive plan (Jenkins 2008, p. 77).

9.6 *Laudato Si:* The Response of the Catholic Church in Ghana in Dialogue with Policymakers

The Catholic Church in Ghana has intensified its efforts to respond to the call of Pope Francis in *Laudato Si* to care for our common home in the wake of the crisis caused by *galamsey*. The Conference of Catholic Bishops of Ghana (GCBC) embarked on sensitization campaigns in parishes of the various dioceses in Ghana. For instance, in 2021, the GCBC, the Ghana Catholic Bishops' Conference, in collaboration with Green Ghana Project, committed to champion the planting of one million trees as part of their response to the Seven-Year *Laudato Sì* Action Platform pronounced by Pope Francis (Samasumo 2021). The Metropolitan Archbishop of Tamale, Archbishop Philip Naameh, who was then the President of the Ghana Catholic Bishops' Conference, made the declaration stating, "We wish to reiterate the Ghana Catholic Bishops' Conference's commitment to plant one million trees during this year's rainy season as announced during our Plenary Assembly in early May" (Samasumo 2021). He further announced that the National Catholic Secretariat would efficiently coordinate and mobilize the efforts and resources in the various dioceses and parishes to achieve the stated aim.

Consequently, there is a serious campaign against *Galamsey* and raising awareness among the lay about the effects of the actions of these artisanal miners. Very prominent in this effort is the Archbishop of Accra Archdiocese, Most Rev. John Bonaventure Kwofie, CSSp., in collaboration with the Conference of Major Superiors of Religious—Ghana, who organized an awareness walk called *Environmental Prayer Walk Against Galamsey (Illegal Mining): Caring for Our Common Home—Laudato Si*. He encouraged the Catholic faithful and all people of goodwill to come out and join in the prayer march. The march concluded with mass at the Christ the King Parish Grotto, Accra, and a petition was sent to the President (Archbishop Kwofie 2024). The petition called on the President to deal with illegal Artisanal Small-Scale Mining decisively for the well-being of Ghanaians (Archbishop Kwofie 2024). It was a success, as the Archbishop, in his thank-you message, appreciated those who honored his invitation, saying, "God bless you for choosing 'eco not ego', indeed, where human capacity cannot reach, prayer can!" (Archbishop Kwofie 2024). This development attests to the impact of *Laudato Si*, not just in Ghana but in the world today.

The Catholic Church has not relented in its efforts to dialogue with political leaders and policymakers; however, the citizens are aggrieved that this menace has not been decisively eradicated. Artisanal Small-scale Gold Mining (ASGM) activities, i.e., *Galamsey,* are ongoing and unsupervised, despite the government's claims that it has criminalized them through several displays of force that appear to have had no effect. According to Blay, in his interview with some locals, the issue of the politicization of *galamsey* came up, since "politicians tread on *galamsey* shores with tact and shrewdness because they fear losing votes" (Blay 2022, p. 145). The Church encourages dialogue with local chiefs and traditional leaders in order to achieve good results in the attempt to ameliorate the worsening situation. Enacting and enforcing laws and commands without collaboration with local chiefs and communities appears to be mitigating the fight against *galamsey* (Blay 2022, p. 147).

It was around May 2013, under pressure from the media, that President John Mahama (now back in power, having been recently elected on December 7, 2024) established a task force of military personnel aimed at fighting illegal small-scale mining, also known as galamsey. According to Gabriel Botchwey et al., the "President was careful to include all illegal mining activities in the task force's remit; however, the subtext was clear—this was a measure aimed at foreign miners, especially Chinese nationals" (2018, p.216). The task force launched a military-style operation, which resulted in the "deportation or 'voluntary' departure of about 4,592 Chinese nationals" (Botchwey et al. 2018, p. 316). Additionally, the Ghana Parliament enacted a legislative instrument known as the Environmental Protection (Mining in Forest and Reserves) Regulations (L.I 2462) in 2022. This legislation aimed to regulate mining activities within forest reserves. Nevertheless, Lucas Binnah, a doctoral student at the University of Mines and Technology (UMAT) in Tarkwa, states that "some foreigners, particularly of Chinese origin, have conspicuously been seen engaging in it, some have been reportedly arrested" (Binnah 2024, p 2).

The Catholic Church has consistently advocated that solidarity in the quest for the good must guide our actions, not solidarity in bad, because it kills the whole community, the same way a lack of solidarity would do (Bujo 2001, pp. 87–88). The Catholic Church continues this awareness campaign to remind citizens that "misuse of the universe through unnecessary exploitation indicates one's attitude toward others, whether ethical or not (Magesa 2020, p. 52). The question is: what kind of world are we bequeathing to new generations? Therefore, let us, in solidarity with the whole of creation, become stewards who care for our environment and one another. Individuals must realize that no one transcends the community, but "thinks and acts in solidarity with the other members of the community (the whole of creation), whilst maintaining a clear sense of personal responsibility to the environment, your neighbor, and the generation to come (Bujo 1990, p. 97).

Finally, despite recent progress and the persistent advocacy of the Catholic Church in Ghana, more effort must be dedicated to grassroots mobilization and empowerment. Many youths involved in mining activities need to be engaged and equipped with skills that benefit them, their local communities, and the environment. For example, in Zambia, "the Council of Churches in Zambia's Rural Initiative for Community Empowerment (RICE) trains farmers in organic and conservation agriculture" (Gichohi 2025, p. 8). Initiatives like this can create opportunities to engage the youth by providing skills that not only protect the environment but also help them earn a livelihood. While advocacy and dialogue with policymakers are essential, developing practical initiatives that offer viable alternatives for the large young population will have a far-reaching impact.

9.7 Conclusion

This paper begins by recounting my pastoral experience as a young Parochial Vicar at a parish in Tarkwa, Western Region of Ghana, where I served for six years. It highlights the impacts of Artisanal Small-scale Gold Mining (ASGM) in Ghana on both the environment and humanity. It demonstrates that environmental degradation cannot be separated from the exploitation of the poor and vulnerable because all of creation is interconnected. Therefore, the call to respond to the cry of the earth is also a call to respond to the cries of the poor. Any approach to solving environmental issues must therefore consider the social dimensions of the ecological crisis. As a result, it emphasizes that Pope Francis's message in Laudato Si' is timely and calls for immediate action and ecological conversion. It affirms the positive effects of the encyclical within the Catholic Church in Ghana, in collaboration with the Conference of Catholic Bishops of Ghana. It recognizes the Church's efforts in promoting dialogue and cooperation with relevant authorities and policymakers to raise awareness about the harmful effects of human actions on the environment and the poor. Through this, we acknowledge God's boundless love for creation and our duty to care for our common home as stewards. We should live in solidarity with all of creation, our collective home, united by God's love evident in creation.

References

Archbishop Kwofie, B.J. (2024) *Environmental Prayer Walk Against Galamsey (Illegal Mining):Caring for Our Common Home—Laudato Si.* Accra, Ghana: Catholic Archdiocese of Accra. At: https://accracatholic.org/thank-you-message-environmental-prayerwalk-against-galamsey/ (20 December 2024).

Basu, N., Renne, E. and Long, R. (2015) 'An Integrated Assessment Approach to Address Artisanal and Small-Scale Gold Mining in Ghana', *Int J Environ Res Public Health*, 12(9), pp. 11683–11698. At: https://doi.org/10.3390/ijerph120911683.

Berry, W. (1977) *The unsettling of America: culture & agriculture.* San Francisco: Sierra Club Books.

Bhattacharya, P. *et al.* (2012) 'Hydrogeochemical study on the contamination of water resources in a part of Tarkwa mining area, Western Ghana', *Journal of African Earth Sciences (1994)*, 66–67, pp. 72–84. At: https://doi.org/10.1016/j.jafrearsci.2012.03.005.

Binnah, L.J. (2024) 'The Galamsey Albatross: Where there is a Will, there is a Way', *unpublished article*, pp. 1–4.

Blay, J.K. (2022) *A Phenomenological Study of Artisanal and Small-Scale Gold Mining in Ghana: Rural Poverty Mitigation or Aggravation?* ProQuest Dissertations & Theses.

Botchwey, G. *et al.* (2018) 'South-South Irregular Migration: The Impacts of China's Informal Gold Rush in Ghana', *International Organization for Migration*, pp. 310–328.

Bujo, B. (1990) *African Christian Morality at the Age of Inculturation.* St. Paul Publications Africa (Christian leadership in Africa). At: https://books.google.com/books?id=6hAvAAAAYAAJ.

Bujo, B. (2001) *Foundations of an African Ethic: Beyond the Universal Claims of Western Morality.* Crossroad Pub.

Clifford, M.J. (2017) 'Assessing releases of mercury from small-scale gold mining sites in Ghana', *The Extractive Industries and Society*, 4(3), pp. 497–505. At: https://doi.org/10.1016/j.exis.2017.05.007.

Deane-Drummond, C. (2017). *A Primer in Ecotheology: Theology for a Fragile Earth.* Eugene, Oregon: Cascade Books.

Deane-Drummond, C. and Koster, H.P. (eds) (2023). *In Solidarity with the Earth: A Multi Disciplinary Theological Engagement with Gender, Mining and Toxic Contamination.*1st ed. London: Bloomsbury Academic.

Francis, P. (2015). *Encyclical letter Laudato si' of the Holy Father Francis: on care for our common home.* London: Catholic Truth Society (Laudato si').

Gibb, H. and O'leary, G.K. (2014) 'Mercury Exposure and Health Impacts among Individuals in the Artisanal and Small Scale Gold Mining Community: A Comprehensive Review', *Environmental Health Perspectives*, 122(7), pp. 667–672.

Gichohi, Paul. (2025). Church's Ecological Stewardship and Sustainability: A Dogmatic Approach to Sensitizing Community on Habit Change in Kenya. International Journal of Professional Practice. 13. 1-12. https://doi.org/10.71274/ijpp.v13i3.596.

Jenkins, W. (2008) *Ecologies of grace: environmental ethics and Christian theology*. Oxford: Oxford University Press.

Magesa, L. (2020) in J.F. Keenan (ed.). *Catholic theological ethics in the world church: the plenary papers from the first cross-cultural conference on Catholic theological ethics*, New York: Continuum.

Obiri-Yeboah, A. *et al.* (2021) 'Assessing potential health effect of lead and mercury and the impact of illegal mining activities in the Bonsa river, Tarkwa Nsuaem, Ghana', *Scientific African*, 13, p. e00876. At: https://doi.org/10.1016/j.sciaf.2021.e00876.

Odozor, P.I. (2022) 'Africae Munus as Christian Policy Statement', in M.A. Agbaw-Ebai and M. Levering (eds). *Africae Munus: Ten Years Later*, South Bend: St. Augustine's Press.

Samasumo, P. (2021). *Ghana: Catholic Bishops to Plant One Million Trees in 2021*. Vatican News. At: https://www.vaticannews.va/en/africa/news/2021-06/ghana-catholic-bishops-to-plant-one-million-trees-in-2021.html (19 August 2025).

Chapter 10
Integral Ecology in Practice: Equity Mining Partnerships with Indigenous Peoples—Opportunities, Risks, Challenges and Contradictions

Raymond Offenheiser

Abstract Indigenous people have received little real social or economic benefit from centuries of mineral extraction on their native lands. Rather, they have historically suffered mass displacement, discrimination, and trauma. Today, they face growing pressure from governments and mining companies to open their lands for mass extraction of strategic minerals. Roughly 75 percent of critical minerals are located on Indigenous lands. Mining companies eager to access these minerals are offering new co-investment and benefit-sharing partnerships to indigenous communities. So, indigenous people must ask: Do they try to protect their lands from the potential social and environmental impacts of industrial mining, or do they negotiate a middle path and what a middle path looks like? Will this partner respect our indigenous culture and practices? Will they listen to and respect our concerns about the environment? These new co-investment and equity partnership opportunities also present challenging questions for Integral Human Development (IHD) and Integral Ecology (IE), as outlined in Pope Francis' encyclical. Laudato Si (Francis. (2015). Laudato Si': On Care for Our Common Home [Encyclical]. Vatican City: Vatican Press). Are such equity partnerships consistent with the core principles of IHD and IE? The objective of this paper is to raise critical questions about the opportunities, challenges, and risks of such new partnerships for indigenous communities, and to generate a wider discussion among proponents of IHD and IE about the implementation of these ideals in complex, real-world environments.

Keywords Mining · Extraction · Business · Private enterprise · Indigenous lands · Development

R. Offenheiser (✉)
Raymond Offenheiser, University of Notre Dame, Notre Dame, United States
e-mail: roffenhe@nd.edu

R. Marcantonio et al. (eds.), *Integral Ecology in the Anthropocene: Africa and Beyond*, The Anthropocene: Politik—Economics—Society—Science 49,
https://doi.org/10.1007/978-3-032-32125-1_10

10.1 Introduction

Integral ecology challenges us to rethink our fundamental relationship with nature and God's creation. It challenges us to move beyond the traditional Anthropocentric conception of homo sapiens as a superior being and master of the natural order. It summons us to a new vision in which we are one with nature and called to care for and steward the gift of nature while applying our human faculties toward advancing the well-being of all mankind within the bounds of what nature can sustainably provide.

Indigenous communities have maintained a close relationship with nature for thousands of years. Their daily lives, culture, and ceremonies are deeply connected to the cycles of seasons, plant growth, and the availability of crucial wildlife and water sources. They have lived sustainably within the limits of what nature provides. They have also learned instinctively to adapt when these natural rhythms are disrupted. Overall, they have proven to be responsible stewards of the environment in their territories.

Colonization and modernization, however, have disrupted their traditional ways of life in many parts of the world. The demands of a global economy have created pressure to access their lands, divided their communities, and upended their traditional lifestyles. While, in some true sense, they might exemplify what living sustainably and practicing integral ecology could mean, their way of life faces challenges from the forces of contemporary market economics and capitalism. How they respond to these market pressures while trying to preserve their cultures, values, and the environmental integrity of their homelands can perhaps teach us important lessons about what living according to the values of integral ecology (IE) requires in practice.

This paper aims to examine the questions that indigenous communities will face in the coming decades as the world's demand for critical minerals increases exponentially, along with the pressure on these communities to access minerals on their lands.

These communities will face difficult choices. They can resist, defending their values, lands, and natural resources. Or, they might see resistance as futile and choose to negotiate strategically with governments and companies. In doing so, they can work to limit land exploitation and ensure their communities have a say in how resources are extracted, including decisions about the scale and cost.

Such partnerships could enable them to take a leading role in protecting their natural resources while also providing benefits for their communities that support increased sustainable and holistic human development. This process will vary among different Indigenous communities worldwide. This paper aims to serve as a guide for communities facing the challenges, risks, and opportunities in the coming decades. It seeks to explore what it means to implement the theological principles in Laudato Si and what Indigenous populations must do to lead in reducing the unchecked exploitation of mineral resources, favoring more balanced and developmentally sustainable methods (Francis 2015).

Centuries of colonization and modernization have taken a heavy toll on Indigenous communities. Despite thousands of years of strong social and cultural connections to their lands and landscapes, Indigenous peoples have faced brutal programs of forced displacement and have been treated as culturally invisible and disposable. Their claims of traditional ownership have been dismissed, and signed treaties have been torn up or ignored (Kaplan 2022). National armies have often been sent to force them to move, leaving them homeless and to struggle on unfamiliar lands and terrains.

In the United States, Canada, and Australia, children were often forcibly taken from their communities and sent away to residential school systems, never to return (Mosby and Millions 2021). With few protections for their rights, limited support from national governments, and few job opportunities, many Indigenous communities remain trapped in poverty and squalor. Drugs and alcohol ravage their communities, breaking families apart. Intergenerational trauma casts a dark shadow that fuels psychological distress and erodes their human dignity, sense of self-worth, and agency.

Only in recent decades have we started to confront this legacy of racism and legally enforced social marginalization. The last Indigenous residential schools in the US and Canada closed from the 1970s through the mid-1990s (Bombay, Matheson, Anisman 2014). The UN Declaration on the Rights of Indigenous Peoples (UNDRIP), which sets the basic framework for protecting Indigenous human rights, was finally adopted by the UN General Assembly in 2007 (UN 2007).

The Declaration is based on four key principles: self-determination, participation in decision-making, respect for and protection of culture, and equality and non-discrimination. It establishes a universal framework of minimum standards for the survival, dignity, and well-being of the world's Indigenous Peoples and builds on existing human rights standards and fundamental freedoms specifically related to Indigenous Peoples. The United States, Canada, and Australia initially voted against the Declaration but later joined the other signatories. These documents closely align with the core principles of Integral Human Development (IHD), which advocate for the holistic development of the entire person.

In June 2011, the UN Human Rights Council unanimously approved the UN Guiding Principles on Business and Human Rights, also known as the Ruggie Principles (UNHRC 2011). These voluntary principles were developed over a six-year period under the leadership of the late Professor John Ruggie, who served as the Special Representative of the Secretary-General on Human Rights and Transnational Corporations and Other Business Enterprises. Building on the UNDRIP, the Ruggie Framework presented a comprehensive, three-pillar approach to business and human rights, including a state duty to protect, a corporate responsibility to protect, and access for stakeholders to effective remedies (Rosenmeier/Neergaard, 2013).

Foundational to the implementation of these new frameworks is the concept of Free Prior and Informed Consent (FPIC). And, nowhere has this principle been more applied and tested than in the extractive industries sector. Free Prior and Informed Consent suggests that every project proposed on Indigenous land should undergo a referendum or a similar collective process agreed upon by the community to determine their support for the project. It assumes that every company must collaborate with Indigenous communities to obtain FPIC for their project.

After centuries of operating with impunity and often with government support within Indigenous territories, the mining industry realized that its operations were vulnerable to public resistance and occasional closures if it did not obtain some form of social license to operate.

With the UN adoption of UNDRIP and later the Ruggie Principles, mining companies realized that public scrutiny and legal challenges would likely grow. The International Council of Mining and Metals (ICMM), the global trade association of the 15 largest mining companies worldwide, issued its first policy statement on Indigenous rights in 2008, which was later updated in 2013 and again in 2024 (ICMM 2024). While this was a significant step for an industry with a long history of mistreating Indigenous populations, it was a cautious initial move. Instead of accepting the idea of "consent," the ICMM chose to frame their policy as an endorsement of Free Prior and Informed "Consultation," showing their discomfort with fully honoring the expressed preferences of the local Indigenous population.

Nonetheless, FPIC has become mainstream in policy discussions when large-scale projects are proposed on Indigenous lands. There have been some cases where the concept has been applied and respected, most notably in Tambogrande, Peru, where Manhattan Minerals withdrew from a major gold mining project after a local referendum voted 96 percent against it (Echave 2005). Despite the company's objections, the Peruvian government respected and enforced the outcome. In another case, BHP Corporation withdrew from a significant iron ore project in the interior of Kalimantan, Indonesia, when faced with strong pressure from local forest-dwelling Indigenous communities. Many companies nowadays demonstrate their recognition of Indigenous rights in their sustainability reports and aim to reach mutually beneficial agreements before exploring Indigenous territories.

International financial institutions have increasingly acknowledged the rights of Indigenous peoples. In 1998, the Inter-American Development Bank adopted a policy requiring prior consent from Indigenous people facing involuntary resettlement as part of a bank-financed project (Deruyttere 2009). The International Finance Corporation (IFC), the private sector lending arm of the World Bank, required in 2007 that it must 'minimize, mitigate or compensate for impacts on Indigenous communities and engage in a process of consultation and "informed participation' with affected populations (IFC,2007). In 2012, it expanded these requirements to include an obligation to obtain free, prior, and informed consent for all projects affecting Indigenous populations (IFC, 2007).

This growing recognition of Indigenous rights and land claims has led to significant changes in public attitudes. In Canada, the Chretien government launched Gathering Strength, a 98-point action plan for reconciliation in 1998, which urges all national, provincial, and local companies and governments to develop their own action plans. And in 2008, Prime Minister Harper issued an apology (Canada.ca 2008) to the First Nations students who attended Canada's notorious residential schools. Similarly, Australia held a national referendum in 2023 to amend the Australian Constitution to recognize full citizenship rights for all Indigenous Australians. Although the amendment did not pass, it received 40 percent of the vote, indicating a notable shift in public attitudes in favor of its Aboriginal population. (Davis 2024)

10.2 Impact of the Just Transition for Indigenous Populations

This shift in public attitudes occurs at a crucial time. With the threat of climate change looming, countries are racing to develop the infrastructure necessary for the transition to a carbon-neutral world. Achieving this goal requires large amounts of minerals, and some speculate that it will necessitate a 400-600 percent increase in global mining activity (IEA 2022). The race is underway to find and secure legal claims and social licenses in regions rich in lithium, copper, and nickel deposits. Besides these essential elements for manufacturing electric vehicle batteries, rare earth metals are scarce and only found in a few locations worldwide. China has the largest deposits, with other significant sources located in the Congo and Canada (Pistilli 2025). Meanwhile, the global mining industry is expanding its prospecting and strategic planning efforts, expecting significant opportunities for growth in investments and project sites.

The race for minerals is also significantly affecting global geopolitics and national security. The strategic minerals needed to power our cars are also essential components of advanced military technology and the progress toward AI as a key element in both offensive and defensive military strategies. Closed-door meetings are taking place in capitals worldwide to determine whether they will have secure and sustainable access to the key metals and minerals necessary to maintain a modern military and national defense system (Vergum 2025).

10.2.1 Global Scale of Indigenous Populations

The pace and scale of this increasing demand significantly impact Indigenous populations, as Indigenous or traditional African tribal groups own 40 percent of the world's lands. Seventy-five percent of the world's strategic minerals are located on or near the lands of Indigenous and tribal communities (Pohl 2025). These groups are starting to feel pressure from governments and global mining companies to allow access for exploration and potential resource extraction. This often puts Indigenous populations against their national governments, which are eager to benefit financially from large-scale industrial mining operations while showing little concern for the possible environmental or socio-economic effects on their Indigenous communities.

Indigenous populations have experienced this before and know that historically, governments and companies have shown little patience or respect for treaties and land claims. Once these projects begin, roads and power lines are built through their territories. Workers migrate from other regions seeking jobs, land, or business opportunities. During this rush, governments usually provide little protection for their rights and interests. Ultimately, they see little or no benefit from such multi-billion-dollar investments in their lands and often face increasing stress on their livelihoods, cultural practices, and family cohesion.

10.2.2 The Opportunity for Equity Partnerships

This growing pressure from major global powers to access reserves of strategic minerals, which are critical for both the clean energy transition and national security, creates potential opportunities for Indigenous communities. Historically, governments and companies have often been hesitant to recognize Indigenous or peasant communities as valuable partners. As a result, Indigenous lands were opened to underground resource extraction without proper consultation. Indigenous populations were displaced, and non-Indigenous workers filled both technical and operational roles in these new mines. Meanwhile, local Indigenous communities were marginalized and received little or no material benefits.

With increasing pressure from business and government to secure access to critical mineral resources, the governments of Australia and Canada have been encouraging companies to consider ways in which Indigenous and peasant communities might participate as equity partners and share in the long-term economic benefits of such projects on their lands.

Although it is still early, more projects are involving Indigenous groups and producing tangible benefits. The momentum for such deals is growing. The First Nations in Canada are building professional institutions, banks, and capacities at both the national and provincial levels to support these efforts. Since 2000, 524 agreements have been signed between exploration and mining companies, Indigenous communities, and governments (NRC 2021).

According to the 2021 census, over 17,300 Indigenous people are employed in Canada's mining and mineral processing industries. About half of these Indigenous workers are in the upstream mining subsector, where Indigenous Peoples make up 11% of the industry's workforce. Additionally, there has been an increase in agreements covering exploration, development, and post-development stages (Natural-Resources.canada.ca 2021): 51 agreements were in place before 2000, and 524 agreements were signed between 2000 and 2021.

The increase results from higher exploration and mining activities, as well as a growing emphasis on building mutual understanding between Indigenous communities and the mining industry. However, fewer agreements across exploration, development, and post-development stages have been made in recent years, leading to a decrease in projects (Natural Resources.canada.ca).

10.3 The Business Case for Equity Partnerships

There is a compelling business case for Indigenous equity partnerships that benefit Indigenous communities. The clearest reason for a community to pursue such a partnership is the long-term economic benefits it can offer. It can support the community's financial sustainability over time by creating jobs, businesses, educational opportunities, and training programs. It can also help develop professional skills and boost confidence and self-esteem among young Indigenous leaders, fostering a spirit of self-determination within the community. Additionally, it can provide essential resources for cultural preservation and strengthen community cohesion. Perhaps most importantly, it promotes self-determination and helps build trust between Indigenous and non-Indigenous leaders and workers.

There is also a strong business case for non-Indigenous companies to pursue such partnerships. The most obvious reason is the opportunity to build an authentic relationship with the Indigenous partner that recognizes the rights and integrity of the Indigenous community and its value as a trusted equity partner in a process of co-creation. Building this kind of relationship from the outset will foster mutual understanding about how future problems and challenges can be addressed. It will contribute significantly to de-risking the project and establish a foundation for a more inclusive and equitable development process.

Finally, Indigenous communities often possess knowledge about a region and its ecology that outsiders may overlook, but can be a significant asset to large multi-year projects with substantial environmental and cultural impacts. Investors and governments are increasingly favoring projects where companies demonstrate a sophisticated approach to engaging with Indigenous populations. In some jurisdictions, governments may soon require Indigenous consent or equity participation. It's therefore increasingly in a company's interest and reputation to show a commitment to reconciliation and partnership.

10.3.1 What Kinds of Deals Are on Offer?

The deals offered today are still in an early developmental stage. They range from simple benefit-sharing agreements, which may or may not include direct financial payments, to more complex arrangements that could involve loan guarantees to help groups secure essential loan capital. Benefit-sharing agreements might not include an equity stake but could ensure employment and training opportunities, participation in project procurement deals, or cultural support.

Once communities achieve financial stability, they may seek direct equity investments involving full or co-ownership. If they want to avoid the burdens and risks linked to full or partial ownership, they can choose a revenue-sharing arrangement or royalty payments based on a project's production or revenue. One option discussed but not yet implemented is creating a sovereign wealth investment fund in the name of a First Nations group, which the group would oversee. The success of this method largely depends on the financial management skills and good governance of the Indigenous organization.

Finally, governments can play an essential role in promoting and supporting such equity deals by providing tax exemptions and incentives to indigenous organizations or businesses, enabling them to benefit from procurement services that mines may need.

10.4 Ensuring Success: Critical Considerations

Entering into an equity partnership is a significant step for most Indigenous communities. It should be approached with careful due diligence and expert guidance on important legal and financial decisions. A crucial first step is to determine whether the proposed partnership will be based on the recognition of Free, Prior, and Informed Consent (FPIC). A key question for decision-makers is whether FPIC will be maintained throughout the entire lifecycle of the mine, not just at the beginning.

Indigenous communities entering such partnerships often face knowledge asymmetries. Companies generally conduct extensive research before making long-term investments. Indigenous groups need to invest in building the capacity of their teams to ensure they have the necessary knowledge to effectively manage and participate in these partnerships.

Communities will have high expectations for any such partnership. Any collaboration must prioritize and adapt to the specific needs and priorities of the Indigenous community involved. Lastly, Indigenous leaders should work to ensure that their partnership is based on a shared interest in the long-term social and economic benefits, as well as the sustainability of both the project and the partnership.

10.4.1 What are the Risks or Challenges of Such Equity Partnerships?

Indigenous peoples have endured centuries of unmet expectations and understand that they may face significant risks when entering into an equity partnership with a mining, oil, or gas company. Beyond the environmental risks, which are costly and unavoidable in any extractive project, there are additional legal, financial, and business risks inherent in such partnerships that can be difficult to manage and will require external expertise and trusted, competent spokespersons throughout the project lifecycle.

10.4.2 Addressing Power Asymmetries

Indigenous communities have historically faced power imbalances. Companies, their national governments, and colonists interested in contesting land claims have frequently questioned the validity of Indigenous rights and land treaties. Naturally, any Indigenous group should approach equity negotiations with external partners cautiously and skeptically. To establish a good-faith partnership, corporate or government partners must set a fair starting point, share vital information, and build trust in everyone's good intentions.

10.4.3 Environmental Damage

Indigenous populations have a deep connection with their land and environment, which is vital for both their livelihoods and spiritual well-being. Extractive projects are often environmentally harmful, often disturbing habitats, water systems, soils, and air quality. When an Indigenous community agrees to an equity partnership, they will want to work carefully with governments and companies to minimize both short-term and long-term environmental damage. Ideally, strong ecological protections and enforcement mechanisms should be integrated into the core of any equity partnership agreement. However, such agreements may not always foresee every potential environmental risk; therefore, Indigenous and corporate leaders must agree seriously on appropriate dispute resolution forums and processes to handle any future issues.

10.4.4 Cultural and Social Impacts

As shown by the loss of ancient cave paintings at Junkan Gorge in northwest Australia, mining projects that focus on accessing high-quality ores instead of respecting cultural sites can destroy sacred locations and weaken cultural norms, social structures, and community bonds. Non-indigenous partners must show a strong commitment to understanding and honoring Indigenous cultural practices and beliefs from the very beginning of negotiations, at all levels of their organization. A company's firm dedication to Free Prior and Informed Consent is crucial for building trust. To protect their rights and interests during a project, Indigenous communities must demand clear assurances and commitments that they will have a meaningful role in decision-making processes during the development and implementation of projects, as well as in establishing important environmental protection measures.

10.4.5 Challenges in Navigating Complex Legal and Business Structures

The idea of equity partnerships is new to Indigenous communities. Unsurprisingly, many of these communities may lack the necessary business, legal, and financial skills to manage the complexities of such negotiations. Governments and companies aiming to adopt this new approach in good faith must show a willingness to help these organizations find appropriate independent counsel and technical expertise.

It is in the best interests of both parties to have all interests on the table and properly represented during every phase of negotiations throughout the entire project lifecycle. Companies should consider making strategic investments in building the skills and capacities of their Indigenous partners. Experience shows that such investments can generate a highly positive return over the course of a 30- to 50-year project. Without such resources, Indigenous organizations will be vulnerable to exploitation, especially in areas with weak regulations, enforcement, and governance. In such cases, the long-term success of the project and partnership could easily be jeopardized.

10.4.6 Short-lived Benefits

A common issue with many past projects is that the benefits left for indigenous communities often do not last long after the project's end. Usually, the original owner sells the site to a company that operates it for a few years to extract remaining value before walking away, neglecting any responsibilities to fix environmental damage or restore the landscape. In a responsible partnership, the community should be left with financial resources and ongoing business activities that can support its future. Equity partnerships, therefore, need to include a clear and realistic plan for long-term economic sustainability from the beginning. Forward-thinking companies today are reevaluating their responsibilities to communities and planning from the start for responsible closure or potentially ongoing site management. Incorporating long-term planning early in relationship building helps ensure both parties understand potential challenges and impacts, as well as how benefits and profits will be shared fairly and equitably over time.

10.4.7 Mutual Accountability

Successful partnerships invest time early in building strong systems and processes to ensure mutual accountability. Indigenous communities and companies should develop suitable tools for providing regular professional oversight, reporting, and consultations on critical issues such as environmental, social, and cultural impacts. These processes should be maintained throughout the entire project lifecycle. Such mutual accountability will allow both parties to make clear assessments of past work and to anticipate potential challenges as a project expands into new areas, where consultation and co-design discussions will be necessary to address potentially complex questions.

10.4.8 Pre-Conditions for Success

Equity partnerships as a new form of benefit-sharing with Indigenous communities present an exciting new frontier of opportunity. However, it's important to include a cautionary note. The reality is that the conditions for success are not in place in many parts of the world. Canada has the highest number of Indigenous equity partnerships. Why is this? Canada has undertaken a bold reconciliation initiative with its First Nations populations, driven by its ambitious 98-point Action Plan. Canadian companies have been asked to join this initiative and demonstrate their commitment to "righting the historical wrongs committed" against First Nations communities.

The government, at both national and provincial levels, has invested considerable time and funding in creating a supportive ecosystem of Indigenous-owned and managed financial institutions. This includes engaging in joint brainstorming sessions with Indigenous leaders on best practices for promoting equity partnerships and encouraging the national mining industry to align its goals with the national reconciliation process. While it is still early to declare these efforts a success, the momentum generated during both the conservative Harper and Liberal Trudeau administrations has persuaded many companies to start exploring various forms of equity partnerships.

Australian mining companies are beginning to move in similar directions. However, the Aboriginal issue in Australia remains complex, as demonstrated by a recent failed referendum that called for full recognition of Aboriginal rights and land claims. It will take time for Australia to develop the social relations, institutional capabilities, and positive narrative that currently exist in Canada, which support such efforts.

Meanwhile, US- and UK-based companies, such as Newmont, are exploring such deals internationally but not within the US, where little effort has been made to develop the supportive infrastructure needed for equity partnerships in mining. Despite this, a mining boom is currently underway in the US. The Southern Ute tribe in southern Colorado is perhaps the only tribe that has successfully built a strong economic foundation for its community through effective resource management, oil and gas property development, methane capture, and a diversified portfolio of private equity investments.

Based on experience, the following is a short list of critical pre-conditions that all groups exploring a possible equity partnership should consider.

10.4.9 Clarity on Indigenous Identity and Territorial Rights

An Indigenous community must start negotiations with any mining, oil, or gas company with a clear understanding of its identity and a strong sense of unity. Leaders should be recognized as legitimate representatives of the entire community. Tribal land titles need to be acknowledged and regarded as legitimate by the relevant public authorities before negotiations can begin. Internal tribal factionalism can undermine any well-meaning project.

10.4.10 A Company with Policy Alignment and Prior Experience

The most dependable corporate partners will be those whose leaders have engaged in a thorough process of reflection on reimagining their relationships with Indigenous stakeholders, have developed new publicly available policies on Indigenous rights and Free Prior and Informed Consent, and have been restaffing their operational divisions with personnel, preferably Indigenous personnel, who are equipped to manage co-equal partnerships with Indigenous communities. Companies with a proven track record of commitment and success in implementing their new policies operationally are likely to be the best corporate partners.

10.4.11 Good Governance and Enforcement Mechanisms

While the main contractual partnership will be with a company, national and provincial governments play a crucial role in initiating and shaping these alliances. They possess specific powers over permitting, land and water use policies, and environmental regulations. Government entities must support from the beginning and foster these partnerships by offering financial or institutional backing. Clearly, a successful extractive project in their province can generate a steady stream of royalties that can be invested in building the institutional infrastructure needed to sustain the project long-term and provide valuable services to both Indigenous and non-Indigenous communities. Indigenous leaders need to trust that both national and provincial legal systems will protect, not weaken, their interests throughout the project's life. To prevent the government from assuming a company will handle the state's responsibilities for public services and other investments in a mining region, companies may sometimes act as advocates, representing the interests of both the company and its Indigenous partner in securing and maintaining essential state support. Although no scenario is perfect, leaders must carefully evaluate their partners and potential opponents before entering negotiations.

10.4.12 Supportive Institutional Ecosystem

Indigenous organizations will require substantial financial support to participate in such partnerships. In Canada, Indigenous-owned and managed banks at the national and provincial levels can be sources of initial capital. There are also university and apprenticeship programs designed to train an Indigenous workforce that could work on-site. Additionally, Indigenous-owned businesses can offer valuable services to the mining operation. This level of institutional capacity is often missing in many other countries and jurisdictions where mining projects are usually located. Companies can play an important role in building these capabilities in collaboration with their Indigenous partners.

10.4.13 Commitment, Transparency, and Accountability

Perhaps the most vital element for a successful partnership is transparency and accountability from both the company and the Indigenous community. A clear commitment to transparency and accountability from the beginning will establish the trust needed to support a project lasting 30 to 50 years. Leaders will come and go, but trust must be transferred with each change in leadership. Transparency and accountability do not happen automatically; systems and processes need to be intentionally designed, embedded, and implemented at every level of both partners' organizational cultures to ensure they become a core part of the ongoing relationship.

10.5 Conclusion

Equity partnerships give Indigenous communities the opportunity to finally benefit financially from the lands they have occupied for thousands of years. They also allow them to have greater control over their lands and manage them according to their traditional beliefs. Companies are increasingly realizing that they can no longer forcibly push or manipulate Indigenous peoples off their lands. In a world where 75 percent of the world's critical strategic minerals are found on Indigenous lands, new ideas and approaches are needed to ensure companies can obtain and maintain social and environmental licenses throughout the life of a project.

We can identify a few such partnerships in Canada, the Pacific Islands, and Australia. However, there are no major cases yet in the US, Latin America, or Africa. The US, with its outdated mining law from 1872 and its failure to recognize the abuses of its Indigenous populations, lags significantly behind. Latin America shows potential, but its institutional support systems and governance practices may not yet be sufficiently strong to sustain these initiatives.

In Africa, the issue of Indigeneity differs from that in the Americas and Australia. Its institutional capacities and governance systems, with some exceptions, would need much more investment both at the tribal level and in remote provinces where most mining takes place. For now, we may see equity partnerships gradually develop in North America, Australia, and possibly later in South America and Asia as social, political, and institutional frameworks become more supportive.

In the meantime, we can learn a lot from the experiences so far. There is fertile ground for university-based research on best practices. A team at the University of Queensland's Center for Social Responsibility in Mining, located in Brisbane, Australia, is currently leading a university-based research program on this topic.

Hopefully, similar research efforts will develop in the US, Canada, Latin America, Asia, and Africa.

As mining accelerates due to the demand for rare earth minerals, lithium, copper, nickel, and cobalt, we can expect new mines to open in areas we hadn't previously considered. Most likely, these new projects will be on Indigenous lands. It would be a tragedy if some Indigenous populations are unable to benefit from these emerging opportunities and secure the long-overdue benefits for their communities. Ultimately, equity ownership is not just about wealth for Indigenous populations; it's about overdue empowerment, justice, and reconciliation for communities recovering from centuries of marginalization and poverty.

The Indigenous peoples of the world are among the marginalized populations that inspired his writing of Laudato Si and its emphasis on Integral Human Development and Integral Ecology. Mining companies worldwide are surveying Indigenous territories to find the metals and minerals that will drive the just transition to a sustainable future. How these Indigenous communities respond to these calls for access to their lands will determine the future of our planet. Will they be motivated by the vision of IHD and IE to resist and defend their territories, seek common ground through sensible partnerships that honor their dignity and environmental values, or will they be coerced and divided by government and corporate pressure to give in to market demands with limited promises of investment in their development or protection of their environmental heritage? It is on this frontline that the theology of Laudato Si will face one of its greatest global challenges.

References

Aboagye Emmanuel Mensah, (2022) *Assessing inclusiveness for Indigenous communities in mineral rights allocation in Ghana, Heliyon,* Vol. 10 Iss. 8, e29710.

Angarova, Galina (2024) *Geopolitical Implications of the "Green Energy" Transition on Indigenous Peoples, Cultural Survival*, Cambridge, MA.

Ash, Gillian (2024) *Social impacts of critical mineral exploration on Indigenous peoples' lands: A case study from the Solomon Islands, The Extractive Industries and Society*. city: publisher

Bobbit, Michael J., *Massachusetts Cultural Council, Native American & Indigenous People's Equity Plan* (FY25-27) Boston: Mass Cultural Council.

Bernauer, Warren and Gabrielle Sloweyz (2020) COVID-19, extractive industries, and indigenous communities in Canada: Notes toward a political economy research agenda, *The Extractive Industries and Society Journal* No.7 pp. 844-846.

Bombay, Matheson, Anisman, (2014) The intergenerational effects of Indian Residential Schools: Implications for the concept of historical trauma, *Transcultural Psychiatry*. Vol. 51, Iss. 3, pp320-338.

Burton, John, Deanna Kemp, Rodger Barnes and John Parmenter (2024) Mapping critical minerals projects and their intersection with Indigenous land rights in Australia, *Energy Research and Social Science.*

Caldecott, Leonie (2016) Towards and Integral Ecology, Humanum: Issues in Family, *Culture & Science*, April.

Cardenas, Daniel, Rick Tallman, Brad Handler, and Morgan Bazilian (2023) *A Pathway to Responsible Mining in Indian Country,* Center for Strategic and International Studies, Nov.

Castillo, Daniel P. (2016) Integral Ecology as a Liberationist Concept, *Theological Studies* Vol., 77, pp353-376.

Canada. Department of Indian Affairs and Northern Development. (1997*). Gathering strength: Canada's Aboriginal action plan.* Ottawa: Minister of Indian Affairs and Northern Development.

Cleghorn, Christine (1999) Aboriginal Peoples and Mining in Canada: Six Case Studies, *Mining Watch Canada*, September.

Deruyttere, Anne (2011), Implementation of the World Bank's Indigenous Peoples Policy*: A Learning Review* (FY 2006-2008) August

Echave, Jose (2005) Canadian Investment Companies Investments in Peru: *The Tambogrande Case and the Need to Implement Reforms, CooperAccion.* Lima: CooperAccion.

Francis. (2015). Laudato Si': *On Care for Our Common Home* [Encyclical]. Vatican City: Vatican Press

Fredericks, Carla f. and Kate R. Finn,Harnessing (2020) *Private Equity for Indigenous People, First People Worldwide*, University of Colorado, Boulder, CO.

Global Witness Press Release (2024) *Rush for critical mineral in the Philippines threatens Indigenous communities and biodiversity*, London, December: Global Witness.

Government of Canada (2008) Statement of apology to former students of Indian Residential Schools.

Government of Canada (2025) *What We Heard: Engagement on the National Benefits-Sharing Framework and Indigenous Loan Guarantee Program.* , Ottawa: Government of Canada.

Haynes, Josiah Equity (2025) Partnerships represent a turning point for Indigenous communities, Indigenous Partnership Success, Investor Advocates for Social Justice, Shifting Gears part II: *Indigenous Peoples' Rights and Mining in the U.S.* , Montclair: Investor Advocates for Social Justice.

Heliyon Journal, Law School, Zhongnan University of Economics and Law, Wuhan, China Digital Report (2022) Geopolitics of the Energy Transition, International Renewable Energy Agency, *Acta Geographica Sinica,* Issue (8) : pp 2050-2066.

Herring, Rachel, Keaton Sandeman and Lyuba Zarsky (2024) Decarbonization, critical minerals and tribal sovereignty: *Pathways towards conflict transformation, Energy and Social Science*, Vol. 111

Hoben, Merrick (2024) To Navigate Conflict, Prioritize Dignity, *Sloan Management Review*, Summer.

Husain, Nyla (2023) *Global Extractive and Industrial Projects Disproportionately Impact Indigenous Peoples,* American Academy for the Advancement of Science.

ICMM, (2024), *Indigenous Peoples: Position Statement*, London, International Council on Mining and Metals.

Indigenous Partnership Success Showcase (2024) *Equity partnerships represent a turning point for Indigenous communities.* , Vancouver: Indigenous Partnerships Success Showcase.

International Energy Agency (2022) *The Role of Critical Minerals in Clean Energy Transitions*: World Energy Outlook Special Report , Paris: International Energy Agency.

International Finance Corporation (IFC) Guidance Note 7: *Indigenous Peoples* , Washington, D.C.: The World Bank.

International Renewable Energy Agency, (2022) *Digital Report: Geopolitics of the Energy Transition*: *Critical Minerals*,. Masdar: International Renewable Energy Agency

Investor Advocates for Social Justice (2025) Shifting Gears Part II*: Indigenous Peoples' Rights And Mining in the U.S.* , Montclair: Investor Advocates for Social Justice.

Jessen, Tyler D, Natalie C. Ban, Nicolas XEMTOLTW Claxton & Chris Darimont (2021) Contributions of Indigenous Knowledge to ecological and evolutionary understanding, *Frontiers in Ecology and the Environment.* Vol. 20, Iss. 2. Pp93-101.,

Kaminsky, Andrew (2024) How Co-Ownership Built a Strong Relationship Between an Indigenous Community and a Mining Company, San Francisco: Triple Pundit.

Kaplan, Genevieve (2022) A Legacy of Negligence: The Historical Mistreatment of Indigenous Peoples in the United States, *Berkeley Public Policy Journal*, April.

Kellner, Elke (2025) Mineral extraction on Indigenous land: employing a relational approach to navigate the convergence of Indigenous and other ontologies and practices, *Energy Research and Social Science*, Volume 125, e104097

Kokoris, Moki (2012) Ensuring Equity for the Indigenous Within Their Nation-States for the Benefit of All, *The Arctic Institute.* , city: publisher

Kung, Anthony, Sarah Holcombe, Joel Hamago and Deanna Kemp (2022) Indigenous co-ownership of mining projects: a preliminary framework for the critical examination of equity participation, *Journal of Energy and Natural Resources Law.* Vol. 40, Iss. 4., pp413-435.

Mosby, Ian and Erin Millions (2021) Canada's Residential Schools Were a Horror, *Scientific American*, August.

Natural Resources Canada (2021) Minerals and the Economy, (https://natural-resources.canada.ca/minerals-mining/mining-data-statistics-analysis/minerals-economy website)

ran (2013) Extractive industries and Indigenous peoples: A Changing Dynamic, *Journal of Rural Studies*, Volume 30.

Owen, John R, Deanna Kemp, Alex M. Lechner, Jill Harris, Rulian Zhang, and Eleanore Lebre, (2023) Energy transition minerals and their intersection with land-connected peoples, *Nature Sustainability*, Volume 6, pp. 203-211

PBS Newshour (2024) How the mineral mining boom endangers Indigenous communities worldwide

Pohl, Wolfhart, Dianna M. Pizarro, Justine Sylvester, (2025) *Protecting Indigenous Peoples rights amid growing demand for land and critical minerals*, World Bank, Washington, DC

Quirino, Maya. (2025) *Criticality and the Securitization of Southeast Asia's Indigenous Mineral Resources,* Heinrich Boll Stiftung Southeast Asia, February.

Rosenmeier, Louise and Neergaard, Peter (2013) "Exploring the Effect of the Ruggie Framework for Human Rights," *The International Journal of Ethical Leadership*: Vol. 2, Article 10.

Ruggie, John (2008), Protect, Respect and Remedy: The UN Framework for Business and Human Rights, Chapter 27 in Protect, respect and remedy: a framework for business and human rights: report of the Special Representative of the Secretary-General on the Issue of Human Rights and Transnational Corporations and Other Business Enterprises, John Ruggie

Stagner, Taylor Dawn (2025) *Most critical minerals are on Indigenous lands, Will miners respect tribal sovereignty*, Grist, March

Teevan, Sinead and Jenny-Lou Campbell (2024) Indigenous Peoples in Mining Regions: From Compensation to Collaboration, OECD Cogito Blog.

The Aspen Institute, Energy & Environment Program (2023) A Critical Mineral Policy for the United States: The Role of Congress in Scaling Domestic Supply and De-Risking Supply Chains, Washington, DC. : The Aspen Institute.

Trahant, Mark (2023) *This Time It Will Be Different,* Anchorage: Alaska Beacon

Tyler Dawn Stagner (2025) Most critical minerals are in Indigenous lands. Will miners respect indigenous Sovereignty?. Seattle: Grist

UN High Commission on Human Rights (2011) "Guiding Principles on Business and Human Rights: Implementing the United Nations 'Protect, Respect and Remedy' Framework", Geneva: Office of the High Commissioner for Human Rights (OHCHR)

U.S. Government Accountability Office, (2024) *Critical Minerals: Status, Challenges and Recovery from Nontraditional Sources*, Washington, DC.: US Government Accountability Office

Zhang, Rulian & Éléonore Lèbre, (2023) Energy transition minerals and their intersection with land-connected peoples, *Nature Sustainability*, Volume 8, February pp 203-211

Zepeda, Carlos (2025) *Integrating the ethics of integral ecologies into global environmental governance*, tandfonline, April.

Chapter 11
Extractive Mining and Popular Movements in Sub-Saharan Africa: An Ethical Consideration of Agency from Below

Mark J. Shimave

Abstract In *Laudato Si'*, Pope Francis emphasizes the need for an integral ecology, arguing that the ecological crisis is inseparable from the social, economic, and political conditions that shape our world. The mining sector in sub-Saharan Africa starkly illustrates this point, where local communities suffer poisoned water, cultural disruption, and deepening poverty—consequences of exploitative mining practices. Meanwhile, consumers far removed from these realities benefit from products made with little awareness of the human and ecological cost. Grounded in the framework of integral ecology, this paper examines the ethical implications of extractive mining in the region. It argues that mining is driven by a logic of extractivism that severs the bonds of human and ecological interdependence, reducing both people and land to mere resources. As an alternative, the paper reimagines integral ecology not as a top-down mandate, but as a dialogical, participatory framework rooted in the agency of the poor and vulnerable. In doing so, it proposes popular movements as a viable form of resistance to extractive injustice in sub-Saharan Africa.

Keywords Mining · Social movements · Politics · Ethics · Agency · Ecology

11.1 Introduction

In his 2015 encyclical letter *Laudato Si'*, Pope Francis powerfully drew the attention of the world to the immensity of the ecological crisis of our time. The world, "our sister," as Francis calls her, "cries out" under the weight of various burdens such as global pollution, water scarcity, loss of biodiversity, and global inequality. In truth, Francis considers the crisis to be double sided with environmental and social dimensions intricately connected. Sustaining this complex crisis is an economic

M. J. Shimave (✉)
Mark Shimave, Boston College, Chestnut Hill, United States
e-mail: shimave@bc.edu

R. Marcantonio et al. (eds.), *Integral Ecology in the Anthropocene: Africa and Beyond*, The Anthropocene: Politik—Economics—Society—Science 49,
https://doi.org/10.1007/978-3-032-32125-1_11

system built on what he regards as a mentality of 'plunder' (Francis 2015, pp.81, 104).

The image portrayed by Pope Francis above finds vivid instantiation across many mining communities in Sub-Saharan Africa today. Given their financial strength, powerful global market networks continue to influence unsustainable mining practices. As a result, local populations are exposed to poisoned water and air, suffer cultural and social disruption, and witness the corruption of local and national governments through revenues generated by extracted commodities. While these communities endure various forms of devastation, consumers living thousands of miles away continue to enjoy the benefits of products made from these resources—often with little awareness of the human and environmental costs borne by those in the extractive zones. The result is a lack of connection between these two ends of the mining chain.

Against this background, this paper argues for the urgent need to build new networks of relationships grounded in agency from below—networks capable of subverting the current global extractive order. The model proposed here envisions a synodal-like network that would connect vulnerable mining communities in sub-Saharan Africa with end-user communities across the globe. Such a network would function as a platform for storytelling and strategic action—a lived ethic of communion and resistance.

This proposal for a synodal-like network of relationships constitutes a fitting response to one of the central hopes expressed in *Laudato Si'*: that faith traditions, particularly Christian churches, would be inspired to embody an ecological commitment capable of disrupting—and ultimately transforming—the exploitative systems sustained by powerful global market forces. Accordingly, this paper demonstrates the Church's capacity to facilitate such a disruptive network through Popular Movements—movements with the potential to amplify and connect the efforts of grassroots and community-level actors working toward transformative change across the global mining chain.

The argument unfolds in four parts. The first part surveys the mining industry in sub-Saharan Africa. The second part provides an ethical analysis of mining practices in sub-Saharan Africa and the global structure of relationships that currently shape those practices. The third part defends the choice of popular movements as a viable response to the current exploitative network of relationships underpinning mining in sub-Saharan Africa. The fourth part demonstrates how the synodal framework can empower the collective agency of popular movements towards transformative resistance to mining extractivism in Sub-Saharan Africa.

11.2 Mining in Sub-Saharan Africa

Sub-Saharan Africa holds approximately 30 percent of the world's critical mineral deposits, 10% of the world's oil and 8% of the global natural gas (Chen et al. 2024, p.1). For instance, estimates indicate that 50 percent of the world's cobalt reserves are located in the Democratic Republic of the Congo (DRC). Zambia is the seventh-largest producer of copper in the world. In addition, Zimbabwe, along with the DRC and Mali holds substantial lithium, nickel, and platinum deposits. As a result of such huge mineral resources, mining is a major sector of the extractive industry in Sub—Saharan Africa.

Whether considered in terms of industrial or artisanal mining, the extraction of minerals in the sub-Saharan region has long attracted international and global markets. The vast global demand for these resources has led to the proliferation of black-market outlets, which often circumvent international trade laws to ensure uninterrupted supply chains. In many cases, these markets are sustained by informal small-scale or artisanal miners—mostly young people—who, due to dire economic conditions, are compelled to sell their limited output immediately at the mining sites for extremely low prices. For instance, Kenyan peace scholar Elias Opongo, conducting ethnographic research in Kenya's coastal regions in 2015, found that some artisanal miners earn as little as $253 per month—or approximately $8 per day—despite working over eight hours a day under hazardous conditions (Opongo 2016, p.3).

On a broader level, these transnational companies who feed the global markets would often insist on purchasing these critical minerals in their raw form. For instance, the DRC, which accounts for 74 percent of global cobalt production, exports 97 percent of its cobalt—mostly unprocessed—to China. Similarly, on a daily basis, over 1,000 trucks are estimated to transport low-concentration, rock-form lithium from Zimbabwean mines to distant African ports, destined for China instead of undergoing local processing (International Energy Agency 2019, p.12). It goes without saying that the easy access to unprocessed raw materials at relatively cheap prices reflects a broader structural challenge rooted in global trade inequalities and the historical legacy of colonial economic systems.

Indeed, such extractive exploitation is also seen in the militarization of most mining regions in sub-Saharan Africa. Owing to their substantial financial power, some transnational companies spend millions of dollars influencing and hiring state security forces and private firms to guard and protect their interests around these regions. For instance, in August 2012, The *Guardian* reported that over a three-year period, Shell spent approximately $383 million on Nigerian security forces for the protection of its staff and installations in the oil-producing Niger Delta region (Hirsch and Vidal 2012, p.7). Evidently, the presence of security forces—as seen in the Niger Delta—ensures that transnational companies maintain easy and steady access to mineral resources.

Such militarization of the mining regions and the complicity of multinationals as well as transnational elites benefiting from the extractive exploitation of minerals is also linked to the recurrence of violent conflicts in such mining regions. Whether it is in the DRC or Central African Republic (CAR), it is common to find non-state actors around mining regions of sub-Saharan Africa employing violence as means of protesting against the exploitation in their communities. Various forms of civil wars in resource-rich sub-Saharan countries such as Sierra Leone, Liberia, Ivory Coast, and the insurgency in Nigeria's oil-rich Delta region pertain to extractive mining.

It is not uncommon to hear allegations of multinational mining companies sponsoring civil restiveness in order to secure easy access to minerals. One of the most cited instances is that of eastern DRC, where some mining companies have empowered rebel groups that had gained enormous influence in the region, with the ADF and the Movement 23 (M23), being some of the deadliest (Nzwili 2025). Unsurprisingly, such collusion between multinational actors and rebel groups is connected with the large cobalt reserves in the region. Ultimately the issue is not simply that communities in sub-Saharan Africa fight because they possess mineral resources, but rather that conflict is, in part, driven by the quest of transnational actors and the structures created by the global markets.

Given their central role in increasing the risk of conflict and serving as a disincentive for peace, mineral resources in sub-Saharan Africa are today largely viewed through the lens of the 'resource curse' rather than as a blessing. These resources play a key role in financing and triggering conflicts. As commodities, these minerals are not inherently a curse; rather, they are 'cursed' by powerful global actors and markets, in collaboration with the ruling elites of sub-Saharan African states (Illesanmi, 2025, p.149). These set of actors place a high premium on these resources, which serve as critical fuel for industrial power and affluent consumerist lifestyles. Thus, while mineral related violent conflicts occur in Sub-Saharan Africa, the triggers or variables in the equation are inescapably due to a globalized supply chain.

Thus far, what is clear is that mining in sub-Saharan Africa is simultaneously local, in how operations are connected to specific places of production, and yet massively global, in how the industry is part of complex structures of global economics, consumption, corporate and political power. Arising from such a complex dynamic is a network of relationships shaped largely by a market structure in which local populations in the mining communities of sub-Saharan Africa continue to suffer the harmful effects of mining, while consumers—particularly those in the Global North—enjoy the benefits of products made from these resources, often unaware of the human and environmental costs borne by communities in extractive zones. Evidently, mining not only has deleterious consequences for the environment and peace in sub-Saharan Africa but also significantly affects relationships between people across the globe.

11.3 Ethical Considerations of Mining Practices in Sub-Saharan Africa

It is difficult to conceive of the human vocation on Earth without some form of mining. This assertion rests on the understanding that, although the Earth's riches and resources are generous gifts from God, they must be extracted and refined to meet essential human needs—ranging from power generation and solar energy to the production of cell phones, computers, buildings, cars, and countless other technologies. While mineral resources may be viewed as gifts, the morality of mining depends on multiple factors: the type of material extracted, its end uses and users, the policies and practices of mining companies, the involvement of affected communities, and the broader socio-economic context. In any case, mining can yield significant benefits when sustainable practices are intentionally adopted and consistently maintained.

Nevertheless, when considered together, the ethical factors outlined above suggest that extractivism, as shaped by the global economy, underpins mining activities in the sub-Saharan region. This form of mining is based on the continual removal of resources from the earth—a non-reciprocal, dominance-based relationship characterized purely by extraction (Klein 2014, p.169). It stands in stark contrast to stewardship, which involves not only taking but also caring for the regeneration of the earth and the continuation of life.

Such extractivism is rooted in a colonial view of Africa as a continent whose natural wealth exists to be exploited. This vision drove the European scramble for Africa and continues to shape postcolonial mining practices in sub-Saharan Africa. American theologian Vincent Miller describes the successive historical assemblages of colonialism, neocolonialism, and neoliberal globalization as a triptych that has established powerful structures, now functioning as default forms of global relationships (Miller 2022, p. 205). What is perhaps most intriguing—and complex—as observed by Ugandan theologian Emmanuel Katongole, are the ways in which this vision is reproduced in the daily practices of local artisanal miners and political elites, who often act as willing collaborators with Western extractive corporations (Katongole 2022, p.158). This complexity reinforces the argument that extractivism, as "colonial plunder," is not only the orienting dynamism of mining in sub-Saharan Africa, but also a dominant expression of sin in the world today.

Like Miller and Katongole, Pope Francis in *Laudato Si'* critiques the sin of extractivism through his concept of the "technocratic paradigm." This paradigm reflects an anthropocentric and androcentric worldview that denies both the intrinsic value of the Earth and the interconnectedness of all life—directly opposing the vision of integral ecology (Francis 2015, p. 79). While integral ecology emphasizes threefold relationships—with God, with nature, and with fellow human beings—the technocratic paradigm isolates humanity from this broader web of life. As a result, Francis argues that extractivism entails a rejection of these relationships, often leading to ecological degradation, social conflict, and human rights violations.

In response to its widespread impact—particularly in the mining industry—Christian ethicists are increasingly identifying extractivism as a social sin of our time. This designation reflects the understanding that social sin includes "any unjust structure(s), distorted consciousness, and collective action(s) that facilitate dehumanization" (Heyer 2024, pp. 59–88).This conception of extractivism as a social sin is effectively reflected in the perspective offered by the 2019 Inter-Religious Working Group on Extractivism, as it maintains that: "Extractivism is a short-sighted model of development seen around the world that exploits natural resources on a massive scale, creating significant economic profits for the powerful few in the short term, but too often resulting in minimal benefits for the communities where these resources are found" (Inter-religious Working Group on Extractivism 2019).

Against this backdrop, the mining industry in sub-Saharan Africa exemplifies a pattern of unjust structures, distorted consciousness, and exploitative practices rooted in a short-sighted developmental model. Today, it constitutes a vicious and sinful pattern due to the systemic injustice and violence it enables. This injustice affects both human communities—especially the most vulnerable—and the environment, as evidenced by some of the harms described in the preceding section. Enabling this pattern of injustice is what some theologians and ethicists have identified as *scotosis*—a form of "knowing ignorance," or the unconscious blocking of understanding. This non-voluntary dimension of social sin reflects the ways in which human beings are often captivated by distorted perceptions that prevent us from seeing rightly (Lonergan 1957, pp.346–52).

Undeniably, such *scotosis* is evident in the mining extractivism in sub-Saharan Africa. This is due to the fact that industrialists and consumers in much of the Global North continue to enjoy the benefits of products derived from these resources—often with little awareness of the human and environmental costs borne by those in the mining zones of the African subregion. Through mechanisms such as externalities, commodity distancing, and pricing systems, global markets facilitate attenuated relationships between producers, consumers, and victims (Miller 2022, p.209). Even when the motivation behind the minerals rush appears benign—such as transitioning from fossil fuels to renewable energy technologies to mitigate climate change—the injustice inherent in extractivism remains undeniably pervasive. Ultimately, the same forces that drive the extraction and production of these mineral resources also perpetuate poverty, environmental vulnerability, and violent conflict in many mining zones in sub-Saharan Africa.

As it were, extractivism—arising from *scotosis*—can be seen as a corollary to what Pope Francis calls a "throwaway culture." This term, as used by Francis, powerfully captures the interconnected evils present in the modern world, particularly as reflected in human attitudes toward nature and the environment. Like the throwaway culture, extractivism is rooted in a disavowal of connectedness; it seeks only to use resource and people often in dehumanizing forms. Extractivism, enabled by global market networks, continues to shape and constrain relationships by obscuring the interconnectedness of people across the globe. In the case of sub-Saharan Africa, it conceals various forms of harm, especially at points of resource extraction.

Unfortunately, global Christianity is also implicated in this web of relationships dictated by the global market. The church as it were is viscerally present at every level of this network: in sub-Saharan Africa, it is present in communities who labor in mines, fields, and factories, and whose land and water are poisoned by extractive enterprises; in the North, it is present among urban professionals supported by the proceeds of these industries, and among consumers—rich and poor—who are fed, clothed, and housed by the products they generate and distribute (Miller 2022, p.16). Indeed, the church is implicated at every level of these global chains, with Christians often engaging one another primarily through the shallow relationships of the market, rather than through properly ecclesial relationships.

Thus, our ethical analysis of mining in sub-Saharan Africa reveals the underlying sin of colonial plunder and extractivism—both of which continue to have profound consequences for the region's people and environment. These mining projects expose the depth of human sin, violence, and systemic injustice. Ultimately, the sin of mining extractivism is not merely a failure of moral will; more fundamentally, it represents a disruption and distortion of the network of human and ecological relationships. It deeply affects how humans relate to one another and to the broader biotic community—in a word, integral ecology.

By naming and analyzing the sin at the heart of the mining experience in sub-Saharan Africa, we can begin to discern a path of repair and restoration—a response grounded in the hope of God's redeeming grace. The question that follows is this: What structures and dispositions must the church cultivate to awaken awareness of the destructiveness of extractivism and to foster networks of relationships capable of transforming those shaped by the global market? Such networks would need to overcome the systems that obscure our interconnectedness and deny the depth of our relational ties—both among human beings, regardless of geographic location, and between humanity and the environment, which continues to suffer violent abuse through extractive practices.

For authentic change to happen in a globalized industry like mining, there is need for authentic and transformative networks of solidarity between producers and end user consumers. As Miller suggests: "those consumers (global north) need to be brought into solidaric empathy with the people in the global South who are bearing the brunt of the consequences for their modern lifestyles". He cites the example of the Central American Solidarity movement in the 1980s that was able to mobilize North American Catholics to support and influence policy change regarding old War–era proxy wars in Central America (Miller 2022, p.212).

A similar solidaric encounter, as cited by Miller, could begin to take shape when the church in sub-Saharan Africa facilitates or empowers networks of relationship and communication that enable deeper forms of communion—connecting the region to the places where the consumers of its mineral resources reside, particularly in the Global North. There needs to be sustained attention among Christians and churches in the Global North to the impacts of extractive mining on vulnerable communities. One way to foster such awareness is by giving greater attention to popular movements in the sub-Saharan region.

Hitherto, local churches in sub-Saharan Africa have engaged in a number of efforts aimed at curbing mining extractivism in the region. A major strategy has been policy advocacy, undertaken by various episcopal conferences that draw national and international attention to exploitative and destructive mining practices through official communiqués. Some of these conferences have also established networks to monitor whether corporate and government actions align with the promises made regarding mining activities. For instance, in the DRC, the Episcopal Commission for Natural Resources has created diocesan "natural resource observatories" to monitor extractive projects for potential human rights violations (Minani 2022, p.47).

While such efforts are commendable, the framework often remains largely elitist, involving only a limited number of clergy and the experts they consult. As a result, a significant portion of the church in Africa—particularly at the grassroots level—is yet to be empowered to actively resist the shallow and exploitative connections enabled by the global economic structure. The church should work toward uniting Christian communities in the mining zones of sub-Saharan Africa with others across the globe in relationships rooted in responsibility and reciprocity. As Miller puts it: "Rather than reading reports only from experts, believers could listen to the experience of fellow Christians from across the globe. This would deepen their sense of connection and obligation. Equally important, by encountering the people at the other end of the supply chain, it would reveal to them how much is ordinarily hidden by our global supply chain" (Miller 2022, pp.184–86).

Miller's insight is not only compelling but also attainable—especially for the Catholic church, whose global reach rivals that of the globalized commodity network. To this end, this essay proposes popular movements as having the potential to become networks that foster authentic solidarity between communities at both ends of the global extractive and economic chain.

11.4 Popular Movements and the Agency from Below

The turn to Popular Movements as a response to mineral extractivism in sub-Saharan Africa is grounded in two foundational principles of Catholic Social Teaching (CST): solidarity and subsidiarity. Solidarity expresses the moral imperative to stand with the marginalized and affirms the intrinsic social nature of human life, rejecting individualism and indifference. Subsidiarity complements this by affirming that social change must emerge from below—through the initiative of local and intermediary communities—rather than through centralized or top-down structures (Verstraeten 1994, p.134). In this sense, subsidiarity gives concrete expression to solidarity by entrusting power and responsibility to those most immediately affected. Together, these principles call for inclusive participation and collective agency as essential to the flourishing of any community.

Popular Movements embody this Catholic vision of social cooperation. Rooted in the Latin American tradition of liberation theology, they are composed primarily of those at the margins of society—those who suffer most from the injustices of extractive economies. These movements typically include three increasingly excluded social sectors: (a) workers who are at risk or lack job security, such as those in the informal sector, the self-employed, migrants, and day laborers—many of whom are unprotected by labor rights or trade unions; (b) landless farmers, small-scale family farmers, and Indigenous peoples; and (c) urban poor communities. The three foundational pillars that continue to galvanize these movements are known as the "three Ls": Land (*Tierra*), Labour (*Trabajo*), and Lodging (*Techo*) (Marshall 2014).

In *Laudato Si'* and his addresses to the World Meetings of Popular Movements, Pope Francis recognizes popular movements as distinct, dynamic, and vital forms of social participation in public life. He considers them capable of creating viable alternatives through the "people's power" they possess, which stems from their position at the grassroots level. According to Francis, their experiences of solidarity grow from below—from the "subsoil of the planet" (Francis 2014).

Notably, around the same time *Laudato Si'* was being written, Francis began sponsoring a series of World Meetings of Popular Movement. His address at the second gathering, held in Bolivia in 2015, has been described as a *magna carta* for popular mobilization aimed at replacing the "globalization of indifference and exploitation" with a "globalization of hope." In his words: "You, the lowly, the exploited, the poor and underprivileged, can do, and are doing, a lot. I would even say that the future of humanity is in great measure in your own hands, through your ability to organize and carry out creative alternatives . . ." (Francis 2015b). By initiating the World Meeting of Popular Movements (WMPM), Francis signals a call to action for coalitions of less powerful yet resilient stakeholders. Today, this forum serves as a space of encounter for grassroots organizations anchored around the needs of the fragilities of the poor and the planet across the world.

Following the lead of Pope Francis, this paper also considers hope in the agency arising from the "subsoil of the planet"—from the network of popular movements in sub-Saharan Africa. At present, these movements are operating as isolated units, lacking regional coordination and synergy. Some of the African movements that participated in the first World Meeting of Popular Movements (2014) include the Informal Economy Association and the National Union of Peasants, both from Mozambique, as well as the South African Waste Pickers Association. However, many other movements remain unidentified, especially those whose experiences with mining extractivism would be essential for developing the kind of solidarity envisioned in this essay.

Although the membership and certain values of these groups are not strictly Catholic, their engagement with the church is rooted in a shared vision of the common good. While the church can facilitate the solidaristic networking of these movements based on its core values, their membership and formation should not be limited exclusively to those who identify as Christians. The inclusion of people from diverse religious traditions, professions, ideologies, and cultures is, in fact, a strength of this endeavor. Such inclusivity is needed given the pluralistic nature of African societies and the complex realities of mineral extractivism, as well as the distorted global economic networks of relationships that sustain it. Indeed, when rooted in CST, Popular Movements can offer an alternative vision of human cooperation—grounded in mutual responsibility, local agency, and the shared pursuit of the common good.

One of the values or structures from its rich patrimony that the Church can offer or mediate through popular movements is synodality. As a dynamic mode of communal discernment and shared responsibility within the Church, it builds on solidarity and subsidiarity by providing a framework for sustained and inclusive collaboration. As an ecclesial framework revitalized by Pope Francis, synodality has the potential to enhance both communication and resistance within popular movements. Through its structure, it can serve as a platform for storytelling and strategy-building. By empowering synodal-like networks of popular movements at both ends of the mining chain, the church can become a space of encounter and exchange—a genuine network of solidarity that uplifts marginalized voices and challenges exploitative systems. Demonstrating why and how the synodal framework can benefit popular movements in fostering authentic solidarity will thus constitute the focus of the final section of this paper.

11.5 Building Networks of Popular Movements Through Synodality

In its ecclesial sense, synodality is a consultative process that seeks to involve the people of God in the life and mission of the church—one that is participatory and appreciative of the gifts and insights of each community that constitutes the church (Moons/Colberg 2025, p.14). In recent times, Pope Francis and the Secretariat for the Synod on Synodality (2021–2024) have described synodality as a 'journey' (*camino*), a 'dynamism' (*dinamismo*), a 'style,' and 'walking together' (Moons 2024, p.16).

By its design, synodality seeks to foster accompaniment, creative communication and action across the global church. Notably, the art of "walking together" begins not at the universal level but at the local level—with parishes, dioceses, national episcopal conferences, continental assemblies, and finally the universal church. The value of this process lies in its potential to create spaces and promote opportunities for listening at the regional level while also linking these various levels in a dynamic, reciprocal exchange between the local, regional, and universal dimensions (Moons/ Colberg 2025, p.166). Thus, practices of synodality prioritize the experiences and agency of those on the margins or peripheries, aligning with the principle of subsidiarity. This bottom-up—rather than top-down—approach is a defining strength of synodality.

Using this structure which synodality offers, Popular Movements can cultivate a solidarity that fosters authentic listening, mutual learning, collective discernment, and coordinated action across the extractive mining chain. If integral ecology embraces relationality and insists on solidarity with creation—recognizing the interconnectedness of social, economic, and human justice—then the synodal framework can connect faith communities in the Global North, where mineral demand is concentrated, with the lived experiences of mining communities in sub-Saharan Africa.

Such a connection would create space for a contemplative gaze on creation, making the consequences of high consumption visible. It would also amplify the voices of those who suffer—voices that, in the words of Pope Francis, "cause embarrassment, with cries that are bothersome" (Francis 2014)—thus prompting solidaric resistance. As he states in the final document of the Synod on Synodality, "Synodality and integral ecology both take on the character of relationality and insist upon us nurturing what binds us together" (Francis 2023, p.18). This entails our solidarity and responsibility toward one another, and toward all human and non-human life.

In arguing for the agency of popular movements in sub-Saharan Africa, and using synodality as a framework for solidaric resistance, this paper envisions a structure similar to the Laudato Si' Movement (LSM). This global movement brings together a wide range of Catholic organizations—approximately 900—and grassroots members from around the world. Their aim is to witness to ecological conversion by walking together in synodality with the universal Church. This is expressed through shared prayer, co-creation with local Laudato Si' Circles, and engagement in global initiatives that bring *Laudato Si'* to life. The movement combines strong local actions with coordinated global campaigns for ecological justice—most notably, leading a record-breaking coalition of Catholic institutions in the global fossil fuel divestment campaign (Laudato Si' Movement, no date).

While the Laudato Si' Movement (LSM) addresses a broad range of issues, the network envisioned in this paper focuses specifically on mining extractivism. Its aim is to create dialogical spaces that connect popular movements, empowered by Christian communities in sub-Saharan Africa, with Christian communities in the Global North—fostering a bond of solidaric resistance. Though deeper solidarity cannot by itself overcome the exploitative networks driving the global mineral economy, it embodies a profound act of moral witness—affirming our shared vocation to care for both the Earth and its most vulnerable communities. It is a crucial strategic step—transforming relationships and awareness needed for broader systemic change.

11.6 Conclusion

This essay has argued for the development of a formidable, synodal-like grassroots movement to resist mining extractivism in sub-Saharan Africa. Drawing on Catholic Social Teaching, it has shown how synodality—understood as a participatory and dialogical process—can serve as a framework for the Church to support popular movements, which have the potential to become alternative models of relationship and ethical agency rooted in lived experience.

Indeed, the synodal framework offers a pattern by which popular movements can massively build coalition against extractivism in Africa. When embraced, synodality would become a lived ethic of communion and resistance for these movements. Thus, the Church—drawing on its long history as a pivotal player in community organizing—should intentionally leverage its network-like structure to accompany and empower popular movements in sub-Saharan Africa. Facilitating the transformation of these movements into networks of solidaristic resistance against the exploitation of land, labor, and lodging represents a renewed way for the Church to fulfill its mission of evangelizing the world.

No doubt, popular movements are emerging as vital expressions of grassroots agency, challenging exploitative systems and articulating alternative models of relationships and development rooted in dignity, solidarity, and care for our common home. Through solidaristic storytelling and organized action, popular movements in the mining communities in sub-Saharan Africa can resist extractives. These forms of ethical agency—rooted in lived experience—embody a bottom-up approach to justice that aligns with the church's call to care for both people and the planet. Ultimately, the paper underscores the transformative power of integral ecology when animated by the experiences and moral claims of those on the margins.

References

Afua, H. and Vidal, J. (2012) 'Shell spending millions of dollars on security in Nigeria', *The Guardian*, 19 August. At: https://www.theguardian.com/business/2012/aug/19/shell-spending-security-nigeria-leak (7 April 2025).

Chen, W., Laws, A. and Valckx, N. (2024) 'Digging for opportunity', in *Regional Economic Outlook: Sub-Saharan Africa—A tepid and pricey recovery*. Washington, DC. At: https://www.imf.org/en/News/Articles/2024/04/29/cf-harnessing-sub-saharan-africas-critical-mineral-wealth (17 March 2025).

Francis (2014) 'Address of Pope Francis to the participants in the WMPM'. At: https://www.vatican.va/content/francesco/en/speeches/2015/july/documents/papa-francesco_20150709_bolivia-movimenti-popolari.html (1 March 2025).

Francis (2015) *Laudato si'*. At: http://www.vatican.va/content/francesco/en/encyclicals/documents/papa-francesco_20150524_enciclica-laudato-si.html (1 March 2025).

Francis (2024) 'For a synodal church: Communion, participation, mission'. At https://www.synod.va/content/dam/synod/news/2024-10-26_final-document/ENG---Documento-finale (15 March 2025).

Heyer, K. (2024) 'Hearts of flesh: Structural sin and social salvation', *CTSA Proceedings*, 78, pp. 59–88.

Himes, K. (2010) 'Peacebuilding and Catholic social teaching', in Schreiter, R.J. and Powers, G.F. (eds.) *Peacebuilding: Catholic theology, ethics, and praxis*. Maryknoll, NY: Orbis, pp. 265–299.

Illesanmi, S. (2025) 'The status and protection of non-human objects in just war tradition: A comparative analysis from an African perspective', in Obadare, E., and Adogame, A. (eds.) *Topographies of African spirituality: Essays in honor of Jacob K. Olupona*. New York, NY: Routledge, pp. 140–165.

International Energy Agency (2019) 'World Energy Outlook'. At: https://www.goldmansachs.com/insights/goldman-sachs-research/direct-lithium-extraction (7 April 2025).

Inter-religious Working Group on Extractivism (2019) 'A just response to extractivism.' At: https://justresponse.faith/node/3. (15 April 2025).

Katongole, E. (2022) *Who are my people? Love, violence, and Christianity in sub-Saharan Africa*. Notre Dame, IN: University of Notre Dame Press.

Klein, N. (2014) *This changes everything: Capitalism vs the climate*. Toronto: Knopf.

Laudato Si' Movement (no date) 'What we do'. Aat: https://laudatosimovement.org/what-we-do/ (Accessed: 7 September 2025).

Lonergan, B. (1957) *Insight: A study of human understanding*. New York: Philosophical Library.

Marshall, J. (2014) 'Challenging the globalization of indifference. At: https://links.org.au/challenging-globalisation-indifference-pope-francis-meets-popular-movements (16 March 2025).

Miller V.J. (2022) 'Mining and the call for solidarity: The networks we have and the synodal network the Church is called to be', in Ceasar, M. and Powers, G. (eds.) *Catholic peacebuilding and mining: Integral peace, development, and ecology*. London: Routledge, pp. 202–220.

Minani, R. (2022) 'The mining industry, conflict, and the Church's commitment in the Democratic Republic of the Congo', in Ceasar, M. and Powers, G. (eds.) *Catholic peacebuilding and mining: Integral peace, development, and ecology*. London: Routledge, pp. 202–220.

Moons, J. (2024) *Witnesses of synodality: Good practices and experiences*. Mahwah: Paulist Press.

Moons, J. and Colberg, K. (2025) *The future of synodality: How we move forward from here*. Minneapolis: Liturgical Press.

Nzwili, F. (2025) 'A silent genocide is unfolding in Congo', *National Catholic Reporter*. At: https://www.ncronline.org/news/silent-genocide-unfolding-congo-church-sources-warn (9 April 2025).

Opongo, E. (2016) *Informal mining and family vulnerability in Kenya, Zambia and Zimbabwe*. Harare: African Forum.

Verstraeten, J. (1994) 'Solidarity and subsidiarity', in Boileau, D. (ed.) *Principles of Catholic social teaching*. Milwaukee: Marquette University Press, pp. 133–148.

Chapter 12
What is Lost Can Never be Found: Climate Change and Loss and Damage from *Laudato Si′* to Tuākoi ‘Lei

Elsa Barron

Abstract Climate change has already led to irreversible forms of loss and damage even as human society continues to increase its rate of greenhouse gas emissions, worsening future climate risks. Scholars have developed moral arguments identifying the injustice of climate-related loss and damage, legal theories and precedent for accountability and enforcement, and policy frameworks to address loss and damage through international climate diplomacy. The principles of integral human ecology also urge consideration for the immediate needs of impacted communities to process the losses and damages they experience, upholding human dignity amidst the interconnected cry of the earth and cry of the poor. In this chapter, I ask, beyond legal and policy pathways, how does the spiritual perspective of Pope Francis's encyclical *Laudato Si'* address the issue of loss and damage for impacted communities? How does faith leadership in highly impacted communities build on this framework and identify needs and responses to loss and damage? Through an analysis of *Laudato Si',* translated "praise be to you," I identify three major forms of loss related to climate change: ecological, social, and spiritual loss. The encyclical invites the urgent development and implementation of holistic solutions to the crisis of climate change, including each of these dimensions of loss and damage. Communities on the frontlines of this crisis are already navigating the realities and response to loss and damage, and offer insights into the action needed to honor the dignity of those most impacted. Through analysis of the Pacific Conference of Churches' (PCC) Tuākoi 'Lei Declaration, translated "good" or "loving neighbor", and an interview with the Secretary General of the PCC, I argue that holistic responses to loss and damage require support for processing anger, grief, and trauma related to climate change and a procces of climate reconciliation. This is a tangible way to operationalize the call to action in *Laudato Si'* and an application of the principles of integral human ecology to the issue of climate-induced loss and damage.

Keywords Loss · Damage · Non-economic · Religion · South Pacific

E. Barron (✉)
Elsa Barron, University of Notre Dame, South Bend, USA
e-mail: ebarron3@nd.edu

R. Marcantonio et al. (eds.), *Integral Ecology in the Anthropocene: Africa and Beyond*, The Anthropocene: Politik—Economics—Society—Science 49,
https://doi.org/10.1007/978-3-032-32125-1_12

12.1 Introduction

Climate change is a multifaceted crisis with devastating impacts on the planetary system and human well-being and health (Adger, W. Neil et al., 2022, Abbasi et al. 2023, Nunez et al. 2019). The impacts of climate change include extreme heat, intensifying natural disasters, biodiversity loss, sea-level rise, and desertification, resulting in death, displacement, and infrastructure and ecological destruction. Loss and damage refers to both these tangible negative impacts on communities today and other less tangible effects such as deteriorating mental health or loss of community, culture, and ways of making sense of the world (Tschakert et al. 2019). These are often referred to as forms of non-economic loss and damage in international policymaking within the United Nations Framework Convention on Climate Change (UNFCCC) (Tschakert et al. 2017, Adger, W. N. et al., 2013, Cianconi, Betrò and Janiri, 2020).

Scholars have long pointed out the injustice of climate change relating to loss and damage, namely that the impacts of climate change are unequally distributed and felt most severely by under resourced communities who have contributed minimally to its causes (Harlan et al. 2015). Recognizing these distributive inequalities, activists have also developed grassroots climate justice demands, evolving from preceding conceptions of environmental justice and emerging from local experiences with the impacts of climate change and demands for voice and sovereignty in decisions impacting their communities (Schlosberg and Collins 2014).

In addition to these grassroots pressures, legal scholars have sought to define frameworks for addressing the injustices of climate change-related losses and damages. One of the most contentious debates regarding loss and damage has focused on whether countries most responsible for climate change should be obligated to compensate for its impacts (Burkett 2014). One of the largest challenges facing legal arguments for compensation within a corrective/ retributive justice framework has been the inability to identify the direct link between an agent's action

Table 12.1 A framework for categorizing loss and damage in *Laudato Si'*. *Source* Author created

Category of climate and environmental losses	Mentions in *Laudato Si'*
Ecological	Loss of forests (8, 24, 32), biodiversity (8, 35), human and environmental health (142, 195), exploited lands (134), and entire species and groups of plants and animals (32, 35, 36).
Social	Loss of lives (25), homes (25), legal protections (25), resources (195), quality of life (184), jobs (128), social capital (128), cultural values (184), social cohesion (46), community living (110), and freedom (142).
Spiritual	Loss of a sense of responsibility (25), identity (46), appreciation (110), and purpose in life (110).

and the resulting harm produced by climate change, the link necessary for establishing causality (Adler 2007). Recently, new advances in attribution science enable end-to-end linkage between the impacts of climate change and the emissions of particular actors such as states or corporations, which may soon solve the challenge of proving causality in the case of climate change-induced losses and damages (Callahan and Mankin 2025). Such attribution science could open new pathways for impacted individuals, communities, and states to seek accountability and compensation by bringing legal cases against major emitters. The International Court of Justice (ICJ) advisory opinion requested by Vanuatu and released by the ICJ in July 2025 further expands the possiblities for climate accountability by indicating that states can seek cessation of climate related harms as well as reparation and compensation for impacts ("Obligations of States," 2025).

The debate around loss and damage liability has also been ongoing through international climate diplomacy within the UNFCCC. The first appearance of a loss and damage concept in these negotiations was in 1991, when Vanuatu, on behalf of small island developing states (SIDS), proposed an insurance pool for addressing sea level rise (Siegele 2023). However, this proposal was rejected by developed country parties and it was not until 2013 when states agreed to the Warsaw International Mechanism (WIM) for Loss and Damage that a structure was developed to research and discuss the topic of loss and damage formally within the UNFCCC. The WIM functions to increase knowledge and dialogue surrounding climate risk and encourage support for states experiencing loss and damage, but it is not a compensatory or accountability-based mechanism. At COP27 in Egypt, negotiators reached decision 2/CP.27 on "Funding arrangements for responding to loss and damage associated with the adverse effects of climate change, including a focus on addressing loss and damage," which established a voluntary fund for addressing loss and damage, without any legal obligation for compensation (UNFCCC 2023). As of 2025, the fund has been established and received its first small contributions, but the process for disbursement is still undetermined.

As the theoretical, grassroots, legal, and international debates surrounding loss and damage continue, the lived experience of loss and damage continues to accelerate. Communities around the world have already had to navigate the loss of homes, communities, and cultures with little way of recourse for their suffering. Given these losses, scholars increasingly acknowledge the role of grief as a natural response to climate impacts, which many argue should be given greater consideration in public discussion and policy responses to climate change (Cunsolo et al. 2020, Cunsolo, Ellis 2018, Pihkala 2024, Varutti 2024). However, intangible experiences such as grief are challenging for policymakers to address and much of the focus of loss and damage conversations within the UN has focused on financial payments for economic losses related to climate change. Moving forward, there is still a need for a better understanding of non-economic forms of loss and damage, including cultural loss, grief, and deteriorating mental health, among others (Menk et al. 2022). Spirituality and faith practices can help facilitate a process of lament for ecological and social loss (Pihkala 2025, De Klerk 2014). Understanding available spiritual frameworks to support communities experiencing climate-related losses and damages through a process of grief is important for fulfilling the aims of integral human ecology, which emphasizes the interconnectedness between environmental, social, and spiritual crises and their responses (Nassar n.d.).

Pope Francis's (2015) encyclical *Laudato Si':* On Care for Our Common Home has been a transformative religious text addressing the climate and ecological crisis through the interconnected framework of integral human ecology. In this chapter, I analyze *Laudato Si'* and identify an ongoing discussion of climate-related loss throughout the text, which I classify into three forms: ecological, social, and spiritual forms of loss and damage. However, I argue that the text stops short of comprehensively engaging with grief as a natural response to the losses it warns of. While the encyclical urges action on the crisis at hand, there remain climate-related losses that are irreversible, and impacted communities need support for processing losses and pursuing justice in addition to action to prevent future harms. More work remains to be done.

In order to better understand the role of emotional processes such as anger and grief in guiding spiritual responses to the crisis of climate change, I analyze the Tuākoi 'Lei Declaration developed by the Pacific Conference of Churches. The Tuākoi 'Lei Declaration builds on *Laudato Si'* by calling out the injustice and even betrayal of climate change and addressing the needs of impacted communities to process anger, grief, trauma, and fear related to this devastation. I argue that this approach to supporting and processing anger, grief, and trauma is an important addition to the framework of climate-related losses in *Laudato Si'* and a key application of the principles of integral human ecology to the issue of loss and damage.

12.2 Methods

In this chapter, I analyze three texts, including Pope Francis's encyclical, *Laudato Si'*, the Pacific Conference of Churches' (PCC) Tuākoi 'Lei Declaration, and an interview with the General Secretary of the PCC, Rev. James Bhagwan. I assess the approach of each of these documents to the crisis of climate change, particularly to climate-induced losses and damages at the community level. I use an emergent thematic analysis to categorize these framings and responses to climate-induced loss and damage (Nowell et al. 2017).

The included interview with James Bhagwan was conducted as a part of a series of semi-structured interviews with approximately twenty policy experts and advocates operating as intermediaries between communities from diverse regions impacted by climate change and international policy development. The interviews were approximately thirty to sixty minutes long and conducted during COP29 in 2024. The single interview with Rev. Bhagwan is included in the following analysis because of its explicit reference to and analysis of the Tuākoi 'Lei Declaration and its themes.

12.3 Loss and Damage in *Laudato Si′*

In his 2015 encyclical *Laudato Si'*, Pope Francis discusses many instances of loss related to climate change and ecological degradation, which I classify into three categories outlined in Table 12.1: ecological, social, and spiritual loss. Each of these losses is raised with concern by Francis, who highlights the shortsightedness with which humanity has accepted the cost of long-term damages for the sake of short-term profit. Francis primarily addresses climate-related losses by focusing on their prevention, including through spiritual and social transformation and climate mitigation and adaptation. For example, he references Patriarch Bartholomew, who "asks us to replace consumption with sacrifice, greed with generosity, wastefulness with a spirit of sharing," emphasizing the importance of inner spiritual transformation to address the outer crisis of ecological degradation (Francis 2015, 9). Francis also makes an appeal for tangible actions through interdisciplinary dialogue, education, and international policy change to push past political opposition and general apathy to act collectively on the climate crisis (Francis 2015, 15). While focusing on transformation, he acknowledges the unlikeliness of full social repentance, yet he does not detail other spiritual responses to great loss, such as anger and grief, that communities impacted by climate change often navigate.

12.3.1 Ecological Loss

Laudato Si' outlines the many devastating ecological losses resulting from climate change, including loss of forests, biodiversity, human and environmental health, exploited lands, and entire species and groups of plants and animals. Francis reflects that,

As long as production is increased, little concern is given to whether it is at the cost of future resources or the health of the environment; as long as the clearing of a forest increases production, no one calculates the losses entailed in the desertification of the land, the harm done to biodiversity or the increased pollution (Francis 2015) 195.

The environment becomes a sacrifice zone for the sake of production and profit, but Francis also emphasizes that the true costs of these losses are far greater than any perceived economic gain.

Where certain species are destroyed or seriously harmed, the values involved are incalculable. We can be silent witnesses to terrible injustices if we think that we can obtain significant benefits by making the rest of humanity, present and future, pay the extremely high costs of environmental deterioration (Francis 2015) 36.

12.3.2 Social Loss

As his reflections suggest, Francis is not only concerned with these ecological losses but also their interconnectedness with human interest, action, and impact. Accordingly, he also overviews the many social costs of climate change and environmental degradation, including the loss of lives, homes, legal protections, resources, quality of life, jobs, social capital, cultural values, social cohesion, community living, and freedom. Most immediately, climate-related disasters and long-term environmental degradation cause death and displacement, yet those on the move as a result of these impacts are not recognized in the international Refugee Convention. Pope Francis notes, "they [climate refugees] bear the loss of the lives they have left behind, without enjoying any legal protection whatsoever" (Francis 2015, 25). The climate-related social losses outlined by Francis are even more expansive than these two frequently-cited realities of death and displacement. He summarizes:

> The social dimensions of global change include the effects of technological innovations on employment, social exclusion, an inequitable distribution and consumption of energy and other services, social breakdown, increased violence and a rise in new forms of social aggression, drug trafficking, growing drug use by young people, and the loss of identity. These are signs that the growth of the past two centuries has not always led to an integral development and an improvement in the quality of life. Some of these signs are also symptomatic of real social decline, the silent rupture of the bonds of integration and social cohesion (Francis 2015) 46.

Not only does Francis argue that the world's ecological losses have devastating effects on communities around the world, but also he is also clear to link these losses to the unraveling of the social fabric that increases the likelihood of all other kinds of violence and harm. He also argues that these interconnected social and ecological crises weaken human institutions, and "Anything which weakens those institutions has negative consequences, such as injustice, violence and loss of freedom" (Francis 2015) 142.

12.3.3 Spiritual Loss

Finally, woven into these social and ecological crises, Francis discusses a spiritual crisis, in which humanity has lost a sense of responsibility, identity, appreciation, and even what he describes as purpose in life. He views climate and environmental degradation as a part of a larger, interconnected system in which "there are a number of symptoms which point to what is wrong, such as environmental degradation, anxiety, a loss of the purpose of life and of community living" (Francis 2015, 110). It is not only the ecological crisis at hand, but our lack of response to it that indicates the spiritual losses of the moment, which Francis argues "points to the loss of that sense of responsibility for our fellow men and women upon which all civil society is founded" (Francis 2015) 25.

He argues that one of the reasons that, as a society, humanity has accepted these losses is that to increase efficiency, assembly-line-style, we have fragmented our knowledge. This fragmentation "often leads to a loss of appreciation for the whole, for the relationships between things, and for the broader horizon, which then becomes irrelevant" (Francis 2015) 110.

12.3.4 Responding to Loss and Damage

Throughout *Laudato Si'*, Pope Francis describes many losses and damages related to the modern society's interconnected ecological, social, and spiritual crises. Because of the gravity and seriousness of these challenges, he acknowledges that "Doomsday predictions can no longer be met with irony or disdain," because,

> "We may well be leaving to coming generations debris, desolation and filth. The pace of consumption, waste and environmental change has so stretched the planet's capacity that our contemporary lifestyle, unsustainable as it is, can only precipitate catastrophes, such as those which even now periodically occur in different areas of the world" (Francis 2015) 161.

What is the appropriate response in light of this grave warning? Francis references Patriarch Bartholomew's call for repentance, acknowledging that the destruction of the biodiversity of God's creation is a form of sinfulness (Francis 2015) 6. Yet, even after everything described in this encyclical, the Pope questions, "Is it realistic to hope that those who are obsessed with maximizing profits will stop to reflect on the environmental damage which they will leave behind for future generations?" (Francis 2015, 190). The implied answer seems to be no, that without radical societal transformation, there can be no true repentance, which would include a change in heart and behavior. In the absence of this transformative repentance, is there a place for anger or grief?

Throughout his writing, Francis maintains a hopefulness, even when describing the dark condition of the present moment. He declares that "No system can completely suppress our openness to what is good, true and beautiful... I appeal to everyone throughout the world not to forget this dignity which is ours. No one has the right to take it from us." (Francis 2015) 205. He doesn't seem to process through anger and forgoes grief for the sake of persuasion that society can still transform. His message is first directed toward those who have bought into the systems of capitalist exploitation that have enabled the climate and ecological crisis at hand as he attempts to nurture society toward transformation.

However, for many members of his church and the greater world at large, the loss related to climate change and environmental degradation is not a hypothetical threat, but a deep and ever-present source of pain. Francis centers his justice-based approach around this interconnected cry of the Earth and the cry of the poor (Francis 2015) 49. For those who have lost homes, family, careers, cultures, and dreams due to the impacts of climate change, there is much emotion to process. Francis explicitly uses the term grief only once and at the end of his encyclical, when he reflects on Mary:

> Mary, the Mother who cared for Jesus, now cares with maternal affection and pain for this wounded world. Just as her pierced heart mourned the death of Jesus, so now she grieves for the sufferings of the crucified poor and for the creatures of this world laid waste by human power (Francis 2015) 241.

If Mary is grieving, are we invited to grieve alongside her? Is this a part of the process of responding to the impacts of climate change? This is a question that *Laudato Si'* leaves unanswered. Addressing loss and damage requires a different approach than addressing climate change. While the latter requires urgency, action, and hopefulness, the former also requires a process of anger, grief, compensation, and repair. I argue that even though he outlines many different forms of loss and damage, Francis does not comprehensively guide impacted communities on how to process such losses, especially those that fall into the social and spiritual categories. In his more recent Apostolic Exhortation, *Laudate Deum*, Francis references the shortcomings of the current state of loss and damage policy (Francis 2023) 51. He follows it with a quote from *Laudato Si'* that "Those who will have to suffer the consequences of what we are trying to hide will not forget this failure of conscience and responsibility" (Francis 2023) 52. How can communities carry and process this betrayal?

12.4 Beyond *Laudato Si'*: Loss and Damage in the Tuākoi 'lei Declaration

Faith leadership in the Pacific encounters an extreme localized crisis of climate change when compared to the global climate crisis Pope Francis discusses. For many living in Pacific island nations, their communities have contributed little to the root causes of climate change, maintaining a close relationship to the land and reverence for the environment that has endured through colonization and resource exploitation. Collectively, estimates suggest that small island states in the Pacific have only contributed around 0.02 percent of global greenhouse gas emissions (WMO 2024). Even so, islands in the Pacific are some of the places most at risk from the impacts of climate change. Five of the fifteen states most vulnerable to the impacts of climate change are island nations located in the Pacific, according to ND-GAIN's vulnerability index (ND-GAIN 2025).

In 2024, The Pacific Conference of Churches published the Tuākoi 'Lei Declaration, twenty years after their first collective declaration on the issue of climate change, Otin Taai, which built a history of ecumenical concerns for social and ecological justice in the Pacific and made concrete recommendations for action on the issue of climate change (Pacific Council of Churches 2024, World Council of Churches 2004). The declaration is intended to take stock of the climate change developments over the past two decades and propose new commitments for Pacific churches and their partners moving forward. I analyze the Declaration together with an interview of one of its creators, Reverend James Bhagwan, for themes surrounding grief and trauma from loss and damage.

Representatives from the Pacific are often called on to share their story and experiences of loss and damage in international fora while simultaneously processing their home community's grief. Rev. Bhagwan compared his community's loss and pain to that of the Israelites in exile in Psalm 137, which recounts the exiled sitting by the river to weep and asking, "How can we sing the songs of the Lord while in a foreign land?" (The Holy Bible: Psalm. 137). For Pacific communities facing climate displacement, this lament is not a metaphor but a reality. Reflecting on his own experience balancing the slow pace of international diplomatic negotiations with the urgent needs of his community, Rev. Bhagwan shared that the biggest question and the largest challenge is, "How do you maintain hope when it seems hopeless?" (Bhagwan 2024).

The impacts of climate change described in Tuākoi 'Lei are extreme: complete submergence of land, forced relocation, leaking nuclear waste, the disruption of entire knowledge systems, and extreme weather disasters. Reflecting on these many impacts, the declaration states, "The stories we have heard this week tell us that the world has not listened to our cries" (Pacific Conference of Churches 2024). Not only does the world not listen, the document also accuses perpetrating states of seeking "to silence Pacific countries on these issues through offers of aid and development assistance" (Pacific Conference of Churches 2024).

In contrast with the earlier declarations of the PCC, which call for increased effort and collaboration and generally maintain a hopeful tone, this declaration expresses greater exasperation and anger regarding the irreverence of the world toward the people, traditions, lands, and oceans of the Pacific.

In our interview, Rev. Bhagwan compared international finance for climate adaptation and loss and damage absent commitments to addressing the root causes of climate change to thirty pieces of silver, referencing Judas's compensation for the betrayal of Jesus (Bhagwan 2024). Without action to eliminate reliance on fossil fuels, the disasters facing communities in the Pacific and around the world will only accelerate, making meager funding today a mere betrayal. Communities in the Pacific express the anger associated with this betrayal and press the urgency of their demands. The Tuākoi 'Lei Declaration states, "We are no longer the 'canary in the coal mine' and time is running out for all of us. Our cry for justice is not only for ourselves but for God's precious biodiversity, of which we are a part" (Pacific Conference of Churches 2024).

This cry for justice resonates closely with the points of urgency in *Laudato Si'*. However, uniquely included in this cry for justice is an acknowledgement of all that is already lost. The declaration recounts:

> We knelt together in prayer on the sands of Togoru settlement and lamented the exploitative and extractive systems of neo-colonialism and enduring colonialism that continue to exacerbate the plight of many of our Pacific brothers and sisters experiencing climate change (Pacific Conference of Churches 2024).

This active lament goes deeper than a mere threat of loss; it acknowledges that great loss is already present in the lives of the communities served by these churches that requires a response that does not attempt a false hopefulness, but recognizes and grieves the irreversible damage at hand.

In their commitment, which serves as the declaration's call to action, the Conference recognizes the "need to address loss and damage through pastoral care, especially the psychological, spiritual and emotional trauma resulting from climate change" (Pacific Conference of Churches 2024). They urge the PCC and its partners to develop a network for counseling to deal with trauma and fear in their communities resulting from the "fear, worry, distress and damage caused by climate change and climate-induced migration" (Pacific Conference of Churches 2024).

In our interview, Rev. Bhagwan elaborated further on the multifaceted needs for pastoral care responding to climate change, reflecting that it is necessary to have spiritual guides to lead communities needing to migrate through a farewelling process of their lands and facilitate their transplantation to a new home with as much care and continuity as is possible. He added that it is also necessary to allow those who choose to stay behind, and receive pastoral care through their gradual process of loss (Bhagwan 2024).

The the Tuākoi ‘Lei Declaration from the Pacific Conference of Churches adds new insights to *Laudato Si"s* foundation for addressing loss and damage. Building on stern warnings of losses and an urgent call to action, Tuākoi ‘Lei offers an expression of anger and lament, and a call to provide pastoral care for communities navigating the grief and anxiety of great loss. The vantage point of communities in the Pacific urges leadership in both climate action and holistic responses to loss and damage. These reflections on anger, grief, and trauma reflect the needs of communities impacted by climate change around the world, and encourage thinking with and beyond *Laudato Si'* to comprehensively address the violence and injustice of climate change.

These texts speak to the imperative that loss and damage responses be more than a small pot of money distributed for sporadic projects in impacted communities. Not only are more robust forms of financial accountability and compensation required, but also a holistic approach to the serious ecological, social, and spiritual dimensions of loss outlined in *Laudato Si'*. Specifically, Tuākoi ‘Lei calls for pastoral and mental health care to help communities navigate their complex experiences of loss and grief. There is also an underlying concern for a justice-based process to be implemented internationally, which could take inspiration from transitional justice and truth and reconciliation processes. Negotiations within the UNFCCC have consistently failed to robustly respond to demands for climate justice, driven by the resistance of nations with high emissions. This failure to explicitly address climate injustice obscures and denies the experience of harm felt by impacted communities and exacerbates the risk of hopelessness and exploitation. Behind the urgency for technical and policy solutions to address the crisis of climate change is a deeper need for reconciliation.

12.5 Conclusion

Pope Francis's encyclical, *Laudato Si'*, published in 2015, explores a wide range of losses relating to climate change, which I categorize into ecological, social, and spiritual forms of loss. It urges systemic transformation to prevent further devastation and repentance of the root causes of the ecological crisis. Ten years after the publication of this encyclical, it is clear that many of the impacts of climate change are not distant threats, but lived realities of communities in the world today. Losses and damages from the impacts of climate change have already occurred, and communities need guidance on how to navigate the new emotions and challenges raised by such devastation within a framework of integral human ecology.

Island nations in the Pacific face some of the largest threats due to climate change, despite having contributed very little to the causes of climate change. Faith leaders in the Pacific have been outspoken about the root causes of climate change and the impact on their home communities. The Tuākoi 'Lei Declaration from the Pacific Conference of Churches builds on *Laudato Si'* by calling out the injustice and even betrayal of climate change and addressing the needs of impacted communities to process anger, grief, trauma, and fear related to this devastation.

Addressing the realities of climate-induced losses and damages today requires the acknowledgement that many of these losses, especially those related to life, land, and cultural tradition, are irreconcilable and irreversible. Impacted communities need guidance through a process of expressing honest anger and grief and support for recovering from climate change-related trauma. Faith communities around the world can learn from the leadership of the Pacific in acknowledging these experiences and evolving beyond simplistic narratives that the world can prevent climate change when it already impacts so many today. We must embrace the much more complex truth that many of the impacts of climate change are already present, devastating, and irreversible by providing holistic support to grieving communities. This includes both financial compensation, non-financial forms of support and care, and a holistic process of international reconciliation for the harms that have been committed.

References

Abbasi, K., Ali, P., Barbour, V., Benfield, T., Bibbins-Domingo, K., Hancocks, S., Horton, R., Laybourn-Langton, L., Mash, R., Sahni, P., Sharief, W.M., Yonga, P. & Zielinski, C. 2023, "Time to treat the climate and nature crisis as one indivisible global health emergency", *Afr.j.prim.health care fam.med.(Online),* vol. 15, no. 1, pp. 1–3.

Adger, W.N., Barnett, J., Brown, K., Marshall, N. & O'Brien, K. 2013, "Cultural dimensions of climate change impacts and adaptation", *Nature climate change,* vol. 3, no. 2, pp. 112–117.

Adger, W.N., Barnett, J., Heath, S. & Jarillo, S. 2022, "Climate change affects multiple dimensions of well-being through impacts, information and policy responses", *Nature Human Behaviour,* vol. 6, no. 11, pp. 1465–1473.

Adler, M.D. 2007, "Corrective Justice and Liability for Global Warming", *University of Pennsylvania Law Review,* vol. 155, no. 6, pp. 1859–1867.

Bhagwan, James, interview by Elsa Barron, November 2024.
Burkett, M. 2014, "Loss and Damage", *Climate law,* vol. 4, no. 1-2, pp. 119–130.
Callahan, C.W. & Mankin, J.S. 2025, "Carbon majors and the scientific case for climate liability", *Nature,* vol. 640, no. 8060, pp. 893–901.
Cianconi, P., Betrò, S. & Janiri, L. 2020, "The Impact of Climate Change on Mental Health: A Systematic Descriptive Review", *Front Psychiatry,* vol. 11, pp. 74.
Cunsolo, A. & Ellis, N.R. 2018, "Ecological grief as a mental health response to climate change-related loss", *Nature climate change,* vol. 8, no. 4, pp. 275–281.
Cunsolo, A., Harper, S.L., Minor, K., Hayes, K., Williams, K.G. & Howard, C. 2020, "Ecological grief and anxiety: the start of a healthy response to climate change?", *Lancet Planet Health,* vol. 4, no. 7, pp. e261–e263.
De Klerk, B.J. 2014, "Enhancing ecological consciousness through liturgical acts of doxology and lament: original research", *Verbum et ecclesia,* vol. 35, no. 2, pp. 1–8.
Francis. 2015, *Laudato Si'* - On Care for our Common Home. *Encyclical.* http://www.vatican.va/content/francesco/en/encyclicals/documents/papa-francesco_20150524_enciclica-laudato-si.html
Francis. 2023, "Laudate Deum." *Apostolic Exhortation.* https://www.vatican.va/content/francesco/en/apost_exhortations/documents/20231004-laudate-deum.html
Harlan, S.L., Pellow, D.N., Roberts, J.T., Bell, S.E., Holt, W.G. & Nagel, J. 2015, "Climate Justice and Inequality" in Climate Change and Society: Sociological Perspectives, Oxford University Press, New York.
Menk, Linda, Thomas Schinko, Veronica Karabaczek, Isabel Hagen, and Stefan Kienberger. 2022, "What's at Stake? A Human Well-being Based Proposal for Assessing Risk of Loss and Damage from Climate Change." *Frontiers in Climate* 4. doi:https://doi.org/10.3389/fclim.2022.1032886.
Nassar, Matilda. n.d., "Concept: Integral Ecology." *Kellogg Institute for International Studies.* https://kellogg.nd.edu/ihd-research-lab-integral-ecology.
ND-GAIN. 2025, "Notre Dame Global Adaptation Initiative Country Index," Notre Dame: University of Notre Dame.
Nowell, L. S., Norris, J. M., White, D. E., & Moules, N. J. 2017, "Thematic Analysis: Striving to Meet the Trustworthiness Criteria." *International Journal of Qualitative Methods,* 16(1). doi:https://doi.org/10.1177/1609406917733847
Nunez, S., Arets, E., Alkemade, R., Verwer, C. & Leemans, R. 2019, "Assessing the impacts of climate change on biodiversity: is below 2 °C enough?", *Climatic Change,* vol. 154, no. 3-4, pp. 351–365.
Obligations of States in respect of Climate Change (Advisory Opinion) I.C.J. Rep. 23 July 2025
Pacific Conference of Churches. 2024, "Tuākoi 'Lei Declaration." *Uniting World.* https://unitingworld.org.au/wp-content/uploads/2024/07/TUAKOI-%E2%80%98LEI-DECLARATION.pdf
Pihkala, P. 2025, "Ecological Grief, Religious Coping, and Spiritual Crises: Exploring Eco-Spiritual Grief", *Pastoral psychology,* vol. 74, no. 1, pp. 69–96.
Pihkala, P. 2024, "Ecological Sorrow: Types of Grief and Loss in Ecological Grief", *Sustainability,* vol. 16, no. 2, pp. 849.
Schlosberg, D. & Collins, L.B. 2014, "From environmental to climate justice: climate change and the discourse of environmental justice", *WIREs Clim Change,* vol. 5, no. 3, pp. 359–374.
Siegele, L. 2023, "Financing for loss and damage under the UNFCCC: Have we come full circle?", *Review of European, Comparative & International Environmental Law,* vol. 32, no. 3, pp. 403–415.
The Holy Bible. Psalm. 137. New International Version (2011). *Biblica, Inc.*
Tschakert, P., Barnett, J., Ellis, N., Lawrence, C., Tuana, N., New, M., Elrick-Barr, C., Pandit, R. & Pannell, D. 2017, "Climate change and loss, as if people mattered: values, places, and experiences", *Wiley interdisciplinary reviews.Climate change,* vol. 8, no. 5, e476–n/a.
Tschakert, P., Ellis, N., Anderson C., Kelly A. & Obeng J. 2019, "One thousand ways to experience loss: A systematic analysis of climate-related intangible harm from around the world", *Global Environmental Change.* Vol. 55, pp. 58-72.

UNFCCC. 2023, "Report of the Conference of the Parties on its twenty-seventh session, held in Sharm el-Sheikh from 6 to 20 November 2022. Addendum. Part two: Action taken by the Conference of the Parties at its twenty-seventh session." https://unfccc.int/documents/626561.

Varutti, M. 2024, "Claiming ecological grief: Why are we not mourning (more and more publicly) for ecological destruction?", *Ambio,* vol. 53, no. 4, pp. 552–564.

WMO. 2024, "Climate change transforms Pacific Islands." https://wmo.int/news/media-centre/climate-change-transforms-pacific-islands.

World Council of Churches. 2004, "Otin Tai Declaration." https://www.oikoumene.org/resources/documents/otin-tai-declaration ,Geneva: World Council of Churches.

Chapter 13
The Garden of Eden of the Dayak People: A Significant Change Study of Empowerment of Villages Through Culture, or How 'Adat' (Customs) Can Bring About Change

Nicolaas Ahmad Heeren

Abstract The land of the Dayak People, the indigenous people of Indonesian Borneo, has been taken over by oil palm plantations. So much so that only very limited original forest is left. What is at stake is thus the ability for the Dayak to continue living in harmony with nature. The question addressed in this paper, is: How can a subjugated people find again, the strength, the know-how, the knowledge and the spirit to gain control over their land and their forest? Laudato Si in point 180 gives a glimpse of the response: "*Forms of cooperation or community organization that defend the interests of small producers and protect local ecosystems from depredation should and can be facilitated.*" And indeed, social actors grown from the Dayak People have tried to empower the Dayak. One of these is the Dayakology Institute. I will explore their approach and will do so through the words of the direct players involved. The methodology we used comes both from project evaluation techniques and development anthropology theory. The study was embedded in several visits over a four-year period, thus enabling us to put the inputs in perspective. We met with 131 persons through 21 Focus Group Discussions (FGDs) from village elders to politicians, from women and youth to program coordinators and forest managers. All these were involved in the program with a long name (reflecting oral traditions) *Natural resource management based on local wisdom as an effort to mitigate and adapt indigenous peoples to climate change*, started in 2015, the year of Laudato Si. The outputs were 4,773 hectares of Land and Forest that will be managed by the customary Adat- leaders according to the customary decision-making processes. This represents almost half of the officialised Customary Forest in Indonesia in that

Nicolaas Ahmad Heeren, Catholic Committee Against Hunger and for Development-Terre Solidaire, nicheeren@gmail.com.

N. A. Heeren (✉)
Catholic Committee Against Hunger and for Development-Terre Solidaire, Paris, France
e-mail: nicheeren@gmail.com

R. Marcantonio et al. (eds.), *Integral Ecology in the Anthropocene: Africa and Beyond*, The Anthropocene: Politik—Economics—Society—Science 49,
https://doi.org/10.1007/978-3-032-32125-1_13

year. The first decision taken was not to allow outsiders buy land for palm-oil plantations. Pope Francis (February 2025): "*the increasing land grabbing of agricultural land by multinationals, large investors and states*" are like "*harmful practices that threaten the right to a dignified life of local communities*".

Keywords Indonesia · Participatory development · Indigenous · Change · Regeneration

13.1 Introduction: Let's Start With a Story…

Pa Vincentius lives in Segumon village of 600 people or so way up north almost on the Malaysian border. Segumon is surrounded by palm-oil plantations. He must be in his eighties (one doesn't ask a respectable village elder and customary leader how old he is), still lean and able to climb a sugar palm-tree. While he is rolling his cigarette, he relates us this story:

"*Back in 2010 while living in Segumon, I wanted outside facilitation to help us change things in our villages. I drafted a letter to Institut Dayakologi (ID). They invited us over. And between 2011 and 2013 we met with ID and we even went to see ID in Pontianak, in the big town, 6 hours away. We went with 4 persons, and we asked for help to set-up a Community Radio in our local language to talk about the Dayak culture and its sacred places. Finally in 2016 we received that radio and in that same year ID started a secretariat in Segumon.*

That's also when we made the participatory map and today we have not only official Communal Land, but even Customary Forest Rights of the most spiritual and holy place (called Tampun Juah) *of what is left of our forest after the Palm-oil Plantations came in. Now we are allowed to decide with our Customary institution how we want to manage our forest. One of the first things we decided was to forbid sales of Dayak land to outside people, especially palm-oil and mining companies. We now have an Adat Police to control the respect of these decisions. When we hear a chainsaw in the forest, we go to check it out.*

In September 2018, Mr. John Kennedy from our village went all the way to Jakarta (on Java, by plane!) and received the decision letter of the Ministry of Forestry from our President Joko Widodo himself.

So we are happy to see these results."

13.2 Where It All Began

As Pa Vincentius mentions, at the source of these activities lies the local Non-Governmental organisation (NGO), Institut Dayakologi (ID) counterpart of the French NGO CCFD Terre Solidaire[1] in a partnership of over 10 years. ID is a special local NGO. First of all, by its choice to work specifically with, and for, the 'indigenous' community of the Dayak. The term in Indonesian is *Masyarakat Adat* or Adat Communities for communities respecting their original customs (or Adat). ID was created by young catholic Dayaks back from university wanting to serve their people. Taking in consideration that Indonesia (285 million inhabitants) is the largest Muslim country in the world, although there are roughly 10% Christians, so still 29,4 million. Luckily inter-religious respect and collaboration do function as Pope Francis' visit to the country in September 2024 showed.

Indeed, the Dayak, of which there are about 200 groups and sub-groups, have long been seen as a 'primitive' peoples whose custom of 'headhunting' has been sticking to their image, rather than their incredible knowledge of the forest or their use of the very rich variety of fruit-trees (the *tembawan* orchards).

Discrimination of Dayaks has been the norm for a long time. During the Suharto years 'modernization' and 'development' were forced upon them. Probably best symbolized by the destruction of the traditional long houses on stilts replaced by individual homes on the ground. This comes back very often, when discussing with people. *This is where the longhouse stood*, so we were told.

Another 'modernization' was the economic 'development' that took place when the Suharto regime changed the status of the Dayak forests from customary managed (as they were under Sukarno, Indonesia's first president) into state-managed forest (see Durand 1994, 2001). This opened the way to ownership by companies to exploit the forest. In the beginning this mainly concerned hardwoods which were cut out of the forest by Japanese companies and exported; followed by plywood production (by Indonesian companies) until the exponential growth of palm-oil plantations over the last decade or so.

When you drive down the Trans-Kalimantan highway from Pontianak, West Kalimantan's capital, everywhere one looks forest is gone and palm oil plantations are the norm. In the District of Sanggau, where ID has been working, 351 thousand hectares have been transformed in plantations and only 217 thousand ha of natural forest are left.

Massive deforestation for monocultural oil palm plantations has tremendous ecological repercussions. It suffices to check on an internet satellite map to be able to count the millions of oil-palm trees planted over the last couple of decades. This translates in loss of captioned CO2 (peatlands are being drained to establish plantations) as 100 thousand ha of tropical peatland stores 228 mega tons of carbon (Keily 2021). It also translates into less biodiversity as plantations are monocultures.

ID did an internal study before the program started and found irregularity and randomness for the rainy season and the dry season making planning agricultural

[1] French for *Catholic Committee against Hunger and for Development—for a World of Solidarity*

cycles much more difficult. This in turn led to bad harvests. Hunting and fishing also became more difficult.

Facing these challenges, a 3-year program called *Natural resource management based on local wisdom as an effort to mitigate and adapt indigenous peoples to climate change* was designed in 2015 by Institut Dayakologi (ID) together with the French Catholic Committee against Hunger and for Development (CCFD Terre Solidaire) and a 700,000 Euros budget funded by the French Development Agency.

13.3 Laudato Si (2015), and the Institut Dayakologi program

2015 was indeed a special year as it was the year of the COP21 in Paris and the publication of the Laudato Si' encyclical letter from the late Pope Francis. Four citations enable us to understand the articulation between Laudato Si (L.S.) and the ID Program (my rendering in bold):

- *LS.180 New forms of cooperation* and *community organization* can be encouraged in order to defend *the interests of small producers and preserve local ecosystems* from destruction.
- LS.179 (…) while the existing world order proves powerless to assume its responsibilities, local individuals and groups can make a real difference. They are able to instil a greater sense of responsibility, a *strong sense of community,* a readiness to protect others, a spirit of creativity and a *deep love for the land.* They are also concerned about what they will eventually leave to their children and grandchildren. These values are deeply rooted in *indigenous peoples.*
- (…) Society, through *non-governmental organizations* and intermediate groups, must put pressure on governments to develop more rigorous regulations, procedures and controls. Unless citizens control political power—national, regional and municipal—it will not be possible to control damage to the environment. *Local legislation* can be more effective, too, if agreements exist between neighbouring communities to support the same environmental policies.
- LS.180 Agriculture in poorer regions can be *improved* through investment in rural infrastructures, a *better organization* of *local* or national markets, systems of *irrigation*, and the development of techniques of *sustainable agriculture.*

The Institut Dayakologi program which ran until 2020 aimed at strengthening solidarity, participation, empowerment and self-esteem of the Dayak culture and its customs (or *Adat*) on two sites, one still in the forest and mountains, the other in the middle of the encroaching palm-oil plantations. The interaction one can observe with Laudato Si is not accidental, but neither is it the local effect of an external cause. No, here we see how the ideas of social and environmental change and the practice of a local NGO and the theory of Pope Francis will interact.

The program had 5 dimensions: A first one around the environment and natural resources protection including mitigation of the impact climate change through awareness building on the origins of climate change. The main idea here was that protection and mitigation needed land-rights for the local people. Through the establishment of participatory maps of villages and their natural resources which in turn were proposed for official recognition by national authorities ID was able to obtain Customary Land and Forest Rights for the local communities.

The second dimension aimed at the local Dayak culture, the Adat and the traditions and how to revitalise them. This sub-divided into various approaches. One was around the sacred site of *Tampun Juah*, in Segumon village in the middle of rows and rows of oil-palms. This cultural dimension was also to be developed in school curricula and through youth groups.

Two important symbolic cultural figureheads also entered in this dimension: the valorisation of the *tembawan*, the traditional mixed fruit-tree orchards in the forest where economic use, biodiversity and natural resource management come together on the one hand; the acquiring of the *Wila*ya *Adat* status (Customary Community Land Rights), on the other.

The third dimension concentrated on organisation and empowerment of local actors: farmers, peasants, women, youth, but also local development organisations which were to make local development plans and organise village assemblies. All this accompanied by training and accompaniment.

The fourth dimensions looked at access to credit but not only for economic investments, but rather as a life-insurance for important moments and steps in the cycle of life: birth, wedding, education, sickness and death.

The last and fifth dimension aimed at alternative Income Generating Activities (alternative IGAs) so as to diminish the Dayak dependency on one economic activity (rubber-tapping). Producer-organisations and training, the stimulation of economically interesting crops (like pepper, red ginger, organic rice) and fish-farming were part of the activities as well as negotiating access to government budget-lines for small projects.

13.4 Program's Results

A programme like this aims at change but does so through concrete activities each having their results or outputs. The actual change these bring about (the outcomes) is another matter. But let's look at outputs first.

Our research in both sites at the end of the programme show that according to the local on-site players, results have been obtained. Some bypassing all expectations and programme objectives, notably the *Hutan Adat* forest rights.

13.4.1 On Mitigation of Climate Change: Changing Mind-sets

Mitigation is a long-term process. The various conscientisation sessions enabled the local population to be more aware. Pa Okta, Program Coordinator in Segumon puts it this way:

> "People have changed their mind-set. They are more aware of what should be and can be done for and with the land. They know what will happen to the land and soil if they do nothing and remain in their habits."

13.4.2 On New Status of Land/Forest: Community Forest Rights

As mentioned by almost all interlocutors, the *Community Land and Forest Rights over 4,775 ha* for the Program's villages Tae and Segumon are a major result. In Okta's words: "Who would have thought at the start of the program that the Village Chief of Tae, who lives in a small house and only owns a motorbike, would go to Pontianak the Provincial capital, fly to Jakarta and be received by the President himself during an official ceremony in September 2018." Or more practically: "the Community Environment Advocacy Committees and the "Adat Police" which are overseeing the implementation and respect of the Adat decision in the forest and in the community. Whenever we hear a chainsaw in the forest, the Adat police will go there".

13.4.3 On Revitalizing Dayak Culture

Giving Dayak culture a new status was part of the program's envisaged results. Probably the inclusion of Dayak culture as an official subject in primary schools is a good indicator. But also the revalorization of certain agricultural practices integrated in forest management. In Pa Okta's words: "Before, the tembawang (fruit tree orchards with sometimes up to 15 different tree-species) were not seen as an interesting activity. Now they are. And on top of that, people feel now that these kinds of activities should be started everywhere".

13.4.4 On Local Development Organizations for Integrated Activities: A beginning of Sustainability

For the local management of social, cultural and cultural activities, two small local organizations were put in place and will take over the role of accompanying local dynamics when the ID presence reduces (after 2020). These local dynamics can be fairly straight forward, like managing the Community Radio in Segumon. Okta: "*But more essentially, especially in terms of sustainability, the local organizations can build trust with the Local Authorities*".

13.4.5 On Credit from the Credit Union FPPK: Steady Beginning, More to Come

Credit is a more material approach to economic empowerment. Results on this activity are encouraging (83% of the capital objective achieved and 70% of the number of new members objective). It is also interesting to see how the FPPK's Peasant Philosophy approach gives a different twist to the Credit Union compared to more classical loans. Indeed CU-FPPK compares its different 'products' to the various stages in Dayak life and culture (preparing land, sowing, reaping, etc.). We find therefore 'products' meant for health, for education next to classical capital-loans for economic activities.

13.4.6 On Alternative Income Generating Activities (IGA): We Need More Time and Accompaniment

In the economic non-credit domain, changes, after 3–4 years of program activities, are less pronounced. Clearly there is need for more time here. Pepper growing was, for instance, proposed as alternative IGA and was taken up by many enthusiastically. In May 2015 pepper prices were relatively high. On ½ ha of pepper a farmer could produce more than 100 kg of pepper and create a supplementary annual revenue of ca. 1.200$. However, since 2015, the drop in commodity prices for pepper by 50%, clearly limited the estimated ultimate effect.

A more critical vision is given by the ID Local Coordinator in Tae, Pa Yohannes. "The Alternative IGAs are still in the starting phase. The CU can indeed give financial inputs, but as the whole value-chain is rarely controlled by the farmers; they remain dependant on middlemen".

13.5 Results, But What Social Change? An Alternative Approach

So far so good. But whilst results were obtained, some of the underlying challenges the program wanted to address might be forgotten. Indeed, the actual purpose and intention of a "Development Program" often tend to subtly slip away while the more operational elements take the upper hand. Indeed, the indicator logic—useful as such—often leads both donor and implementer to concentrate on the fulfilment of these purely quantitative indicators rather than the long-term change also envisaged. I.e. the question "What are you doing to change what?" becomes the "Did you do the training session for the target group and how many participated?".

This is of course a classical development project dilemma (output versus outcome, or short-term results versus long-term change). Both are important, but the latter is challenging. Behind that challenge also lies a methodological (and budgetary) one. As NGO, the answer to "Are there ways to measure social or qualitative change the programs are obtaining or enabling?" is often: socio-anthropological studies. That is a good approach, it is also true that they remain hard to operationalise (expertise, duration, cost,…). For those reasons, post-project impact studies (see Gaugnard/ Heeren, 1999) are often not done by NGOs, even though its usefulness is underscored by all players in development projects...

Instead, we opted for an approach which has its shortcomings, but which has the advantage of being feasible. An attempt to better understand the changes a 3-year program can bring about as "Told by the players/actors themselves" (Crozier/ Friedberg 1977).

The approach consisted of a 2-week mission of a small team of two near the end of the program to discuss with as many stakeholders as possible about the important changes (according to them) which have taken place in their environment (be it social, ecological, economic or governance) as a result (or not) of the program. The methodology consisted mainly of focus discussion groups (FDG's) with small groups of about 4 to 6 people.

The leading question was "What has changed over the last 5 years in your village which you find important?" from where we spiralled off to other questions either quantitatively (how many?) or qualitatively (why is that important to you?). There were also site visits to broach the same or other subjects in informal settings but also observe the players in their relation to each-other, to nature and to religious sites.

Of course, anyone can "tell you a story". And it is clear that there can be easily biases in this method. Admittedly this approach is subjective and leaves a series of questions out in the open, i.e. Who defines what is 'change'?, What is significant? Over which duration, with what impact? Etcetera.

But by increasing the number of interlocutors, one can start to establish patterns and realities which are more than just a story. A total of 21 FDG's were held plus some individual interviews with a total of 130 persons (42 women, 88 men).

The multi-actor approach has, I believe, the advantage of the quantity of players involved and the therefore whatever is 'change' and 'significant' is indeed what the interlocutor feels is important in his or her actual life. It also helps to have points of view and counter-points of view because of the number of players and the variety of type of players (women, men, young, old, traditional leaders, politicians, teachers, etcetera) allowing to compensate bias.

13.6 Significant Changes in a Complex Environment

In general, and even more in a forest-environment with indigenous people, conceptual spheres such as environment, (political) governance, social, culture or economy, are difficult to separate as all are inter-connected. In the words of Pa Marcelus in charge of the local organisation set-up in Desa Tae, and an Adat leader, "Adat equals Naturality/Spirituality/Solidarity". With Pope Francis we could say "an integral ecology".

However, this integrated approach renders the program also much more complex for all dimensions must be considered, and all dimensions will probably have activities planned and implemented. This complexity also applies to the outside factors which influence on, and in, the different spheres whether these are of private nature (plantations, mines) or public (Province and Central State) or again economic (companies) or cultural (customs, church, etc.). We have opted in the full study for a description of the changes as they are felt and told by the different players in roughly four spheres of influence: Environmental, Socio-cultural, Political & Governance and Economic. Even though boundaries between these spheres are blurred.

Nevertheless, when discussing, one of the first things that came up is "the customary forest and how it became ours again". The forest for the Dayak is probably the best metaphor for the four spheres above: it is simultaneously nature, livelihood, culture, ownership and management.

> It also echoes the words of Pope Francis:
>
> LS.137 Since everything is closely interrelated, today's problems call for a vision capable of taking into account every aspect of the global crisis.

For the sake of briefness, we will not develop all the dimensions which the original study embraced but concentrate on those which seem to reflect in the best way Laudato Si's ecological ideas.

13.6.1 Environmental Changes: A Garden of Eden

Without a doubt, the most significant change is the obtaining of the Customary Forest Rights (*Hutan Adat)* for 2,840 ha from the hands of the Indonesian President Jokowi and 4,773 ha of Customary Land Rights (*Wilaya Adat*). All this is based on participatory mapping executed by the villagers themselves supported by a local NGO bringing in the necessary technical skills.

The Dayak live in the forest. They take its fruits, grow food, use its resources and believe in the power of creation within its mighty tress. This Customary Forest status allows for these economic activities to be managed by the local community and governed by Adat institutions. Contrary to the Protected Forest status which is State-owned and does not allow for economic activity thereby excluding Dayak community from their habitat.

The decisions of the Customary Adat Assemblies to prohibit sales of land to outside parties (notably Plantations) is helping to reduce the risk of loss of forest.

This is no doubt the major and essential change the program has obtained in the lives of people and was confirmed time and time again in the group discussions.

Interestingly indeed, the fact that other Dayak groups not targeted by the program heard about it through their ethnical connexions (not in the least because the rehabilitation of the Dayak customary sacred site *Tampan Juah*), enabled the Program to go beyond its original territories.

In Segumon, the extent of the territory was actually an unknown given. Pa Darius, an Adat Advisor, said as much: "before we had no idea of our own territory. Now we know it is 2,327.38 ha (He insists on the details after the decimal point) because of our participatory map".

Direct effects of this evolution in forest status are that it is again possible to have a small farm (maximum 2 ha per household) in the forest. The traditional slash-and-burn approach is forbidden and the starting of a small farm is regulated by the community via the Adat system but also evaluated by the Ministry of Forestry and Environment in order to verify that the new regulations are respected.

Also, the ID's success in obtaining Community Forest status let to interest from outside Kalimantan. 28 directors from local administrations from elsewhere in Indonesia came to visit to study the *Hutan Adat* decision process.

13.6.2 Agro-Forestry: Forests are Orchards

Why is the *Hutan Adat* so important? In the words of one woman from Padang village "*The Hutan Adat is for the next generation*". One can almost hear Pope Francis saying "They are also concerned about what they will eventually leave to their children and grandchildren. These values are deeply rooted in indigenous peoples" (LS179).

The best illustration of this, are the *tembawang* or orchards where nature and culture are combined over generations in an original version of agroforestry as it is protecting biodiversity while reaping economic benefit. For a *tembawang* one should imagine a forest-like slightly opened-up woodland of about 2–3ha where more than 15 different species of fruit trees are growing, are (re)planted and have been growing for years as some are 50 m or more tall. Over time, each species might be there in important numbers. Pa Napis, for example, says "I have more than 30 big durian-trees on my 2 ha tembawang. The tree-gardens are in the family for generations. Or, in some cases in the hands of the community collectively (Common Tembawang)".

The richness in types of fruit-trees is just astonishing: different varieties of durian, *nangka* (jackfruit), rambutan, starfruit, blood-starfruit and many other indigenous fruits with other economic possibilities, such as the *tengawang* which can be transformed into butter, *nyalit* which can be turned into soap (to clean gold, no less).

Pa Napis explains how it works: "According to the season, fruits are collected and consumed locally or sold. The income is for the family and in case of the Common Tembawang for the village as a whole". The place of the Common Tembawang is also where the old Longhouse stood when the Dayak still lived collectively. "Before it was a village, this place was one huge tembawang".

Indeed, the old maps from South-East Asia, collected by Suarez (1999), from the 16th Century, when what laid beyond Burma and Thailand was still unknown to Western seafarers (it was not unknown to their Asian counterparts), map-makers used to draw Adam and Eve in their Garden of Eden eastwards beyond the Malaysian peninsula.

It is what I felt, when walking in the tembawang under the canopy high up, one could imagine that the Garden of Eden where so many fruits are just for the taking is here in Indonesian Borneo. Indeed, just to the east of Malaysia.

13.6.3 Community Decision Making: Cultural Revitalization

With the change in management status of the forest from Protected Forest to Community Forest, the forest governance changes from governmental authorities into the hands of the Adat leaders. Before the Adat leaders were not involved with the authorities; both spheres, administration on one hand and traditional ceremonies on the other, were separated. The *Adat* decision-making goes beyond just the forest. Other aspects of community life are now again in the hands of the *Adat*.

This transfer asks for a real synergy between administration and Adat leaders. It is important both for the formal recognition and for the sustainability of the program results. But "there is also more critical thinking in the dialogue between the community members and the Local Authorities", so says Pa Heri, a local farmer.

The Customary Adat Assemblies are part of the cultural revitalization. Indeed, these Assemblies have not been held for many years. It is only when the program started in 2016 that this social and cultural heritage was picked up again. Decisions on regulation in terms of governance were taken. It is interesting to look at some examples and see how they intervene in the social-cultural but also in the ecological and economic spheres of influence. Note that Adat Assembly decisions are formalised by the District head. This being a sort of legalisation of decisions in the traditional system by the 'modern' system. Again, a synergy between Adat and Local Authorities.

Some examples of these *Adat* decisions are given by the Local leaders. "It is now forbidden to sell land to outside parties (i.e. plantations companies) in Tae". "The distance between residential areas, water sources and crops is now also regulated" (spatial planning). More in general, "Local crops and their varieties are very much stimulated using local knowledge as it is much more environmentally friendly". Before the program, mainly rubber trees were planted, now there is a variety of local crops (diversification).

Women are also benefitting from the Adat decisions. The Padang village women group, Lidi Kuwie, mentioned that "before the program they did know much about the local government and the village administration. Since then we know and we are invited to the Local Authority meetings". This could lead to more political involvement of women. Ibu Linda and Ibu Yuliana of the Peasant-women organization in Maet village also underline this benefit for women: "We now know and understand more about village regulations".

The Adat management system also foresees sanctions which are feared by the local population. Indeed, it has been seen that immaterial and spiritual sanctions have more effect than simply monetary ones: "If one of the Adat rules is not respected by a large number of people, the Catholic Priest will not come on Sunday to celebrate mass", so says one interlocutor.

In Segumon, so the Adat Leaders tell us, "the decision was taken to forbid the cutting of trees in the small forest that is left around the Sacred Site of Tampun Juah. We set-up special committees to oversee those decisions". In the words of Pa Dalang (Head of the Environmental Committee), "*If we hear the sound of the chainsaw, we go and step on them*".

13.7 Qualitative Changes: Cultural and Spiritual "Now We Feel Empowered"

Although it was already the custom to welcome strangers into the villages through certain rites by the Adat-leader and prayers by a Shaman, this practice has now strengthened itself. In order to enter certain parts of the forest, these rites are necessary too. It can happen, as it did to me, that while walking through the forest, at a certain point (often a big and high tree), prayers and offerings have to be made for the visitors before continuing their way.

The cultural revitalisation is expressed by our interlocutors in other words: For Pa Karia, the highest Adat Leader "Adat is customs, law and a way of life to ensure peace among the community".

This idea of a mental shift (or internal empowerment) is also mentioned by Pa Burhan, village head, when he says that "we are proud of the change. Now we feel more empowered. Before we were powerless. The Decision letter for the Hutan Adat (from the hands of the President) symbolizes that power."

And indeed, one could think about the Hutan Adat status it as a "transfer of power" in social terms when the President decides to bestow upon (or return to[2]) a forest-community the power to manage their own natural resources.

[2] The decision to make forest State-property was taken in the Suharto era. Under Sukarno these forests were 'commons' of the Dayak peoples.

The Adat Forest status exists because of the 28 participatory mapping exercise that was carried out in the village. Way above the 7 planned. The map, co-signed by all in the village, is seen by many as a sacred map. It is guarded as treasure. And in actual fact, after the signing public ceremony, the map is kept in a special place in the house of one of the Adat Leaders, most probably of shaman level. But the sacred Adat ritual is performed in the central place close to the village Catholic Church with a copy of the Lourdes Grotto with the Virgin Mary overseeing the ritual…

The preservation of the Holy Site of *Tampun Juah,* surrounded by massive Palm Oil plantations, for the 2 main ethnic Groups and 116 ethnic sub-groups is another major change. The site is also a Tembawang with trees more than 25 meters high and indeed very impressive as a small dense forest. Ancestor and customary anthropomorphic objects of both main ethnic groups are positioned at the entrance of the Holy Site. Statues made of wood and of stone. Although we are later told that the real sacred objects are much deeper in the forest and no-one not initiated is allowed to visit them. In the words of one Adat Leader: This is just the cover of a book; the real message or story is inside…

The 116 ethnic sub-groups state that their origins lie here in the Tampun Juah Home Ground from where they migrated to Sarawak (Malaysian Borneo) and even all the way to the Sultanate of Brunei (more than 600 km to the extreme North of Borneo).

"Before the program started in Segumon", so tells us Pa Johan, "people did not come here often. There was a vague idea about the site but no real orientation and the actual site was almost forgotten except its name". Which is surprising as so many sub-ethnic groups see their origins in Tampun Juah.

Pa Kennedy notes that "the younger generation will need to explore more, to discover more about the Adat. It can motivate youth to study tradition. However, it should not be a burden". He confirms that "change is not easy, but the revitalization is in the end about ownership: the right to (manage our) land and to water. Are they the Dayak's property or can they be transferred to a 'company' just for the sake of it?"

13.8 Conclusion: Significant Change? Integral Ecological Empowerment

Perhaps that is the conclusion of the story of the Institut Dayakologi's program to defend the Dayak culture, its habitat and environment, that Change is not easy, but revitalization is (also) about ownership".

Two more citations from Pope Francis on World Food Day 2021 connect with this program:

- The principles of subsidiarity and solidarity should be considered as the foundation of development programs, giving priority to the needs of workers, farmers, the poor and the hungry.

- "Each of us has a role to play in transforming food systems for the benefit of people and the planet"

The first one reflects on the solidarity work of Institut Dayakologi in its development program starting from the needs of the Dayak people to repossess their land, their forest and being able (again) to manage those themselves.

It is an act of empowerment. It's both political and spiritual. The fact that so many other Dayak ethnic sub-groups feel part of the action as they feel that their roots have been reconfirmed through the *Tampun Juah* sacred site also shows that the program has been able to touch another significant demand, not material, but rather immaterial and spiritual.

And finally, we all have a role to play. I tried by supporting this program when working for the CCFD Terre Solidaire and by putting the words of the Dayak people on paper echoing their way to protect the planet.

References

Crozier, Michel and Friedberg Erhard, *l'Acteur et le système*, Points Essais/Seuil, Paris, 1977

Durand, Frédéric, *Les forêts en Asie du Sud-Est, recul et exploitation, le cas de l'Indonésie,* L'Harmattan, Paris, 1994. Paris : Droit, Politique, Espace Agraire au Bresil

Durand, Frédéric, *La jungle, la nation et le marché*, Atalante, Nantes, 2001

Graugnard, Gilbert and HEEREN, Nicolas, *Guide d'indicateur d'impact*, Catholic Committee Against Hunger and for Development-Terre Solidaire-Université Catholique de Lyon/F3E, Paris, 1999

Francis, Holy Father, *Encyclical Letter Laudato Si'*, Holy See, Rome, 24 May 2015

Suarez, Th., *Early mapping of Southeast Asia*, Periplus, Singapore, 1999

Chapter 14
Socio-Ecological Interventions for Addressing Mental Health in Uganda

Godfrey Ddungu and Deena Shelton

Abstract Integral Human Ecology (IHE), highlighted by Pope Francis's *Laudato Si'*, presents a comprehensive framework that integrates environmental stewardship, human dignity, and psychosocial well-being. This paper explains Uganda's escalating mental health crisis, which is exacerbated by structural poverty, historical conflict, social stigma, and ecological degradation, and offers existing mental health benefits of IHE-based models using The Bethany Land Institute (BLI) is presented as a model for demonstrating how land-based education, agroecology, and spiritual formation can contribute to mental health. Suggestions are made that IHE-aligned interventions such as narrative and creative storytelling, empathy and active listening training, and movement-based therapies could provide added benefits and aid in addressing the ongoing mental health concerns in Uganda. This paper explains how IHE provides a valuable theoretical and practical lens for designing mental health interventions that are ecologically grounded and culturally sensitive.

Keywords Socio-ecological systems · Mental health · Interventions · Storytelling · Narrative

Godfrey Ddungu, Fresno State University, goddungu@mail.fresnostate.edu and Deena Shelton, Le Tourneau University, deenashelton@letu.edu.

G. Ddungu (✉)
Fresno State University, Fresno, California, USA
e-mail: goddungu@mail.fresnostate.edu

D. Shelton
Le Tourneau University, Longview, Texas, USA
e-mail: deenashelton@letu.edu

R. Marcantonio et al. (eds.), *Integral Ecology in the Anthropocene: Africa and Beyond*, The Anthropocene: Politik—Economics—Society—Science 49,
https://doi.org/10.1007/978-3-032-32125-1_14

14.1 Socio-Ecological Interventions for Addressing Mental Health in Uganda

Though the concept of mental health is broadly recognized as a universal concept, sociocultural factors influence the conceptualization and responses to problems associated with mental and relational health issues (Asiimwe et al. 2023). Mental health is often characterized by issues in cognitive, emotional, and social functioning that challenges an individual's ability to function effectively in their lives and relationships (Opio et al. 2022). However, the concept of mental health is effectively understood as a state of well-being that provides people with the ability to function well in terms of internal well-being, good work, and connections to others (WHO 2025). It was seventy years ago that Dr. Brock Chisholm, the first Director-General of the WHO, said, "without mental health, there can be no true physical health" (Kolappa et al. 2013).

Despite the decades-long call for global recognition of the importance of mental healthcare in relation to the overall functioning of individuals and suffering communities and positive responses from broader systems (e.g., the Ugandan government), follow-through for provision of such services have not been realized (Nalukenge et al. 2019). Though many mental health issues can be treated without an exorbitant fiscal burden added to communities, health systems all over the world struggle to provide adequate and quality services (WHO 2025). In addition to the difficulty accessing services, people living in countries with comprehensively lower resources are often exposed to more adverse experiences and stigma leading to higher risks of mental health problems.

14.2 Uganda's Mental Health Crisis

Uganda is currently grappling with a mental health crisis, with approximately one-third of the population suffering from mental health disorders, yet less than half receiving any form of care (Murray et al. 2025; Shah et al. 2017). This gap in care is attributed to limited resources, pervasive stigma, and systemic challenges in the healthcare infrastructure. As the population continues to grow and socio-economic pressures mount (World Bank Uganda 2016), the need for innovative, integrated mental health solutions becomes urgent.

Ugandan citizens face high rates of poverty and severe mental health impacts from the history of long-lasting armed conflicts in the country (Iverson et al. 2021). The long-lasting ramifications of the conflicts include mental health issues due to high rates of children were orphaned, abducted and used as soldiers, abused, and left living in poverty. These experiences result in emotional distress, high rates of post-traumatic stress disorder, and social stigma issues that impact long-term functioning as adults. Recent prevalence studies suggested that 22.9% of Ugandans under the age of 18 and 24.2% of Ugandan adults suffered from mental health disorders (Opio et al. 2022).

Social stigma remains one of the most formidable barriers to mental health care in Uganda. It is deeply entrenched in both public and professional spheres and is especially harmful to vulnerable groups including people living with HIV/AIDS, survivors of violence, and the poor (Ssebunnya et al. 2009; Quinn et al. 2014; Liebling et al. 2016). These marginalized individuals often experience not only social exclusion but also internalized shame, leading to depression, anxiety, and withdrawal from social life (Ddungu 2023; Kinyanda 2019). Ugandans experiencing multiple marginalized identities often experience profound pain related to stigma and discrimination, especially from close family members, causing mental and emotional distress. For example, in a study exploring the experiences of people living with HIV/AIDS in rural Uganda, participants described intentional social isolation from family, which can hinder the desire to seek mental health support (Ddungu 2023). One participant described being treated "like [she] was already dead, something with no value" (Ddungu 2023, p. 88).

14.3 Laudato Si' and Integral Human Ecology

In his second encyclical, *Laudato Si'*, Pope Francis (2015) addressed what he considered to be the most pressing concerns of our global existence related to the issues in how humans relate to God, the earth, and one another. This encyclical draws on Biblical insight and wisdom to define the earth as a gift to humans to steward and treat with care (rather than dominate and plunder) and identifies the breakdown in this relationship as it relates to the global reach of a technocratic paradigm. Pope Francis suggested that the broad prioritization of possession and control of resources has created a profound ethical gap between rapid technological progress and the development of moral responsibility, with critical implications for mental health particularly among vulnerable populations. He directly addressed the deteriorating state of the world (the physical strain on our land and resources) and the people inhabiting it (deteriorating quality of life, breakdown of society, and global inequality).

Pope Francis (2015) diagnosed the issue as modern society's glorification of human power expressed through the pursuit of control and domination embodies what he terms "modern anthropocentrism," claiming that despite immense technological advancements, there has been no corresponding evolution in human conscience, ethical values, or social responsibility. This imbalance also reflects a disrupted relationship between humanity and the natural world—particularly the land (Pope Francis 2015). Reducing nature to a mere resource for exploitation results in a brokenness to human's spiritual, cultural, and communal ties to land and each other. The answer presented focuses on opening dialogues and moving toward a complete change in lifestyle, rejecting consumerism and self-centered living. One sign of human flourishing in this is recognition of the power of the individual to impact their broader systems and environment. When individuals experience themselves influencing their broader systems, it can create capacity-building effect, lending to the desire to seek more areas of positive influence and create more positive change (see Shelton/Bridges 2021).

Since its publication, *Laudato Si'* has been used by individuals, organizations, and broader economic groups to address issues of sustainability in economic systems and in the human spirit. Efforts like the Laudato Si' Action Platform (2025) provide guidance to equip institutions and individuals to foster initiatives by offering guiding goals and a framework to guide effective development and implementation for Integral Human Ecology (IHE) based programs and initiatives. The seven goals provided by the platform are: (1) response to the cry of the earth, (2) response to the cry of the poor, (3) ecological economics, (4) adoption of sustainable lifestyles, (5) ecological education, (6) ecological spirituality, and (7) community resilience and empowerment.

14.4 Laudato Si' Based Program Example: Bethany Land Institute

Laudato Si' based programs designed to answer the needs of the poor in rural Africa provide strategies address individual, community, and economic issues to create sustainable living conditions for all living things. One such program is the Bethany Land Institute, which provides training to students in farming, agroforestry, agribusiness, focusing their programs to address the dual problem of social and economical crisis in Uganda (Katongole 2022). Their goal is to train economically disadvantaged young people from rural Uganda, to care for the land and creation, while simultaneously learning to care for themselves and one another. Education programs based on IHE models like BLI serve populations with increased risk of mental health concerns due to experiences with trauma and economically disadvantaged status. By fostering inclusive, land-based communities where people see land as taking care of them and offering dignity-affirming work, BLI helps counteract the psychological damage caused by exclusion (Katongole 2022). The community model reduces isolation and promotes belonging which are key components in mental health recovery. Moreover, BLI's spiritual formation reconnects individuals with sources of hope and meaning, helping to rebuild coping mechanisms that are often eroded by trauma and discrimination (Rothert et al. 2020).

BLI's initiatives in land restoration, sustainable agriculture, education, and spiritual formation intentionally engage vulnerable populations, especially youth, in regenerative work that nurtures psychological resilience, dignity, and hope (Katongole 2022). This integration of ecological restoration with psychosocial healing addresses the root causes of mental distress in Uganda, such as land degradation, displacement, and disconnection from cultural and spiritual identity (Ashukem/Ngang 2022).

14.5 Integral Human Ecological Interventions and Mental Health Impact

While IHE-centered models do not directly emphasize mental health interventions, the overarching philosophy of IHE provides proactive mental and emotional benefits to those involved through peripheral health-centered practices known to enhance emotional and mental well-being. Pope Francis urged people to deny consumerism and the drive for power to embrace the joy and peace, experiencing a "deep enjoyment free of the obsession with consumption" (2015, p. 144). This countercultural call for a change in the focus from extrinsic life goals (i.e., emphasizing possessions, image, status) versus intrinsic life goals (i.e., connection with others and personal growth) is a move that has positive mental and emotional outcomes for any who listen (Soares/Moniz 2023).

Overconsumption and materialism are associated with reduced life satisfaction and lower levels of reported happiness, poorer moods, and increases in psychological problems (Soares/Moniz 2023). However, nature plays a crucial role in promoting human health, and it deeply nourishes our sense of happiness and contentment (Kuo 2015). Just spending time in natural environments whether in forests, parks, or even tree-lined streets significantly improves not just physical health, but emotional well-being. In addition to a boost in mood, nature experiences seem to have a positive impact on human cognitive processes including memory, attention, concentration and impulse inhibition (Bratman et al. 2012).

The emphasis in IHE on connection and response to God and other living things emphasizes a focus on empathy and social support as basic foundations of human existence (Pope Francis 2015). Individuals who report higher levels of social support, or the security of having closer relationships with others with a mutual exchange of love and care, report fewer mental health and addiction issues and greater recovery from mental health issues experienced (Bjørlykhaug et al. 2022). Collectivist African contexts emphasize the necessity of horizontal (interpersonal) and vertical (spiritual) relationships for meaning, purpose, and a healthy life (Wissing et al. 2020).

The lack of employment opportunities and chronic financial insecurity in Uganda intensify psychological distress that ultimately leads to mental health issues, particularly among youth (Rothert et al. 2020). By offering education, training, and a clear path to meaningful livelihood through ecological stewardship, programs like BLI empower young people to regain agency, purpose, meaning, hope, direction which are key factors in combating anxiety, depression, and emotional numbness. Responses that encompass an integral ecology prioritize healing of the relationships between people, community, and nature by reclaiming land not only as an economic asset but as a resource for psychosocial and spiritual renewal (Katongole 2022).

14.5.1 Mental Health Interventions for Integral Human Ecology-Based Programming

Programs designed under Laudato Si' guided models inherently incorporate proactive measures for protecting and enhancing mental health experiences by those served. However, because the populations served have significantly increased likelihoods of traumatic stress and other mental-health concerns (Opio et al. 2022), it could be beneficial to include more direct work related to mental health care to enhance the relationship to self, thereby enhancing the options for relationships to others. This section includes suggestions for mental-health grounded interventions, education, and/or therapeutic engagement that would support IHE-based programming.

14.5.2 Storytelling and Narrative Skills

The power of storytelling in overall health and wellbeing is a multidisciplinary concept. From the full therapeutic interventions under the theory of narrative therapy to creative writing structures, creating understanding through crafting the narrative of our life experiences holds significant implications for personal mental health and healing (Ricks et al. 2014). Narrative therapy suggests that many people struggle to express difficult thoughts, feelings, and emotions which can become a barrier to when they need to process difficult life experiences. However, narrative techniques allow therapists to aid clients in learning to externalize their problem, present it as a tangible storyline, and make determinations on how the story can be changed or altered. While these techniques are used in therapeutic processes, they are skills that can be used outside of therapy for deeper journaling processes and personal exploration or discovery.

Beyond traditional writing for process, storytelling through representative means (art, music, drama, etc.) may also be effective for facilitating internal healing processes. When working with communities or groups that have experienced historical or intergenerational trauma, exploring community healing tactics or collectivist-based exercises may be more beneficial than individual-based interventions (Chioneso et al. 2020). Adapting processes to accommodate for other cultural or group preferences including (but not limited to) oral culture traditions and historically significant art or music representations are important considerations for anyone implementing these types of interventions.

Adding focus on narrative storytelling for healing would be beneficial for programs using IHE-based models for intervention (Ricks et al. 2014). Whether using therapists to help program participants narrate their story through therapeutic means or creating general individual or shared exercises to narrate the individual and community stories, creating structure to facilitate processing pain and creating more opportunities for hope and vision for their future. These exercises would be especially helpful for participants who struggle to discuss their emotions and experiences without pre-process and could promote stronger connections between them and others.

14.5.3 Empathy and Listening Skills

In *Laudato Si'*, Pope Francis emphasized listening as a foundational act of compassion and ecological awareness by and urged readers to "listen to both the cry of the earth and the cry of the poor" (Pope Francis 2015, p. 34). The ability to effectively communicate and deeply connect to others is vital to living an embodied witness of the connection described by Pope Francis. Active listening is a foundational element to communication and connection, can practically be taught in skill-based training (Tustonja et al. 2023). Listening is a key component in empathy, which is strongly associated to healthy relationships and social functioning (Teding van Berkhout/ Malouff 2015). While the definition of empathy can vary to empathize the cognitive or affective processes associated, processes to engage cognitively with another's experience and generate compassion or emotional connection to their experience can be learned through training. For people who have difficulty connecting to others or who may show deficits in their skills related to listening and understanding others' experiences, empathy training may aid in preventing interpersonal safety issues, especially for those who have histories of trauma.

14.5.4 Brain/Body Practices

Mind–body interventions, which leverage the connection between mental and physical processes to enhance well-being, are especially valuable in resource-limited settings like Uganda, where access to formal mental health care is scarce (Guendelman et al. 2017). These culturally relevant, low-cost, and group-based techniques—such as dance, movement, breathing exercises, mindfulness, and storytelling—support emotional regulation, social connectedness, and psychological healing through community engagement (Pandey et al. 2018; Vaisvaser 2021). The Expressive Therapies Continuum provides a framework for categorizing such practices, from sensory experiences like drumming and dance to symbolic activities like art and writing, aligning well with Uganda's rich cultural traditions (Hinz 2019).

Evidence shows these interventions improve brain function, emotional regulation, and neurotransmitter activity, particularly benefiting trauma-exposed youth (Grasser/ Marusak 2023). While meta-analyses affirm their effectiveness in reducing mental health symptoms, more research is needed in African contexts. Overall, mind–body practices present scalable, holistic strategies for promoting resilience and mental health in Ugandan communities.

14.6 Conclusion

Addressing Uganda's mental health crisis requires a transformative approach that transcends conventional biomedical models and embraces a more holistic, socio-ecological paradigm. As this article has argued, the integration of Pope Francis's (2015) vision of IHE with community-based, culturally resonant interventions presents a compelling framework for promoting psychological well-being in resource-limited contexts. Programs like the Bethany Land Institute illustrate the potential of land-based education, ecological restoration, and spiritual formation to rebuild dignity, social cohesion, and resilience among vulnerable populations. The established programming aligned with IHE models inherently promotes many aspects of healthy mental and emotional functioning. Suggestions for practical practices such as narrative practices and creative storytelling, empathy and listening training, and movement-based healing are considerations for targeted mental-health additions to current models. As Uganda continues to navigate complex challenges related to poverty, conflict, and environmental degradation, a shift toward integrated, community-empowering, and ecologically attuned interventions may offer a path to sustainable and inclusive mental health care.

References

Ashukem, J.N., and Ngang, C.C. (2022). 'Land grabbing and the implications for the right to development in Africa', *African Human Rights Law Journal*, 22(2), pp. 1–23. https://doi.org/10.17159/1996-2096/2022/v22n2a4

Asiimwe, R., Nuwagaba-K, R.D., and Dwanyen, R.K. (2023). 'Sociocultural considerations of mental health care and help-seeking in Uganda', *SSM-Mental Health,* 4(15), 100232.

Bethany Land Institute. (2025). *Our work* [Online]. At: https://www.bethanylandinstitute.org/our-work (30 April 2025)

Bjørlykhaug, K.I., Karlsson, B., Hesook, S.K., and Kleppe, L.C. (2022). 'Social support and recovery from mental health problems: a scoping review', *Nordic Social Work Research,* 12(5), pp. 666–697. https://doi.org/10.1080/2156857X.2020.1868553

Bratman, G.N., Hamilton, J.P., and Daily, G.C., (2012). 'The impacts of nature experience on human cognitive function and mental health.' *Annals of the New York Academy of Sciences,* 12, p. 118–136. https://doi.org/10.1111/j.1749-6632.2011.06400.x

Chioneso, N.A., Hunter, C.D., Gobin, R.L., McNeil Smith, S., Mendenhall, R., and Neville, H.A. (2020). 'Community healing and resistance through storytelling: a framework to address racial trauma in Africana communities', *Journal of Black Psychology,* 36(2-3), pp. 95–121. https://doi.org/10.1177/0095798420929468

Ddungu, G. (2023). "An Exploration of the Lived Experiences of Individuals Living With HIV/AIDS in Rural Uganda", *Walden Dissertations and Doctoral Studies*, 14079.
https://scholarworks.waldenu.edu/dissertations/14079

Grasser, L.R., and Marusak, H. (2023). 'Strong mind, strong body: The promise of mind–body interventions to address growing mental health needs among youth', *Mental Health Science,* 1(2), pp. 58–66. https://doi.org/10.1002/mhs2.16

Guendelman, S., Medeiros, S., and Rampes, H. (2017). 'Mindfulness and emotion regulation: Insights from neurobiological, psychological, and clinical studies', *Frontiers in Psychology*, 8, 220. https://doi.org/10.3389/fpsyg.2017.00220

Hintz, L.D. (2019). *Expressive therapies continuum: a framework for using art therapy.* New York, NY: Routledge.

Iverson, S.A., Nalugya, J., Babirye, J.N., Engebretsen, I.M.S., and Skokauskas, N. (2021). 'Child and adolescent mental health services in Uganda', *Journal of Mental Health Systems,* 15, 66. https://doi.org/10.1186/s13033-021-00491-x

Katongole, E. (2022). 'Mission as integral ecology: doing theology at Bethany', *Mission Studies, 39*, pp. 163–185. https://doi.org/10.1163/15733831-12341845

Kolappa, K., Henderson, D.C., and Kipshore, S.P. (2013). 'No physical health without mental health: lesson unlearned?', *Bull World Health Organization,* 91(1), pp. 3–3A. https://doi.org/10.2471/BLT.12.115063

Kuo, F.E. (2015). 'How might contact with nature promote human health? Promising mechanisms and a possible central pathway'. *Frontiers in Psychology,* 6, 1093. https://doi.org/10.3389/fpsyg.2015.01093

Laudato Si' Action Platform (2025). *Who we are* [Online]. At https://laudatosiactionplatform.org/laudato-si-goals/ (1 May 2025)

Liebling, H., Davidson, L., Akello, G.F., and Ochola, G. (2016). 'The experiences of survivors and trauma counselling service providers in northern Uganda: Implications for mental health policy and legislation', *International Journal of Law and Psychiatry, 49*(Part A), pp. 84–92. https://doi.org/10.1016/j.ijlp.2016.06.012

Murray, J.S., Ainslie, L., Alpough, M., Schramm, C., and Showalter, C. (2015). 'The scope of Mental Illness and Straus of psychiatric Care in Uganda', *Issues in Mental Health Nursing,* 36(11), pp. 877–883. https://doi.org/10.3109/01612840.2015.1049311

Nalukenge, W., Martin, F., Seeley, J., and Kinyanda, E. (2019). 'Knowledge and causal attributions of mental disorders in HIV-positive children and adolescents: results from rural and urban Uganda', *Psychology, Health & Medicine,* 24(1), pp. 21–26. https://doi.org/10.1080/13548506.2018.1467021

Opio, J.N., Munn, Z., and Aromataris, E. (2022). 'Prevalence of mental disorders in Uganda: a systematic review and meta-analysis. *Psychiatric Quarterly,* 93, pp. 199–226. https://doi.org/10.1007/s11126-021-09941-8

Pandey, A., Hale, D., Das, S., Goddings, A.-L., Blakemore, S.-J., and Viner, R.M. (2018). 'Effectiveness of universal self-regulation–based interventions in children and adolescents: a systematic review and meta-analysis', *JAMA Pediatrics,* 172, 566–575. https://doi.org/10.1001/jamapediatrics.2018.0232

Pope Francis. (2015). *Laudato Si': on care for our common home.* Huntington, IN: Our Sunday Visitor.

Quinn N, and Knifton L. (2013). 'Beliefs, stigma and discrimination associated with mental health problems in Uganda: Implications for theory and practice', *International Journal of Social Psychiatry.* 60(6), pp. 554–561. https://doi.org/10.1177/0020764013504559

Ricks, L., Kitchens, S., Goodrich, T., and Hancock, E. (2014). 'My story: the use of narrative therapy in individual and group counseling', *Journal of Creativity in Mental Health,* 9(1), pp. 99–110. https://doi.org/10.1080/15401383.2013.870947

Rothert, J., VanDerwerken, D., and White, E. (2020). 'Socioeconomic factors and happiness: evidence from self-reported mental health data', *Empirical Economics,* 58, pp. 3101–3123. https://doi.org/10.1007/s00181-019-01655-y

Shah, A., Wheeler, L., Sessions, K., Kuule, Y., Agaba, E., and Merry, S.P. (2017). 'Community perceptions of mental illness in rural Uganda: An analysis of existing challenges facing the Bwindi Mental Health Programme', *African Journal of Primary Health Care & Family Medicine,* 9(1), pp. e1–e9. https://doi.org/10.4102/phcfm.v9i1.1404

Shelton, D., and Bridges, C.W. (2021). 'A phenomenological exploration of adoptive parent experiences of support during the adoption process', *Adoption Quarterly,* 25(3), pp. 212–243. https://doi.org/10.1080/10926755.2021.1976336

Soares, L., and Moniz, S. (2023). 'Overconsumption and the effects on mental health and well-being: a review', *Current Research in Diabetes & Obesity Journal, 17*(2), pp. 1–5. https://doi.org/10.19080/CRDOJ.2023.17.555957

Ssebunnya, J., Kigozi, F., Lund, C., Kizza, D., and Okello, E. (2009). 'Stakeholder perceptions of mental health stigma and poverty in Uganda', *BMC International Health & Human Rights, 9*, pp. 1–9. https://doi.org/10.1186/1472-698X-9-5

Teding van Berkhout, E., and Malouff, J.M. (2016). 'The efficacy of empathy training: a meta-analysis of randomized controlled trials', *Journal of Counseling Psychology,* 63(1), pp. 32–41. https://doi.org/10.1037/cou0000093

Tustonja, M., Topić Stipić, D., Skoko, I., Čuljak, A., and Vegar, A. (2024). 'Active listening – a model of empathic communication in the helping professions', *Medicina Academica Integrativa,* 1(1), pp. 42–47. https://doi.org/10.47960/3029-3316.2024.1.1.42

Vaisvaser, S. (2021). 'The embodied-enactive-interactive brain: bridging neuroscience and creative arts therapies', *Frontiers in Psychology*, 12, 634079. https://doi.org/10.3389/fpsyg.2021.634079

Wissing, M.P., Wilson Fadiji, A., Schutte, L., Chigeza, S., Schutte, W.D., and Temane, Q.M. (2020). 'Motivations for relationships as sources of meaning: Ghanaian and South African experiences', *Frontiers in Psychology,* 11, 2019. https://doi.org/10.3389/fpsyg.2020.02019

World Bank Uganda (2016). *World Bank open data* [Online]. At: http://data.worldbank.org/country/uganda?view=chart (1 May 2025)

World Health Organization. (2022). *Barriers to mental health care in Africa* [Online]. At: https://www.afro.who.int/news/barriers-mental-health-care-africa (29 April 2025)

World Health Organization. (2025). *Mental health* [Online]. At: https://www.who.int/health-topics/mental-health#tab=tab_1 (27 March 2025)

Chapter 15
Nonviolence and Integral Human Ecology: A Kenyan Case Study

Sheila Kinsey and Benard Islanda

Abstract This study explores the application of Integrity for Nonviolence (IFN), a values-centered process, in Kenya and demonstrates its direct connection to Integral Human Ecology as articulated in Pope Francis's encyclical, *Laudato Si'*. The collaboration with the Justice, Peace, Development, and the Integrity of Creation (JPDIC) program addresses the "*cry of the poor*" and the "*cry of the earth*" by fostering care for the environment and educational opportunities for students and communities in partnership with the Getiongo Nyabigege Self-help Group. The project implemented aspects of integral ecology, importantly the belief that "*everything is interconnected*", by using IFN's core values (creating respect, practicing honesty, promoting justice, and nurturing wholesome relationships) to cultivate integrity at the personal and environmental levels. Workshops were successfully delivered to participants, providing a non-threatening space to confront violence and nurture nonviolent alternatives. The training emphasized that caring for the community is inseparable from caring for nature. Over the last two years, the program has helped 27,550 people: 24,510 students, 3,000 parents, and 40 educators/facilitators in 30 schools and 4 settlements. The program has plans to continue and expand. A key practical component was an organic farming project, where students applied the values of IFN to their work with the land as well as their interpersonal studies. This initiative showed that a "*profound interior conversion*", the journey of the integrity of the heart, which mirrors the external actions required for a healthy environment. By integrating the development of nonviolent behaviors with practical skills in land stewardship, the program affirmed that unifying and promoting the essential dignity of human life while protecting diverse and fragile ecosystems is a challenge, but also a source of

Sister Sheila Kinsey, Integrity for Non-Violence, skinsey.fcjm@gmail.com and Brother Benard Islanda, Justice, Peace, Development, and Integrity of Creation Commission, benardisandacmm@gmail.com.

S. Kinsey (✉)
Integrity for Non-Violence, Vatican City, Rome, Italy
e-mail: skinsey.fcjm@gmail.com

B. Islanda
Justice, Peace, Development, and Integrity of Creation Commission, Vatican City, Rome, Italy

R. Marcantonio et al. (eds.), *Integral Ecology in the Anthropocene: Africa and Beyond*,
The Anthropocene: Politik—Economics—Society—Science 49,
https://doi.org/10.1007/978-3-032-32125-1_15

hope. The success of this collaborative framework is providing a model for ongoing projects in sustainable, peace-building action rooted in life-affirming values.

Keywords Peacebuilding · Non-violence · Consultation · Gardening · Creation · Care

15.1 Introduction

Integrity for Nonviolence® (IFN) (Kinsey 2025), is a values-centered process dedicated to the resistance of violence and the promotion of nonviolence by providing strategies and interventions for change within ourselves, society, the church, and communities. This collaboration was with the Justice, Peace, Development, and the Integrity of Creation (JPDIC) program, an initiative of the Getiongo Nyabigege Self-help Group. The group was comprised of 30 families who, in the year 2019, came together in order to improve livelihoods and to care for the environment in response to Pope Francis's 2015 encyclical *Laudato Si'* and its call to listen to the "*cry of the poor*" and the "*cry of the earth*" by addressing endemic issues of poverty, unemployment, economic instability, insufficient educational opportunities, and inadequate healthcare, including the historical and ongoing effects of the HIV/AIDS pandemic (Odeny 2013, p. 1–10).

One focus area is the creation of an environment of trust whereby the students and community members affected by violence were able to deepen their understanding of the causes and effects of violence and how to embrace the values of nonviolence. This reflection elaborates on the Kenyan experience, highlighting the influence of IFN® and the development of a viable program to help Kenyan students and communities, all explored in light of the Gospel mission. In *Laudato Si'*, Pope Francis reminded us that "*the teachings of the Gospel have direct consequences for our way of thinking, feeling and living,*" and "*how such a spirituality can motivate us to a more passionate concern for the protection of our world*" (Francis 2015, §215, §216).

The collaborative spirit unites efforts for shared responsibility and collective actions of nonviolence in our care for the world, fostering positive change through the interaction of personal and environmental integrity. The team spirit has encouraged involvement by persons who are interested in supporting the program. Finally, we celebrate how this collaborative framework creates future possibilities.

15.1.1 Integrity of the Heart

Integrity for Nonviolence® is a model for change. Pope Francis said that "*In our spiritual tradition, desire is seen as evidence of our innate yearning for God and the peace that he alone can give, as a kind of compass directing our lives to their ultimate goal*". (Francis 2022) In *Laudato Si'* he warned us not to lose our compass so that we keep in sight *"the great motivations, which make it possible for us to live in harmony, to make sacrifices and to treat others well."* (Francis 2015e, f, g, h, i, j, k, l, m, n, o, p, q, a, b, c, d, §200) Integrity is at the core of our being, uniting who we are with our actions, unifying what we say and do, and harmonizing all significant areas. Through integrity, we become who we are meant to be. It is the place where we are alone with God, and where our reflections encourage us to integrate our contemplative and active lives. The Catholic Social Principles and the significance of the Social Justice Methodology are explored in light of the Gospel mission for planned change.

15.1.2 Uniting Values of Nonviolence in Action

The integral values of IFN® are *creating respect, practicing honesty, promoting justice,* and *nurturing wholesome relationships.* These values suggest behaviors that actualize them and are interrelated and developmental in their connectedness between ourselves, others, and our places of work. These corresponding behaviors can support one another to maintain the spirit of nonviolence. Nonviolence has a listening presence that is open and receptive to the challenge of the moment. By sharing experiences, we further develop a common understanding of our values, providing a basis for right action. Nonviolence is a way of life, and the first action is to do no harm (Fig. 15.1).

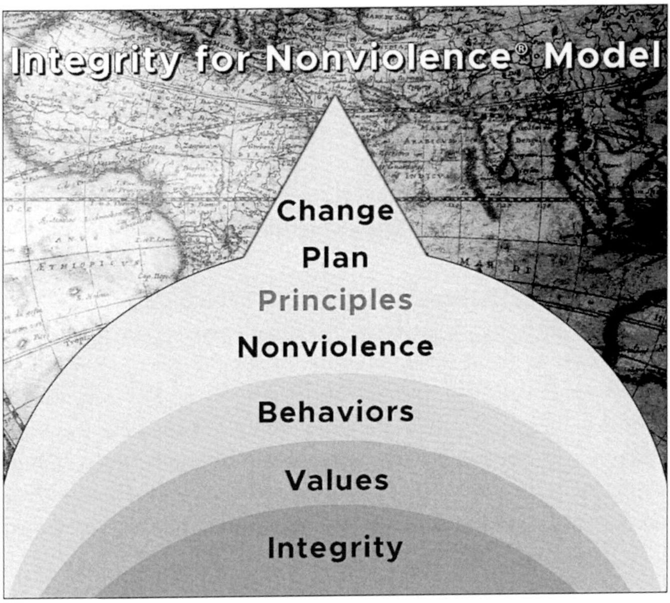

Fig. 15.1 The model utilized by Integrity for Non-Violence. *Source* Author created

15.1.3 Ecological Experiences and Integrity for Nonviolence

The experience of Integrity for Nonviolence® recently in Kenya explored the human and environmental implications of integral ecology. Pope Francis defined integral ecology as *"... everything is interconnected, and that genuine care for our own lives and our relationships with nature is inseparable from fraternity, justice and faithfulness to others."* (Francis 2015e, f, g, h, i, j, k, l, m, n, o, p, q, a, b, c, d, §70) The program looked at ways that actions for personal and ecological growth may be cultivated, such as the core values of IFN®, by considering both the human and the environmental dimensions of any issue. The heart of nonviolence seeks to be attentive to evaluating situations where there may be the active presence of violence. When we become conscious of what we say and how we live, we can foster confidence in knowing what is ours to do.

15.2 Methods

The idea for this collaboration was conceived by Sr. Sheila Kinsey and Br. Benard Isanda to develop workshops using her program, Integrity for Nonviolence®, and was implemented in 2024 in Kenya. To carry out this program the JPDIC program selected and trained 20 facilitators for one week in the values of IFN®. In turn, they would give workshops on those values to 5,510 students in 10 schools and 3,000 parents from 4 settlements. This year, 2025, 20 additional facilitators were trained in the IFN® program with the same values. Together, these 40 facilitators reached out in workshops to the 20 selected primary and secondary schools with an attendance of 20,000 students (19,000 of the students completed the program).

15.2.1 Integrity for Nonviolence® and Integral Ecology

Pope Francis spoke of the challenge of integration of both the personal and environmental concerns: *"...it is no longer enough to speak only of the integrity of ecosystems. We have to dare to speak of the integrity of human life, of the need to promote and unify all the great values."* (Francis 2015e, f, g, h, i, j, k, l, m, n, o, p, q, a, b, c, d, §224) These values are considered here in reference to integral ecology, taking into consideration both people and the environment in all decisions. The program's students considered the environmental, social, economic, and cultural conditions. They were challenged to look inside themselves to see their interior connections to doing what they know is necessary. In *Laudato Si'* Pope Francis wrote that *"The external deserts in the world are growing, because the internal deserts have become so vast. For this reason, the ecological crisis is also a summons to profound interior conversion"*. (Francis 2015e, f, g, h, i, j, k, l, m, n, o, p, q, a, b, c, d, §217) This is a journey of the integrity of the heart, uniting the values of nonviolence in actions that mirror a conversion within with what happens externally in the outside world (Fig. 15.2).

The compass is IFN's metaphor for guidance. The four values are *creating respect, practicing honesty, promoting justice,* and *nurturing wholesome relationships.*

Creating respect recognizes the dignity of each person and sustains a non-threatening place. The students contributed to the welcoming and receptive atmosphere. The issues addressed included all the manifestations of violence at the personal, social and environmental levels such as physical, verbal, spiritual, bullying, gender-based violence, illiteracy, poverty, unemployment, care of the environment, and climate change among others. The intent was to raise awareness of potential areas of violence so the participants could engage with a healthier perspective.

How does this experience enable nurturing specific responses to the environment? The destructive conditions of the environment are evidence of possible climate

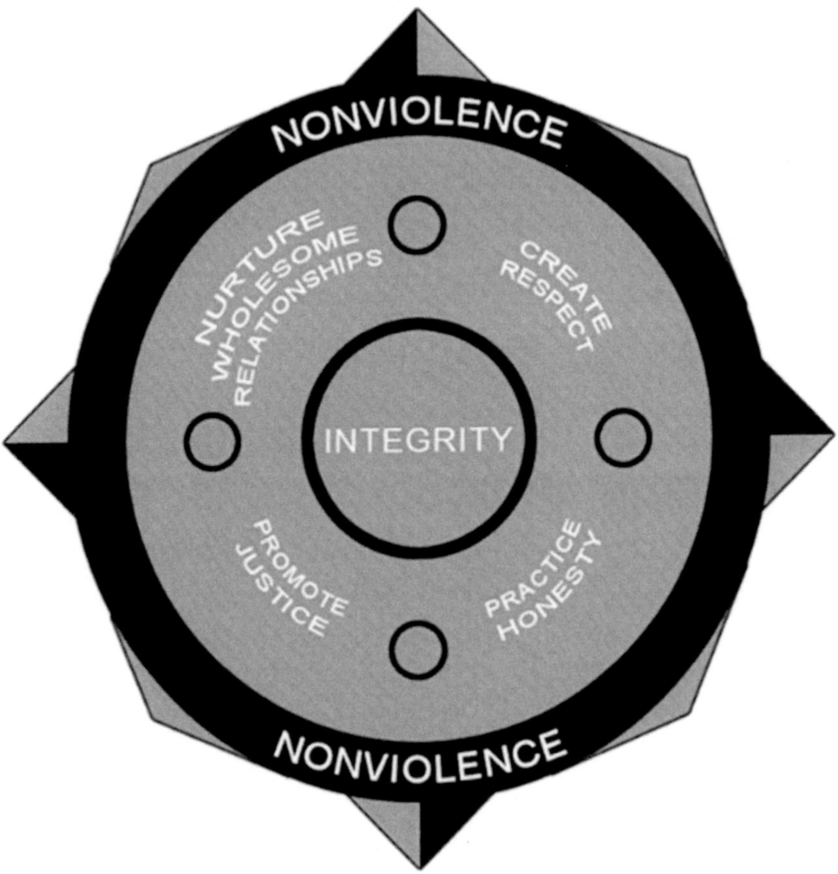

Fig. 15.2 The Non-Violence Compass. *Source* Author created

change such as changes in the growing seasons, floods, droughts, and erosion of soil. (Nyandiko 2020) Valuing the dignity of the person and having organized actions, the program poses the question : What care does the earth require of us? This included an understanding of the needs of particular plants and finding ways to restore the devastated or uncared for land. (Kuria et al., 2024) As Pope Francis wrote, *"Living our vocation to be protectors of God's handiwork is essential"* (Francis, 2015e, f, g, h, i, j, k, l, m, n, o, p, q, a, b, c, d, §217) The dignity of the land includes an environmental perspective that considers the importance of protecting biodiversity and diverse ecosystems. The students worked in cultivating land where they planted vegetables and applied organic practices such as the use of compost, poultry and cattle manure as well as mulching. In this way, respect for the land was cultivated by developing a healthy garden that integrated with the local ecosystem.

The value of **practicing honesty** entails open and truthful communication. *"We are able to take an honest look at ourselves, to acknowledge our deep dissatisfaction, and to embark on new paths to authentic freedom."* (Francis 2015e, f, g, h, i, j, k, l, m, n, o, p, q, a, b, c, d, §205) This call for honesty includes acknowledging the dissatisfaction of behaviors such as cheating, gossiping, stealing, distorting the facts, and hypocrisy. The freedom to discover the true self is based on honesty and strengthens good character development in becoming who we are called to be by God. By attending to the importance of accurate information, students were attentive to learning the appropriate skills to help themselves grow academically and spiritually.

To include consideration of the environment, this would necessitate open and truthful communication so that credible and accurate information is always available. *Laudato Si'* supports *"the honesty needed to question certain models of development, production and consumption"*. (Francis 2015e, f, g, h, i, j, k, l, m, n, o, p, q, a, b, c, d, §138) When making their plans, the students took into consideration their needs and those of the land. In this process, they had the freedom to discover their true selves and to choose what they valued doing. This Gospel mission entails a deeper sense of why we care for the land.

Promoting justice is the value which focuses on accountability and responsibility for all that has been entrusted to us by God that is ours to do. It is important to cultivate the knowledge that there is the possibility that everyone can have the essentials for a good life if everyone agrees and works towards harmony, righteousness, equity and fairness. In the group discussions among the students, fairness was always at the front of the discussion and equality between genders, classes, ages, and all other groups.

Regarding the environment, accountability and responsibility extend to the integral caring for the land, there needs to be clarity about the contribution of work that goes into the use of the land, but also in how the harvest is to be dispensed. Sharing the harvest with family and those in need in the community and potentially making a profit are the goals of the organic farming project. Since the land for the farming project was donated, negotiations and fairness are significant elements to consider. It is important that agreements are in place to adequately address potential conflicts over the use of shared resources such as water, tools and access to land. Basing this requirement on justice creates a healthy sense of shared responsibility among the students.

The value of **nurturing wholesome relationships** includes demonstrating trust and support and having the necessary personal and financial resources. The project was developed with the hope of offering a path to food and economic sustainability for the community. The students were eager to participate and demonstrated that they were trustworthy and supportive of the wellbeing of the land and interested in holistic gardening methods. The students were encouraged to count on each other and to contribute to the community. As the students worked with the land, they developed their skills with gardening as well as social interaction, resulting in four healthy gardens and vital social bonds *"to promote a sustainable and equitable development within the context of a broader concept of quality of life."* (Francis 2015e, f, g, h, i, j, k, l, m, n, o, p, q, a, b, c, d, §192)

15.2.2 The Process of the Integrity for Nonviolence® Model

The IFN® process is centered on the integrity of the person. It helps students build character and commit to a way of life based on the values of nonviolence. The values and ideas of IFN® provide a model for the students to discover their gifts, potential for success, and a foundation for social contributions rooted in justice and equitability. Personal behaviors were discussed at the individual level, and the areas of violence were carefully described through stories and pictures. Several categories and descriptions of abuse were explored and discussed. These included all forms of abuse: physical, verbal, psychological and spiritual. Financial and online abuse as well as gender-based violence were explored. (CCGD et al., 2024) Following this discussion, the program emphasized the components of character, defining personal integrity through a consistent commitment to nonviolent behavior as a way of life. Building on this personal development, the program transitioned to environmental activities. The students were given instruction and training in gardening and caring for plants, cultivating the land, planting crops, using fertilizer correctly, and how to irrigate the land to nourish their crops. (Merti Integrated Development Programme [MID-P], 2025) Local plants were chosen for viability and yields, such as greens like kale, amaranth, sweet potatoes, okra, and squash. (Wamalwa, et al. 2023) Additionally, their work included planting 3,000 trees and caring for the chickens they were provided. They were able to put the values of the program into practical operation while caring for the garden. The skill-building instructions helped the students develop confidence and recognize the importance of further training. By working with the land with the knowledge and tools they needed, they experienced a sense of well-being that came from a productive and caring relationship with their environment. A feeling of stewardship pervaded the project, not only from the value of the work and potential rewards, but also how it enhanced the resourcefulness and interconnectedness with the larger community.

15.3 Catholic Social Principles

These ideas and actions are guided by the Catholic Social principles. The principles of social justice are also characterized in the Integrity for Nonviolence® model. Pope Francis underlines the foundational significance of the Common Good as *"the central and unifying principle of social ethics"* (Francis 2015e, f, g, h, i, j, k, l, m, n, o, p, q, a, b, c, d, §156) in relation to the other principles. In a special way the relational connections of the dignity of the person are emphasized: solidarity with one another, subsidiarity where students were encouraged to make decisions, common good for all, and the careful stewardship of creation were woven into the experience. Pope Francis has a *"particular concern for distributive justice; whenever this is violated, violence always ensues."* (Francis 2015e, f, g, h, i, j, k, l, m, n, o, p, q, a, b, c, d, §157) From this Catholic perspective, we gain wisdom to clearly and objectively assess reality in order to discern what is right to do.

15.4 Plan: Significance of the Social Justice Methodology of See, Judge, and Act

These ideas and activities are guided by Catholic Social Teaching (CST) principles, which puts the values into practice, primarily the common good as it applies to our context of integral ecology. (Christie et. al. 2019, p. 896-906). To articulate a plan, the influence of social justice is outlined in the methodology of See, Judge, and Act. Seeing assesses the situation; Judging comes through the meaning of our faith along with evidence-based observations; and Acting involves determining specific actions.

Within the human family there is the strength to address the ecological crisis as a community. In *Laudato Si'*, Pope Francis reminded us that "the ethical and spiritual roots of environmental problems, which require that we look for solutions not only in technology but in a change of humanity; otherwise, we would be dealing merely with symptoms." (Francis 2015e, f, g, h, i, j, k, l, m, n, o, p, q, a, b, c, d, §9).

15.4.1 See

Pope Leo XIV, spoke in Verona that *"History and practical experience have taught us that authentic peace takes shape from the ground up"* (Leo XIV, 2025), which supports the importance of the organic and authentic human spirit's potential that can grow into a powerful movement. The students in the Kenyan project were from disadvantaged backgrounds and had endured significant poverty. There were few opportunities for employment and little in the way of resources to cultivate the land for the betterment of their social condition. The opportunity to become involved in IFN® created a safe place to confront violence in their society and to become aware of unhealthy characteristics in themselves. They also were taught important skills in how to care for the land and animals. This experience gave hope to the students and their families that more was possible. Learning to deal with issues of violence placed them in the realm of their consciences, where they are alone with God and where their integrity dwells. The skill of careful listening, along with the teaching and sharing with the facilitators and other participants in the program encouraged the desire for a commitment to the common good of one another and all of creation.

15.4.2 Judge

Pope Leo XIV remembered Pope St. John Paul II's words that *"peace is an indivisible good; it is either everyone's or no one's"* (John Paul II 1987), which acknowledges that all humanity needs to be included in the movement to justice and peace. The remedies to violence were discussed in the workshop in the context of nonviolent values and behaviors, recognizing stages of growth and personal experiences. The facilitators took to heart these words from St. Paul to the Romans 10: 14-15, "How then shall they call on Him in whom they have not believed? And how shall they believe in Him of whom they have not heard? And how shall they hear without a preacher? And how shall they preach unless they are sent? As it is written: How beautiful are the feet of those who preach the gospel of peace, who bring glad tidings of good things". The students and community members are only able to apply the principles of nonviolence insofar as they have knowledge. The prayer of St. Francis of Assisi was often sung, calling forth its themes of peace, love, and humility. The prayer shows that the culture of nonviolence joyfully embraces all life that is threatened by violence.

15.4.3 Act

The process entails concrete actions starting from where they are and uses interactive dialogue sessions and songs based on a culture of nonviolence that builds peace. They discussed ways to act in a healthier manner by attending to values and corresponding behaviors of IFN® as a way of life. "If you want peace, prepare institutions of peace" said Pope Leo XIV and "peace is possible when disagreements and the conflicts they entail are not set aside, but acknowledged, understood and surmounted." (Leo XIV 2025). He asks that realistic projects are developed and acted upon with the cooperation and support of all involved, for a sustainable and more just conclusion. The program embraced education and promoted teaching skills for the care of the land. These are the seeds of hope that are encouraged to take root in the spirit of the community. From the perspective of Pope Francis, it can be realized that: "It is good for humanity and the world at large when we believers better recognize the ecological commitments which stem from our convictions." (Francis 2015e, f, g, h, i, j, k, l, m, n, o, p, q, a, b, c, d, §64) Due to peer pressure, these youths are able to influence each other, and a greater societal impact can be created using these transformed mindsets. Society is influenced by ideas, and with good ideas embedded in the values of IFN®, the fruits are peace, inclusivity, dignity, accountability, respect, sufficient resources for sharing, informed decision-making, and personal and social development. This training is meant for metanoia—a profound transformation in beliefs, perspective, perceptions, and spirit. The welfare of the planet with its diverse ecosystems and human societies will benefit if all people were to make ecological commitments based on the spiritual conviction that everything is interconnected.

15.5 Change: Process for the Experience

The movement for conscious change creates energy when preparatory work is translated into a developed program. The transition process is critical for promoting sustainable change through advocacy, addressing the important key components. This comprehensive approach ensures that the project's values are assimilated into a sustainable way of being, integral to a healthier way of life for all participants.

15.5.1 Change with Program Development

Creating the transformation necessary for a program entails having a vision, necessary skills, and incentives to bring meaningful actions, and a prepared plan for the specific change to occur.

15.5.1.1 Vision

Integrate the care of the Earth and care for the person. The community garden project was implemented in different phases and has now been operating for two years. There were thirty schools and 4 settlements whose students and community members were involved during the 5-day training sessions of 2024-2025. The trained facilitators included married laity, religious priests, brothers, sisters, and diocesan priests who offered spiritual and development input. The program trained 20 school principals as facilitators before the students. Additional expertise was sought from external sources. The sessions were led with a good collaborative spirit which arose from the personal strengths of individuals happily participating for the common good.

15.5.1.2 Skills

Although issues were addressed before IFN®, the program was conducted with a more integrated approach on a wider scale and for a longer period of time. The holistic development of the participants helped increase their knowledge of the importance of behavior change and responsible stewardship of the planet for sustainability and equitability. The mission of the Catholic Church is to promote the Gospel of Jesus as a way of life. During contemporary times, the Church's social doctrine has helped us to make contemporary applications. The students and other participants were given information and training on the facts and causes of violence and on the different ways nonviolence can be integrated into their lives. They learned how to cultivate the land, to use organic fertilization methods, water-saving irrigation methods, and how to use natural pesticides. The group discussions were empowering and reflective in establishing a training environment that encouraged personal insight into the experiences of nonviolence. This process helped build empathy and respect with the participants and increased their self-determination. A central concept of IFN® is that the program's training is based on the principle that violence destroys, while nonviolence and the power of love enhance sustainability of the world for all life.

15.5.1.3 Incentives

The IFN® Program raises awareness that a choice can be made, and it offers training in the skills needed to choose nonviolence over violence. As Pope Francis reminds us in *Laudato Si'* that "shared identity, with a story which can be remembered and handed on. In this way, the world, and the quality of life of the poorest, are cared for, with a sense of solidarity which is at the same time aware that we live in a common home which God has entrusted to us." (Francis 2015e, f, g, h, i, j, k, l, m, n, o, p, q, a, b, c, d, §232)

15.5.1.4 Resources

The facilitators and the students deeply appreciated the IFN® program and training. They found it to be empowering, enlightening, transformative, and timely. Due to the success of this ongoing program, more land is to be donated for the productive use of the students. This commitment from donors will enhance further skill development for the students, thus helping them and their families and communities in caring for the environment and for themselves as well. By cultivating growth and life inside of themselves, they cultivate growth outside also, as a source of creating an interconnection of the sacredness of life and spirit

15.5.1.5 Plan

The IFN® program in Kenya is moving forward with future projects. There have been many calls asking for it to be implemented in other schools, which is a very welcome but daunting challenge. These requests signal that the success and transformational impact of the initial programs are demonstrating that the values and methodology of nonviolence are resonating with communities seeking positive change. The challenges include securing the resources to sustain an ongoing coordinated effort. The process would include sustaining the message of Integral Ecology to integrate the care of people and communities with the care of the Earth, finding and training more facilitators for new project sites, commitments from donors, in addition to land which is already being donated, and continuing to address the ongoing issues of food insecurity, unemployment, and inadequate healthcare options. The requests for expansion of the program will need resources to support the nonviolence model as it grows.

15.5.1.6 Advocacy Efforts

When the change is being promoted through advocacy efforts, there are vitally important areas that the transitions process should address. These include naming conflict, dealing with rationalizations, clarifying the reality, sorting through proposed outcomes, testing and then implementing the studied options, sharing the meaning of the work with others, and assimilating the advocacy into a sustainable way of being, which is integral to a healthier way of life. There is much energy around these conflicts that needs to be addressed. Besides the sustainable use of the land, they want to work on the critical issues of food insecurity, unemployment, and insufficient income. This would entail the work of advocacy in another phase of the program. The current initiatives from the facilitators are encouraging ongoing nonviolence workshops and reaching out to new people.

15.6 Results

At the end of the 5-day workshop in 2024, 5,510 students, both boys and girls, 3,000 parents, men and women, and 20 facilitators (8,530 people in total) across 10 schools and 4 settlements, received training in the values of IFN®. In the second program of 2025, 20 schools participated, and 19,000 students and 20 additional facilitators were trained (19,020 people in total). To date, the two-year program has helped 27,550 people.

The creation of an environment of trust through active listening, honesty and open relationships during the workshops by the facilitators empowered the young people and their parents and the community to share first-hand experiences about their broken relationships in need of healing. They spoke of violence and poverty, depression and loneliness, unemployment and desperation, divorce, and sometimes the loss of life. The program workshops brought good news and hope to people suffering from poverty and sickness, and to the environment burdened with abuse and neglect. The commitment of the facilitators who imparted the values of IFN® to the participants through their dedicated time, spirit, listening, and encouragement were acts of generosity that awakened a will in the participants for a different way of thinking, a methodology and approach that is liberating, enabling, energizing, empowering, and most importantly, full of hope.

15.7 Conclusion

The success of the collaboration has grown, with 27,550 people helped: students, parents, and educators. The program will continue to integrate the practice of nonviolence by offering training in interpersonal skills and organic farming, but plans are in place to offer classes in viable employment skills along with a certificate of completion for the students, in order to help them find and secure jobs. The project itself is also creating job opportunities and is helping to stabilize economic uncertainty in the area.

Where does integrity fall relative to values and methodology? In *Laudato Si'* Pope Francis spoke about his concern for integrity when he wrote *"…it is no longer enough to speak only of the integrity of ecosystems. We have to dare to speak of the integrity of human life, of the need to promote and unify all the great values."* (Francis 2015e, f, g, h, i, j, k, l, m, n, o, p, q, a, b, c, d, §224) This is the challenge: to interconnect the work of fostering the integrity of ecosystems with the personal integrity of our lives. This is possible through the conscious practice of behaviors that actualize the values of creating respect, practicing honesty, promoting justice and nurturing wholesome relationships. *"An integral ecology is also made up of simple daily gestures which break with the logic of violence, exploitation and selfishness."* (Francis 2015e, f, g, h, i, j, k, l, m, n, o, p, q, a, b, c, d, §230) Consequently, it is the integrity of the heart that is able to unite, unify and harmonize the values of nonviolence in action. We need to be concerned and realize the connectedness of these actions in creation. As Pope Francis reminds us that in our Judeo-Christian tradition, creation *"has to do with God's loving plan in which every creature has its own value and significance."* (Francis 2015e, f, g, h, i, j, k, l, m, n, o, p, q, a, b, c, d, §76) He realized that *"Inner peace is closely related to care for ecology and for the common good because, lived out authentically, it is reflected in a balanced lifestyle together with a capacity for wonder which takes us to a deeper understanding of life."* (Francis 2015e, f, g, h, i, j, k, l, m, n, o, p, q, a, b, c, d, §225) In summary, the program was blessed with the conception of a collaboration between the ideas and values of Integrity for Nonviolence®, the mobilization of human and financial resources, the involvement of the beneficiaries and their overwhelming participation, and the commitment of the facilitators and their concern for the way forward on the issues raised. This has brought about opportunities for a new understanding between the transformed self and the environment with the implementation of the values. This process has been a celebration of the framework of collaboration. The passion of the majority to have more of the same raises more hope that these ideas, like a much-loved garden, will continue to grow!

References

Christie, I. P., Gunton, R., & Hejnowicz, A. (2019). "Sustainability and the common good: Catholic Social Teaching and 'Integral Ecology' as contributions to a framework of social values for sustainability transitions." *Sustainable Development*, *27*(5), 896–906.

Collaborative Center for Gender and Development (CCGD), University of Nairobi Women's Economic Empowerment (WEE) Hub and United Nations Population Fund (UNFPA) (2024) *Technology Facilitated Gender-Based Violence (TF GBV) in Tertiary Institutions in Kenya.* Nairobi, Kenya: UNFPA Kenya. At: https://kenya.unfpa.org/sites/default/files/pub-pdf/tf_gbv_report_web_1.pdf.

Francis, P., 2015, §200.

Francis, P., 2015, §70.

Francis, P., 2015, §224.

Francis, P., 2015, §217.

Francis, P., 2015, §205.

Francis, P., 2015, §138.

Francis, P., 2015, §192.

Francis, P., 2015, §156.

Francis, P., 2015, §157.

Francis, P., 2015, §9.

Francis, P., 2015, §64.

Francis, P., 2015, §232.

Francis, P., 2015, §230.

Francis, P., 2015, §76.

Francis, P., 2015, §225.

Francis, P. (2015). *Laudato Si': On Care for our Common Home.* [online] The Holy See. §215, §216. At: https://www.vatican.va/content/francesco/en/encyclicals/documents/papa-francesco_20150524_enciclica-laudato-si.html.

Francis, P., (2022). *General Audience of 12 October 2022 - Catechesis on Discernment: 5. The elements of discernment. The desire.* [online] At: https://www.vatican.va/content/francesco/en/audiences/2022/documents/20221012-udienza-generale.html [13 Oct. 2025].

John Paul II, P. (1987). *Sollicitudo Rei Socialis.* [online] www.vatican.va. Available at: https://www.vatican.va/content/john-paul-ii/en/encyclicals/documents/hf_jp-ii_enc_30121987_sollicitudo-rei-socialis.html.

Kinsey, S. (2025). *Integrity for Nonviolence Home Page.* [online] At: https://integrityfornonviolence.org/ [15 Oct. 2025].

Kuria, A., Bolo, P., Ntinyari, W., Orero, L., Adoyo, B., Korir, H., Syano, N., Ngethe, E. and Nduru, S. (2024) 'Understanding farmer options, context and preferences leads to the co-design of locally relevant agroecological practices for soil, water and integrated pest management: a case from Kiambu and Makueni agroecology living landscapes, Kenya', Frontiers in Sustainable Food Systems, 8(1456620 At: https://www.frontiersin.org/journals/sustainable-food-systems/articles/https://doi.org/10.3389/fsufs.2024.1456620/full.

Leo XIV, P. (2025). *To the Members of the Popular Movements for Peace (30 May 2025).* At: https://www.vatican.va/content/leo-xiv/en/speeches/2025/may/documents/20250530-movimenti-pace.html [13 Oct. 2025].

Leo XIV, P., 2025.
Nyandiko, N.O. (2020). Devolution and disaster risk reduction in Kenya: Progress, challenges and opportunities. *International Journal of Disaster Risk Reduction*, 51, p.101832. https://doi.org/10.1016/j.ijdrr.2020.101832.
Odeny, T.A., Penner, J., Kulzer, J.L., Leslie, H.H., Shade, S.B., Adero, W., Kioko, J., Cohen, C.R. and Bukusi, E.A. (2013). Integration of HIV Care with Primary Health Care Services: Effect on Patient Satisfaction and Stigma in Rural Kenya. *Aids Research and Treatment*, 2013, pp.1–10. https://doi.org/10.1155/2013/485715.
Merti Integrated Development Programme (MID-P) (2025) *Promoting School Gardening for Sustainable Food Production in Isiolo County*. Isiolo, Kenya: MID-P. At: https://www.midpkenya.org/promoting-school-gardening-for-sustainable-food-production-in-isiolo-county/.
Wamalwa, E., Okeyo, A.A. & Kinyanjui, H. (2023). Feasibility of commercial production of amaranth leaf vegetable by small- scale farmers in Kenya. *CABI Digital Library*. At: https://www.cabidigitallibrary.org/doi/pdf/10.5555/20133232453 (10 October 2025).

Chapter 16
Living Together, Growing Together: Just Agroecological and Energy Transitions in Uganda's Farmlands

Vincent Ferrer Bukenya, Joseph Ssekandi, and Jimmy Spire Ssentongo

Abstract Renewable energy and agroecology live in contested spaces. Deep agroecologists argue for 'going slow on energy.' Agroecology is then the timely challenge to, among others, unsustainable agricultural models that prioritize yield over sustainable farmlands. Uganda is instead way below acceptable energy thresholds for household and productive use of energy in farmlands. A key question is how to balance energy needs with sustainable practices. This study places at the centre of energy justice households of refugee and host communities at the margins in the populous Bidibidi, northwestern Uganda. It unpacks realties and lived experiences of energy deprivation, with respect to energy access, source protection, and alternatives. Using a qualitative case study, data was gathered through a household survey, focus group discussions, key informant interviews, and spatial analysis. Findings reveal four overarching aspects: Shrinking sources of wood; dirty access to dirty fuel; low appetite for energy source protection, and; dicey access to alternative energy sources. Renewable energy calls for integrating energy in agroecological pursuits to optimize farming systems. This is echoed in Laudato Si given that agroecology is a foundational operationalization of integral ecology. It is a call for alternative worldviews or rethinking processes that underlie present-day policy and practice.

Keywords Agroecology · Energy · Cookstoves · Migrants · Farming

Vincent Ferrer Bukenya, Uganda Martyrs University, bukenya.vincent@stud.umu.ac.ug, , Joseph Ssekandi, Uganda Martyrs University, sekjoseph@gmail.com, and Jimmy Spire Ssentongo, Uganda Martyrs University, jsssentongo@gmail.com.

V. F. Bukenya (✉) · J. Ssekandi · J. S. Ssentongo
Uganda Martyrs University, Nkozi, Uganda
e-mail: bukenya.vincent@stud.umu.ac.ug

R. Marcantonio et al. (eds.), *Integral Ecology in the Anthropocene: Africa and Beyond*, The Anthropocene: Politik—Economics—Society—Science 49,
https://doi.org/10.1007/978-3-032-32125-1_16

16.1 Introduction

Renewable energy and agroecology live in contested spaces (Adriaensens et al. 2025; IDS/IPES-Food 2022), shaping how these paradigms are understood, framed, financed, and implemented. On the one hand, deep agroecologists argue for 'going slow on energy' (Wezel et al. 2020; Altieri/Nicholls 2012). 'Slow on energy' entails a deliberate shift away from high-energy, industrialized farming systems toward low-input, ecologically attuned practices (Caporali 2015; Altieri/Nicholls 2012). Energy is thus argued against as a disconnector from the layered agenda on low-external-input systems to be ecologically attuned, localized incremental transitions that run bottom-up and socially embedded, on-farm circularity, among others. Agroecology is then established as the timely challenge to, among others, unsustainable agricultural models (Shiva 2023) that prioritize yield over sustainable farmlands, downplay local knowledge and agroecological practices, and complicate land degradation and pollution.

On the other hand, the renewable energy agenda demonstrates that clean energy is the quiet backbone between ecological harmony and social justice (Dash 2024; Hercher-Pasteur et al. 2020). Clean energy can be an enabler and a disconnector of productive agricultural systems and human life. It is instead the type or source of energy, its sourcing, and use that need to be checked on applicable scales. Agroecology instead looks at the checks on conventional agriculture. A critical question is how to align human energy consumption with responsible, sustainable practice.

This is a call echoed in Laudato Si on alternative worldviews or rethinking processes that underlie present-day policy and practice (Gocko 2024; Pasquale 2019). For example, integral ecology demands that the burden of proof (ecological debt) falls on those who would act against the common good (LS § 51). Integral ecology, as articulated in *Laudato Si* (Ch. 4), is the idea that how we treat the environment is deeply connected to how we treat each other. We cannot fix nature without fixing society—and vice versa. This is summed up in paragraph 49, saying: The cry of the earth and the cry of the poor are one and the same (LS § 49). [Energy] poverty, inequality, consumerism, etc., are one interwoven crisis that demands a shared solution.

Integral ecology is, as such, a framework that embodies ecological alignment, social justice, cultural preservation, and economic resilience. Whelan (2024) called for that alignment, namely: a shift towards an agricultural model that respects and works with ecological principles and processes, aligning closely with the approach of agroecology. Maier and Monti (2023) underlined how Laudato Si's vision resonates with agroecology's plural rationality, interdisciplinarity, and people engagement in the agri-food chain(s).

The study thus draws directly from Laudato Si and its bold, integrative reframing of how all people of goodwill think about ecology, economics, and ethics in the 21st century (LS §217). It finds root within Laudato Si given that agroecology is a foundational operationalization of integral ecology. Far from a farming method, agroecology is a systems approach that mirrors the very architecture of integral ecology.

The principle of equity needs to be taken into consideration in all environmental decisions" (Francis 2015; §51 & §186).

Uganda, for example, is instead way below acceptable energy thresholds for household and productive use of energy in farmlands (NDF 2025; Munro/Bartlett 2019). Energy access remains a critical justice issue in low- and middle-income countries—such as in Uganda where 90% use traditional fuels for basic survival (Elasu et al. 2025; Gianvenuti et al. 2022). 'Slowing down' energy inputs, thus, risks dangerously conflicting with livelihood targets. Energy sufficiency is instead called for, not a universal red-card. That's why terms like 'transformative,' 'context-sensitive,' or 'adaptive' modalities (Lage et al. 2023) find fit.

Integration of renewable energy into agroecological pursuits is a strategic use of energy towards optimizing farming systems (Osabohien et al. 2025). This study, consequently, places at the centre of energy justice the households of refugee and host communities at the margins in Uganda.

16.1.1 Methods

The topic and results explored here are part of a broader study on the question of household access to energy for cooking and energy deprivation among refugee and host communities at the margins in Uganda. The study employed a mixed-methods approach to an explanatory case study design, particularly the *mixed-methods sequential explanatory design.* Under this two-phase research approach, the study begins with the collection and analysis of quantitative data, followed by the collection and analysis of qualitative data to help to deepen insights from statistical findings or trends (Creswell/Clark 2017). Following Creswell's mixed methods approach (Creswell/Creswell 2005), the study conducted in-depth data collection through individual interviews, focus group discussions, and a household survey.

Spatial analysis was conducted on how much runway left towards catastrophic depletion of the above ground biomass (Muhe/Argaw 2022) for the refugee settlements in Uganda, given the protracted nature of the use of wood energy for cooking—examined using variation in the Sentinel -2A optical imaging.

Taking an in-depth case-study of Africa's most populous single location refugee settlement called Bidibidi in northwestern Uganda, the study unpacks the realities and lived experiences of energy deprivation, with respect to energy access, source protection, and alternatives. Data was gathered through a household survey, focus group discussions, key informant interviews, and spatial analysis. Figure 16.1 shows the spatial distribution and selection of the study area.

A total of 403 households (N=403) were surveyed using questionnaires. Sixteen (N=16) focus group discussions (semi-structured group interviews) were conducted. Fourteen (N=14) key informant interviews were adopted as community interviews. These covered a) refugee and host community structures plus informal authorities and b) sector authorities (government and non-governmental actors, such at the Ministry of Agriculture).

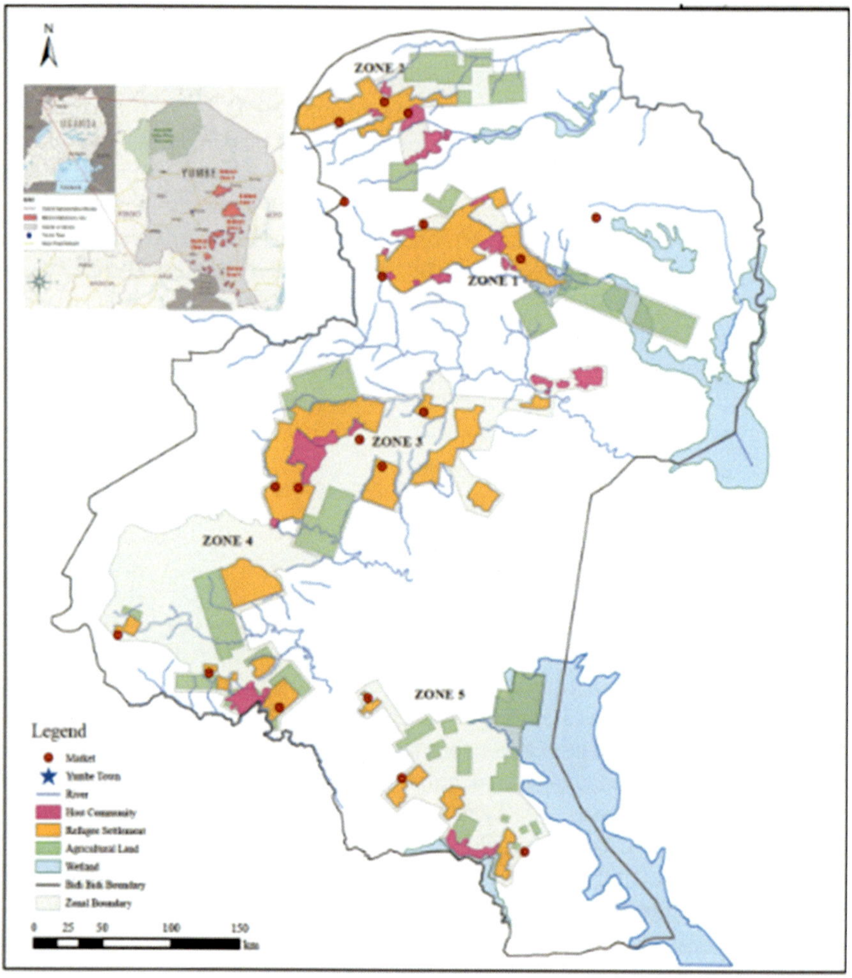

Fig. 16.1 Map of the study area. Notice that 'Zones' are the major geographical sub-divisions of a settlement—for example, Bidibidi has five (5) Zones, comprising of 75 villages in total. These are distinct from but akin to 'regular', Sub Counties in host communities. *Source* Author created

For the ethical approval, this study sought and was granted by the Research Ethics Committee of Uganda Christian University (approval number UCUREC-2024-885). As required, the research was cleared by the Uganda National Council for Science and Technology (UNCST) under registration number A527ES. Confidentiality was upheld throughout the research as provided for in the approved protocol.

The study engaged participants using a structured checklist comprising ordinal and categorical indicators reflecting their lived experiences in sourcing and using wood energy as well as perceptions of (re)source protection. Responses were transformed using ordinal scaling to enable principal component analysis (PCA) in Statistical

Package for the Social Sciences (IBM SPSS Statistics), with assumptions tested for sampling adequacy and factorability. The PCA aimed to identify latent dimensions underlying community trade-offs between immediate energy needs and long-term nature care. Table 16.4 shows the results.

16.2 Findings

Findings reveal four overarching yet intertwined aspects: Shrinking sources of wood energy for cooking (dirty fuel); dirty access to dirty fuel; low appetite for energy source protection; and, dicey access to alternative energy sources.

16.2.1 Shrinking Sources of Wood Energy for Cooking

Source of data used for analysis: GEE/Sentinel -2A. Note: There is a significant decrease in biomass density; as such, in 2024, it appears more fragmented and reduced in intensity compared to 2015.

Trends in biomass stocks or above ground biomass distribution in the Bidibidi refugee settlement area are presented in Fig. 16.2 (above), for 2015 and 2024. In 2015, large patches of dark green and medium green in the eastern and southern parts of the study area indicate higher biomass. Scattered low biomass zones (yellow/very light green) are minimal. In 2024, a noticeable reduction in dark green areas is observed, especially in central and southwestern parts. Expansion of light-yellow zones (zero biomass), particularly in the west due to deforestation or land use change. Biomass density in 2024 appears more fragmented and reduced in intensity compared to 2015. In people's words: forests are running away.

16.2.2 Dirty Access to Dirty (Wood)fuel

Households live a contested choice for energy. Refugees and hosts grapple daily with dirty access to dirty fuel. The phrase 'dirty access' means that the way people obtain fuel is problematic: that is, unsafe, unhealthy, or degrading. They walk long distances to gather wood from woodlands or forests that are "fast running away" from homesteads. They face physical danger such as sexual harassment. There are gendered burdens—as often women and children bear the tasks—that mean hazardous and inequitable supply chains.

Dirty fuel refers to wood for cooking being polluting, inefficient, and unsustainable. It produces smoke, soot, and harmful gases like carbon monoxide. Modalities used for cooking in households of refugee and host communities make the wood burn poorly, requiring more fuel for less heat. That wood (firewood, charcoal, etc.) are

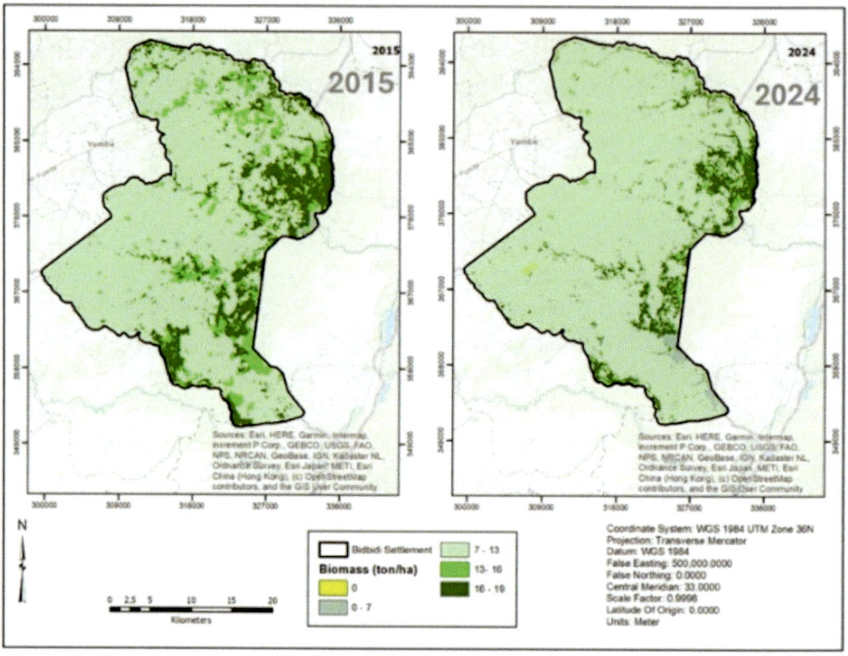

Fig. 16.2 Shows "forests running away from the people" between 2015 and 2024. *Source* Author created

often harvested unsafely, contributing to deforestation. This results into precarious reliance on unsustainable wood energy for cooking.

The overreliance on wood energy has led to immediate resource strain, with most participants reporting reduced availability of wood to be found and increasingly difficult access to firewood. It's harder to reach or collect, due to distances, unregulated local spaces, or competition tendencies between refugee and host communities. This points to ecological stress and socio-spatial deprivation in fuel access.

Take for example, asked if, given consumption patterns they live daily, household survey participants think that the (wood) energy sources are sustainable for the next 10 or more years, 55.8% (225) were not sure, 35.7% (144) said an outright no. Only 8.4% (34) were hopeful, with a clear yes.

Focus group discussions with female and male, refugee and host communities, indicated lived deep concerns over 'forests running away.' For example, women in Village 2 in Zone 3 decried, thus:

> "Previously, we did not have to travel far for firewood, but with people increases, we now have to go deep into the bush, facing dangers from wild animals and potential attacks. To guard against these risks, we travel in groups." FGD_Female_Refugees_V2_Z3

A participant's voice decries the receding woodlands due to the high demand for fuel exerted on them, highly limiting access to wood, and the attendant exhaustion of trekking long distances to get firewood.

"[People] go a distance ... because the surrounding environment is now lacking. Because one, the population which is using it is bigger than the resources around. And due to this [population] being here more than now seven years, I think the rate of consumption has gone high. And that's why in the nearby surroundings, it's very difficult to have access to wood." KII Z3 (Zone 3).

Other voices decried the exposure to risks like rape and its terrible effects especially among women, and particularly where circumstances cannot enable them to go in sizeable groups to fetch firewood from distant places. It is a situation one bad thing after another that have happened.

Social friction has also emerged between hosts and refugees over the available woodlands and woodlots—within the limited access to wood and wood resources.

Given the high population density refugee settlement, competition over natural resources was observed, particularly wood for fuel and shelter construction as quarry stones. Nearly seven in ten (68.73% or 277) participants in the household survey indicated having had a real issue in trying to access wood energy for cooking. Twenty (20) forms of social friction were identified (reported by participants are non-monetary costs), which ranged from low-intensity, often routine or symbolic actions with minimal coercion to severe forms (Table 16.1).

As a consequence, an overwhelming 87.45% (272) participants indicated that it is the landlords that hold the key. Landlords are a category of layered, unstructured and often informal authority that wield a lot of power of access to land and its resources.

Participant's voices thus indicated more than sporadic disputes or tensions between the two communities over the available wood resources, within the context of severely unclear and or communal land tenureFig. 16.3.

Constructive engagement was, however, reported 19 (4.56%) times in terms of receiving individual advice (2), entering into some form of agreement (2), receiving

Table 16.1 Moderate to severe indications of social friction between households and 'landlords.' *Source* Author created

Situation	Frequency	How the situation was operationalized
Labour	122	Low to coercion, depending on context
Threatened	001	Psych. coercion, potential escalation
Chased	003	Elements fear, exclusion, or hostility
Stopped	008	Denial of access or activity
Caned	004	Dispensing corporal punishment
Beaten	015	Meting out physical violence
Abused	004	Physical, emotional, or sexual harm
Rebuked	018	Social or verbal disapproval
Warned	042	Verbal caution, no formal consequence
Sent back	001	Forced return to the homestead
Imprisoned	001	Locally institutionalized punishment

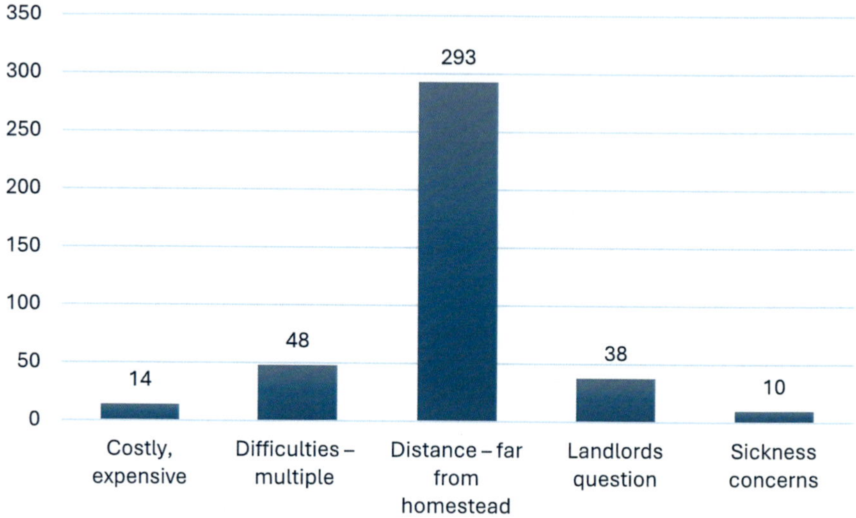

Fig. 16.3 Challenges in accessing receding wood. *Source* Author created

awareness (2), and getting involved into some form of community dialogue (13 times) in the household survey conducted.

As wood becomes scarce (Fig. 16.2), households have to venture farther from settlements, exposing especially women and children to risks including violence.

> "...the challenge here we have is population we have over 1 million people in the district including refugees and hosts. As a result, we have conflicts over these resources. I am here to look at all the natural resources not only forest resources; people are fighting over poles, over firewood, even those stones for construction, people still struggle for them because the number is much." KII [area authority] Barakala Town Council, Zone 1.

Social conflicts, as indicated in Table 16.1 above, are expressed in intra- and inter-household bickering over wood energy, as well inter-community tensions between refugee and host communities. Additional perspective is provided below.

> "Actually, seeing the population here and the influx and the demand … it was realized that people are fighting all over, scrambling for wood. People have to cook. There is no way; they have to go and cut trees. [People] have now cut all the natural trees around because they have to cook. People have been fighting, chasing each other. But to promote coexistence and sustainability, we just thought of [supporting] them to plant trees." KII Kululu Sub County, Zone 3

16.2.3 Low Appetite for Energy Source Protection

Sourcing modalities are odd hunter-gatherer extraction of wood energy for cooking. Sourced clandestinely, wood energy for cooking provides low incentive to protect wood sources. Even where there are local regulatory structures, things are being implemented.

> "Secondly, people ask: 'When peace comes, we are going back to Sudan, why should I plant these trees?'" KII (area authority).

> "…even if people are willing to plant the trees, there are some people who still have the mindset that even if they plant this tree, they will remain here just for people who are here (hosts). So, they have no time of planting the trees ... So, if you call community members to come for a meeting and you start a topic of maybe planting trees, no one can put their attention." KII Refugee Welfare Council III

Public conversations—especially leaderships—tend to shift away from the deeper questions of energy deprivation and justice, and instead assign blame to the forcibly displaced populations and wood cutting. This distracts from the real issues and those truly responsible for sustainable access to (wood) energy for cooking. Participant voices say it best:

> "…as a sub-county, we have some bylaws, but when it comes to real implementation, there are gaps. Take bush burning. Despite our resolution, many places have been set ablaze. Mostly they say are set up on fire by refugees, as if ... refugees are the only people to set fire! No, refugees may have their part there. Hosts are the people who like looking for these small, small animals (rats) just for consumption; and in the process often set fire. This is a two-way traffic. Leaders need to stand in the middle and tell people that bush burning has a lot of dangers." KII Barakala Town Council, Zone 1.

Sustainable access to biomass is a clean energy solution when it is well sourced and used. Risks related to sourcing are in unsustainable harvesting complicate land degradation, biodiversity loss and ultimately deforestation. Risks related to use are that some biomass conversion methods, like traditional firewood direct burning are inefficient and unclean. Overwhelmed, despite the Working Group on Environment and Energy (WorkGrEEn), people restlessly frame the threat as: 'Forests must not run away from us.'

> "When it comes to implementation of particular laws or ordinances, it is sometimes very hard. Because the ownership is based on the communities and to bring these minds together for a common goal, it is sometimes very hard. We engage key landlords and the elders. We have not engaged with them so much … most times what we do is to target the general community … and we just talk to them as a community without targeting those elders." KII Romogi Sub County.

People often talk about returning home, but the reality is uncertain. Many of us arrived in 2016, expecting to stay for no more than two years, yet we have now been here for over seven years with no clear end in sight. (FGD male refugees, village 4, Zone 1).

16.2.4 Dicey Access to Alternative Energy Sources

Refugee and host communities at the margins, women-led farmlands, and youth-led innovations have unequal energy access amidst biting energy deprivation and

Table 16.2 Household attempts at newer ways of cooking. *Source* Author created

Newer ways of cooking attempted	Response	
Improved cooking (energy-saving) stove	14	3.36%
Pour water on used wood after cooking	4	0.96%

Table 16.3 Use of dry grass for cooking ≥ once a week. *Source* Author created

Use of dry grass for cooking	Freq.	Percentage
Every day	33	8.2
3–4 times a week	42	10.4
Once a week	108	26.8
Never	220	54.6

economic poverty (Elasu et al. 2025; Gianvenuti et al. 2022). This impacts their capacity to adopt and adapt to agroecological methods. Results from the household survey indicate a daunting set of realities.

When asked if they have attempted newer ways of cooking, less than 4% households reported yes. They named the modalities: Improved Lorena stoves (Table 16.2).

When asked whether they had every used dry grass for cooking, over 45% households indicated that they had used some form of grass at least once a week (Table 16.3).

Current assistance, such as solar cooking set with runaway unit cost (Tran et al. 2021), drives the adoption agenda, amidst biting economic poverty and the issues of hard household choices that painstakingly deprioritize sustainable energy practices. People's voices have re-echoed the similar sentiments.

Table 16.4 PCA results showing participant lived experiences on wood sources/resource protection. *Source* Author created

Variable	PC1	PC2	PC3	PC4
Household preparedness	0.211	0.083	**−0.52**	−0.09
Attempt saving energy	0.084	−0.15	0.153	**0.739**
Engage with newer ways of cooking	−0.1	0.257	−0.2	**0.65**
Seek advice from actors	−0.13	**0.623**	0.068	0.023
Educating children on responsible wood harvesting	0.159	−0.02	**0.646**	0.038
Support others that are in need of wood	−0.04	0.393	**0.465**	−0.13
Attend meetings between hosts and refugees	0.206	**0.549**	−0.12	0.015
Personal participation in tree planting	**0.627**	0.185	−0.08	−0.04
Village possess tree growing community action agenda	**0.678**	−0.16	0.112	0.054
Eigen Value	*1.895*	*1.272*	*1.242*	1.074
Explained Variance	*0.172*	*0.159*	*0.149*	0.129
Cumulative Variance	*0.172*	*0.331*	*0.48*	0.609

> "We have raised our issues, but the responses we receive are not convincing. Despite numerous meetings, no action is taken when our concerns reach the Prime Minister's office." (FGD mixed host Ujorokunga village, Kululu Sub County (Zone 3).

As such, principal component analysis (PCA) was applied to consistent indicators on wood energy access—such as household participation, community action, support to the other, varying methods of cooking for saving energy –to identify latent dimensions shaping access patterns. Component scores (the individual transformed observations) and loadings (the weight by which each standardized original variable is multiplied to get the component score) were extracted. A varimax rotation, a form of orthogonal rotation, used to enhance interpretability by concentrating high loadings within distinct or dominant components.

Retained components (based on eigenvalue and scree criteria) revealed thematic clusters such as social collaboration, energy-saving behaviour, wood resource stewardship, commitment to tree-growing, as well as institutional presence offer insights into dominant pressures that influence wood energy access in refugee settlements in Uganda. These results were validated through various qualitative triangulation, robustness checks, et cetera, to enabling onward use and framing to inform policy and interventions on wood energy for cooking.

A biplot visually integrated the variables and observations—in a single 2-dimensional space—to allow for interpreting how community behavior and engagement patterns cluster and relate to principal components (Table 16.4).

In Figure 16.4 (below), the biplot clearly provides salient indicative linkages in the lived experiences:

- Proactive community structures such as village agendas for tree planting are enablers of household action on energy-saving and cooking practices;
- Socially-embedded practices like scheduled community dialogues or avenues to seek information on energy issues enable even the households resistant to energy-saving to open to newer cooking methods;
- Household preparedness is dear to many. Despite the circumstances, there were indications of attempts to engage in energy-saving modalities.
- Households showed minimal attention to passing on 'tested' forms of adaption to their young ones or neighbors. This reflected the constraining realities of other forms of deprivation which have led families to deprioritize energy conversations, particularly making women and youth suffer triple layers as frontline actors on household cooking.

Data viewed through multinomial logistic regression indicated high uncertainty about the future of wood use. Both refugee and host communities were apprehensive, even though the former were higher with an odds ratio of $e^{4.509} \approx 90.9$, and a highly significant p-value = 0.000 (Table 16.5).

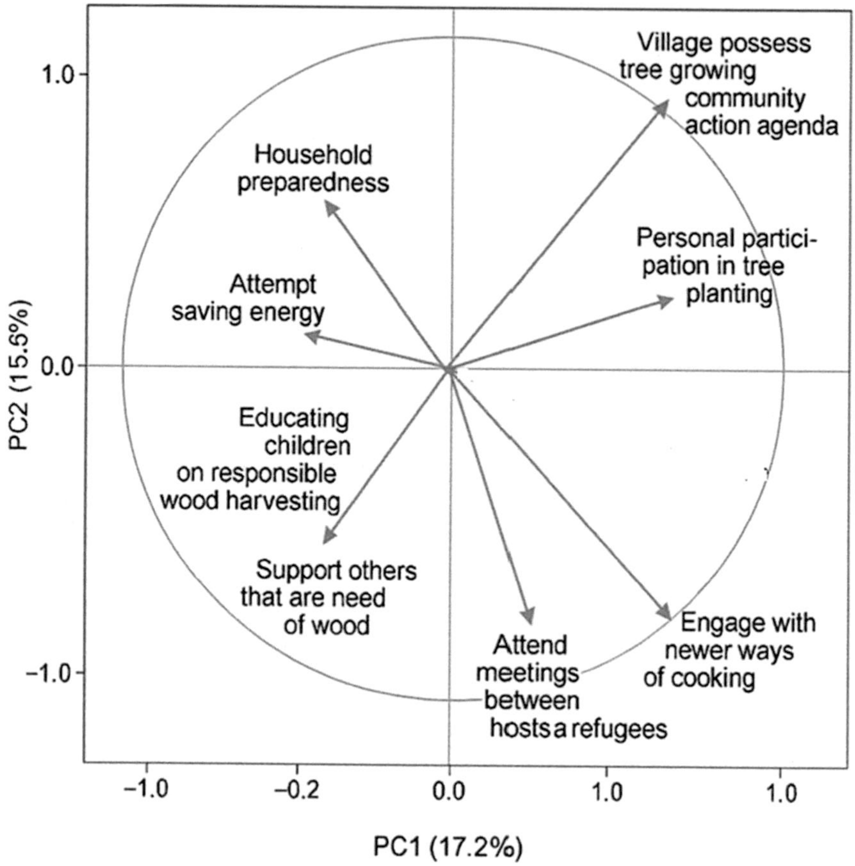

Fig. 16.4 Shows visualization of PCA results on lived experiences on wood sources protection. *Source* Author created

Table 16.5 Shows multinomial logistic regression on participants' perception of the future of wood use. *Source* Author created

			Coef.	St. Err.	t-value	p-value	[95% Conf	Interval]
No	Respondent type	Host	1	.	.	.	.	.
	Refugee	Refugee	4.509	1.837	3.7	0	2.029	10.022
	Constant		1.632	0.475	1.68	0.093	0.922	2.888
Not sure	Respondent type	Host	1	.	.	.	.	.
	Refugee	Refugee	3.881	1.492	3.53	0	1.827	8.244
	Constant		3	0.795	4.15	0	1.785	5.042

16.3 Discussion

The study demonstrated that small choices at the household and community levels contribute to large-scale construction or restoration or destruction, when considered in aggregate terms across time and space. For example, findings revealed through principal component analysis that community and individual actions related to sustainable sourcing and social engagement are key drivers of variation in the populations engaged. The study underlines that energy choices are not isolated technical decisions; they are instead deeply embedded in economic, social, and ecological realities—from singular households to national and global spaces.

16.3.1 Shrinking Sources of Wood Energy for Cooking

Participants' voices in the metaphor of "forests are running away from us" capture lived experience of energy deprivation, ecological degradation and financial constraints. It underscores the call for periodic tracking, monitoring, and reviews of the threats of depletion of biomass resources. This essentially echoes Laudato Si's call to hear both the cry of the earth and the cry of the poor, raising the stakes of the issues from technical scarcity to moral urgency. It is the question of equity that enables the framing of threats of biomass depletion as both an environmental and ethical crisis. Area authority (in the decentralized government structures) says it loudly:

> "So, as a case of Bidibidi, the settlement, it was a forested area. With a very thick tree cover. But now, because of high demand for wood energy, that's why you see most of the natural trees have gone." KII Yumbe District Local Government

16.3.1.1 Unsustainable ('Dirty') Access to the 'Dirty Fuel'

The study reveals and questions the survival-driven logic of accessing wood energy through harmful practices. While such access may be economically rational, especially in the short-to-medium runs, it is ecologically and socially destructive.

It is hardly surprizing that firewood harvester, a charcoal burner or a logger who accesses wood energy through indiscriminate tree cutting or half-baked briquettes makes some economic sense today; yet threatens to deplete humus and puts their existence at crossroads (Dal Ferro et al. 2019). The very survival necessity and economics today edge populations closer to a looming ecological collapse. The charcoal burner and firewood harvester are emblematic of this dilemma, whereby households are caught between immediate needs and long-term harm.

As such, decisions about the energy system—from micro to macro considerations—have moral implications and cannot be irresponsibly and casually made anymore without considering the questions of justice, in the tradition of Plato's notion of wellbeing—equity, fairness, and intellectual virtue or virtue with reason (Sovacool et al. 2016).

Despite periodic tracking of Sustainable Development Goal 7 that occurs through several key mechanisms—stakeholder engagement, data-making, policy-making, grants-making, and projects-making (Marzouk 2025). However, local priorities in communities at the margins hardly reflected in energy strategies since the decentralized governments, non-government entities, and private sector are weakly engaged. Technical data alone tells some story, but when organically woven with empowered local agency, it becomes a richer set of energy realities— deepening our understanding of electricity access, clean cooking, efficiency, and locally-appropriate renewable sources.

People are deeply bothered that, after all, the runway left—before depletion—may not be that long, given an unsustainable rate of using wood energy for cooking and other needs. When asked about the sustainability of wood use for the next 50 years, household respondents were unequivocal about their worries.

This study demonstrates that communities at the margins, particularly refugee and hosts in settlements deserve sustainable biomass sourcing and use. As transitional energy sources, wood energy for cooking needs sustained integration into renewable energy solutions such as biogas and solar. When handled sustainably, wood energy and services can provide beneficial solutions, advantages, and safeguards (Barasa et al. 2022; Bamwesigye et al. 2020). Voices from the field say it out loudly:

> "Humanitarian organizations and the government need not to limit support to only visibly vulnerable groups. All refugees [and hosts], regardless of physical appearance, face various challenges and need receive equitable assistance. Regular feedback to the community … on the progress of reported issues and initiatives. This will help build trust … to address our concerns," (FGD, mixed gender, refugees, village 8, Zone 3).

16.3.1.2 Low Appetite for Energy Source Protection

The study revealed an acute lack of integrated stakeholder and or multi-sectoral engagement, on wood energy for cooking, which points to making weaker the regulatory function of the decentralized governance structures. Households are left to muddle through; pitting the weaker or infirmed against harsh realities. Presence of robust local energy strategies will address systemic gaps in energy source protection. There is an elevated need for a shift in people's restorative appetite—from passive tolerance of unsustainable wood use to active stewardship. It is a call for justice and equity in energy decisions.

A female, area leader (refugee welfare council), among refugees, spelt it out most succinctly:

> "... but the challenge we have here ... there are some people who still have the mindset that even if they plant a tree, they will remain here just for people who are here (hosts); refugees think of going back to their country after all). So, they have no time of planting the trees... The bad things, which I see, that is still happening here in the community. ... this time, we are also lacking partners, and then also community mindset is also… the of category of wood...

So, if you call community members to come for a meeting and you start a topic of maybe planting trees, no one can put their attention." KII RWC3 Zone 1

16.3.1.3 Dicey Access to Alternative Energy Sources for Cooking

The study revealed the loud call for integration of wood energy into affordable transitional solutions like biogas and solar. Participants' voices acknowledged that while alternatives are starting to exist, access is uneven and fragile. Deep gaps in equitable support and feedback mechanisms reinforce the precariousness of alternative energy access, as FGDs underscored:

Authorities [must] take prompt action on reported cases, such as rape and corruption, ensuring that offenders are punished to set an example and deter future crimes, particularly when collecting firewood, to mitigate risks from wildlife and other dangers. FGD_Mixed_Refugees_V8_Z3

16.3.1.4 Narrative Synthesis

Sustainable biomass sources, sourcing and use in refugee settlements constitute a critical intersection of environmental justice, energy equity, and durable solutions. This is nested within the lived realties that, beyond the foreseeable future, households therein will unfortunately continue to use biomass and biomass-derived energy sources for cooking, among others. When handled sustainably, wood energy and services can provide beneficial solutions, advantages, and safeguards (Barasa et al. 2022; Bamwesigye et al. 2020).

Rapid population growth, has indeed long outpaced infrastructure and resource planning. Realtime evidence is needed to assess the scale, drivers, and governance responses to such tensions.

Consequently, as part of transitional energy sources, wood energy for cooking needs measured and sustained integration into renewable energy solutions such as biogas and solar. This thus contributes to the race towards the global SDG7 commitment to "safe access to sustainable, reliable, affordable, and modern energy services for all (displaced) people by 2030" (Marzouk 2025).

16.4 Conclusion

While the cost on populations is high, affecting economic growth, social health, education outcomes, among others; the biomass energy sector continues to grapple with balancing survival necessity and aspirational nature care. Household live contested choices for energy for cooking and productive use. Sourced clandestinely, there is low incentive to protect wood sources. People's voices cried out loud, as they haplessly framed the threat as: 'Forests are running away from us.' This provides valued lived experience of ecological degradation, echoing Laudato Si's call to hear both the cry of the earth and the cry of those at the margins.

Beyond the binary framing—of 'energy versus agroecology'—there's a web of pathways that call for nuanced co-creative engagement of stakeholders on household and productive use of clean energy, with the farmlands at the centre. Remote sensing data on above ground biomass change, qualitative insights from focus groups or key informants, and much more, provided robust evidence. Farmlands are indeed a living interface where energy justice, agroecological regeneration, and community resilience intersect and are co-designed.

This is a call that is reechoed in Laudato Si on rethinking processes that underlie present-day policy and practice. Participants too cried out for incentives for alternative energy sources (clean cooking), formalized permissions on access to wood energy for cooking (rather than unclear muddling through, genuinely inclusive action. For, refugee and host communities—girls, boys, women, men, among others—deserve just agroecological and energy transitions to live together and grow together. It is to rethink the systems and decisions that drive people to unsustainably ransack woodlands.

References

Adriaensens, J., Dessein, J. and Adam, J., 2025. Science as a territory in dispute: an analysis of power and paradigms in the conceptualization of agroecology. *Agriculture and Human Values*, pp.1-15. https://doi.org/10.1007/s10460-025-10723-2.

Altieri, M.A. and Nicholls, C.I., 2012. Agroecology scaling up for food sovereignty and resiliency. In *Sustainable agriculture reviews: Volume 11* (pp. 1-29). Dordrecht: Springer Netherlands.

Berke, T., and L. Larsen. Using Land to Promote Refugee Self-Reliance in Uganda. *Land 2022, 11, 410*. 2022. https://doi.org/10.3390/land11030410

Caporali, F., 2015. History and development of agroecology and theory of agroecosystems. In *Law and Agroecology: A Transdisciplinary Dialogue* (pp. 3-29). Berlin, Heidelberg: Springer Berlin Heidelberg.

Creswell, J.W. and Creswell, J.D., 2005. Mixed methods research: Developments, debates, and dilemmas. *Research in organizations: Foundations and methods of inquiry, 2*, pp.315-326

Dash, A.K., 2024. Regenerative Sustainability: Homo Ecologicus and the New Enlightenment. In *Regenerative Ecosystems in the Anthropocene: A Transdisciplinary Ecosystemic Framework for Regenerativeness* (pp. 33-59). Cham: Springer Nature Switzerland.

Duguma, L.A., Ariani, C., Watson, C.A., Okia, C.A. and Nzyoka, J., 2019. State of biomass resources in refugee-hosting landscapes: the case of Rhino Camp and Imvepi Refugee Settlements in West Nile, Uganda. https://doi/full/https://doi.org/10.5555/20220286691

Francis. (2015). *Laudato Si': On care for our common home [Encyclical letter]. The Holy See*. https://www.vatican.va/content/francesco/en/encyclicals/documents/papa-francesco_20150524_enciclica-laudato-si.html

Gianvenuti, A., Bedijo, N.G., Jalal, R., Hitimana, L., Walter, S., Linhares-Juvenal, T. and Xia, Z., 2022. Woodfuel consumption in Refugee Hosting Areas and its impact on the surrounding forests—The case of Uganda. *Forests, 13*(10), p.1676. https://www.mdpi.com/1999-4907/13/10/1676

Gocko, J., 2024. Ecology and Justice: From Environmental Justice to Integral Ecology of «Laudato si'». *Studia Ecologiae et Bioethicae, 22*(1), pp.75-82. https://doi.org/10.21697/seb.5799

Hercher-Pasteur, J., Loiseau, E., Sinfort, C. and Hélias, A., 2020. Energetic assessment of the agricultural production system. A review. *Agronomy for Sustainable Development*, *40*(4), p.29. https://doi.org/10.1007/s13593-020-00627-2

International Panel of Experts on Sustainable Food Systems (IDS & IPES-Food), 2022. Agroecology, regenerative agriculture, and nature-based solutions: Competing framings of food system sustainability in global policy and funding spaces. https://coilink.org/20.500.12592/38x5rx

Lage, J., Thema, J., Zell-Ziegler, C., Best, B., Cordroch, L. and Wiese, F., 2023. Citizens call for sufficiency and regulation—A comparison of European citizen assemblies and National Energy and Climate Plans. *Energy Research & Social Science, 104*, p.103254. https://doi.org/10.1016/j.erss.2023.103254

Maier, R. and Monti, P., 2023. Integral ecology as critical principle of environmental sustainability in the agri-food chain: Epistemological and ethical inputs from Laudato si'. *Current Opinion in Environmental Science & Health*, 31, p.100415. https://doi.org/10.1016/J.COESH.2022.100415

Marzouk, O.A., 2025. Summary of the 2023 report of TCEP (tracking clean energy progress) by the International Energy Agency (IEA), and proposed process for computing a single aggregate rating. In *E3S Web of Conferences* (Vol. 601, p. 00048). EDP Sciences. https://www.e3s-conferences.org/articles/e3sconf/abs/2025/01/e3sconf_icegc2024_00048/e3sconf_icegc2024_00048.html

Muhe, S. and Argaw, M., 2022. Estimation of above-ground biomass in tropical afro-montane forest using Sentinel-2 derived indices. *Environmental Systems Research, 11*(1).

Munro, P.G. and Bartlett, A., 2019. Energy bricolage in Northern Uganda: rethinking energy geographies in Sub-Saharan Africa. *Energy Research & Social Science*, 55, pp.71-81.

Nordic Development Fund (NDF). *26.01.2025. 600 million Africans without energy are facing a crisis.* https://www.ndf.int/newsroom/600-million-africans-without-energy-are-facing-a-crisis.html

Osabohien, R., Adeleke, O.K., Osabohien, E.U., Khalid, A.A., Aragaw, A.D., Onyesom, M., Agene, D., Ihayere, O., Al-Faryan, M.A.S., Ifere, E.O. and Igharo, A.E., 2025. Energy access, entrepreneurship and gender dynamics in food security in Sub-Saharan Africa. *Discover Sustainability*, *6*(1), p.413.

Pasquale, F. ed., 2019. *Care for the world: Laudato Si' and Catholic social thought in an era of climate crisis.* Cambridge University Press.

Shiva, V., 2023. Agroecology and Regenerative Agriculture: Environmental activist and food sovereignty advocate Vandana Shiva presents sustainable solutions for hunger, poverty, and climate change. *Mother Earth News*, (318), pp.38-41.

Tong Zoran. Agroecology: Integrating Ecological Principles into Farming Systems. *J Agric Sci Bot.* 2025;9(1):283 J Agric Sci Bot 2025 Vol. 9 Issue. https://www.alliedacademies.org/articles/agroecology-integrating-ecological-principles-into-farming-systems.pdf

Wezel, A., Herren, B.G., Kerr, R.B., Barrios, E., Gonçalves, A.L.R. and Sinclair, F., 2020. Agroecological principles and elements and their implications for transitioning to sustainable food systems. A review. *Agronomy for Sustainable Development*, *40*(6), p.40. https://doi.org/10.1007/s13593-020-00646-z

Whelan, M.P., 2024. Agroecology's moral vision. *Agriculture and Human Values*, *41*(2), pp.413-426.

Part III
Theoretical Reflections of Integral Ecology

Chapter 17
Liberation Theology as a Foundational Path to Indigenous Populations' Preferential Protection in Natural Disasters

Portia Ozioma Chigbu

Abstract Does liberation theology offer a moral and ethical basis for prioritizing the rights of Indigenous populations during natural disasters? Answering this question involves analyzing the core principles of liberation theology to understand its teachings and applying them to situations where these communities face disruptive environmental events. Indigenous populations consider their cultural identity central to their livelihoods, as their traditional lifestyles often lead them to live in remote areas. These locations isolate them from mainstream society and expose them to specific human rights issues such as limited access to healthcare, inadequate political representation, housing insecurity, and poor living conditions. From the perspective of liberation theology, I argue that Indigenous populations are among the most vulnerable groups during natural disasters and therefore warrant special protection. By examining case studies of the Oromos and Sengwer peoples, I analyze their social and political circumstances to illustrate their living conditions, emphasizing the need for prioritized protection in disaster scenarios. This research highlights the interconnectedness of these people's living conditions and the environmental challenges they face, as articulated by integral human ecology. By linking liberation theology to the idea of giving preferential protection to Indigenous populations, this paper offers the principles and methods within the theology to morally and ethically justify prioritizing Indigenous communities during disasters.

Keywords Liberation theology · Kenya · Ethiopia · Farming

Portia Ozioma Chigbu, University of Notre Dame, pchigbu@nd.edu.

P. O. Chigbu (✉)
University of Notre Dame, Notre Dame, IN, USA
e-mail: pchigbu@nd.edu

R. Marcantonio et al. (eds.), *Integral Ecology in the Anthropocene: Africa and Beyond*,
The Anthropocene: Politik—Economics—Society—Science 49,
https://doi.org/10.1007/978-3-032-32125-1_17

17.1 Introduction

As the impact of climate change intensifies, natural disasters such as earthquakes and floods are expected to worsen, affecting millions worldwide. The consequences of natural disasters are among the most severe hazards of this century (Southard 2017, p. 6). One of the groups most affected is indigenous populations, who make up a large portion of the poor and vulnerable. Defined broadly as a social group sharing culture, identity, social struggles, and purpose, indigenous populations account for 19% of the global population living in extreme poverty (International Labor Organization 2019, p. 13). With more than 73.4% of the world's indigenous populations living in rural areas, it is estimated that they are three times more likely to face extreme poverty compared to non-indigenous populations (Southard 2017, p. 8).

When it comes to disasters, poor people are disproportionately impacted by natural disasters. Termed pre-disaster vulnerabilities, they live in extremely deplorable conditions with poorly constructed shelters, making their environment unsafe. As a result, disasters worsen existing shortages because resources are scarce, forcing people to compete for them, which leads to conflict, inequality, and power struggles among different groups trying to gain control (See Mazdiyasni/AghaKouchak 2020, p. 1).

Because indigenous people make up the low-income population, their cultural, social, political, and economic backgrounds influence how they are affected by natural disasters, especially given their integral role in the ecosystem. Their lifestyle is typically described as traditional and cultural, often locating them in remote areas away from other population groups and developmental infrastructure. As a result, they are more vulnerable to natural disasters and are the most severely impacted social group in the aftermath.

This is the situation of the Oromos, Sengwer, and several other indigenous populations in Africa and globally. Although they come from different social and political backgrounds, they share similarities, such as living traditional lifestyles and supporting their livelihoods in customary ways. As a result, they are often marginalized and excluded within their respective communities. These exclusions have led to oppression, discrimination, and human rights violations, stripping them of their lands, cultural and social ties, and threatening their survival. In the face of climate disasters, these issues are becoming even more critical.

Although States and international organizations have established guidelines for distributing disaster relief to different population groups, these groups typically include men, women, and children, often excluding indigenous populations (See UNHCR 2023). This exclusion occurs because most indigenous populations are generally excluded from mainstream government services before disasters; thus, the same tendency persists during relief efforts after disasters. These guidelines reference human rights, international, and humanitarian laws to identify violations and specify corresponding rights and standards. A fundamental law and core principle in disaster relief is that affected populations must enjoy the same rights and freedoms as others, and discrimination should not occur (See UNHCR 2023). Therefore, all disaster-affected groups must receive aid during natural disasters, and no group should be left out.

Considering the exclusions faced by indigenous populations, practically applying liberation theology means prioritizing the most vulnerable groups. Indigenous communities should be given priority for protection because they are a distinct social group with a low baseline of vulnerability and marginalization before a natural disaster occurs. Since their starting point is low, natural disasters tend to increase their vulnerabilities, leading to unfair consequences for them. Therefore, in accordance with the principle of equity in human rights, they must be prioritized, and their specific needs and circumstances must be identified and addressed in disaster situations.

Prioritizing their protection is particularly important in light of the economic, social, and legal emphasis on protecting people's rights in Africa, as exemplified by the African Charter and other regional human rights documents.

The task of this paper is to demonstrate that there is a justifiable reason to prioritize protecting indigenous populations affected by disasters due to their living conditions.

As an ethical and moral justification for prioritizing indigenous populations, I rely on the Christian theory that centralizes the poor and vulnerable in society as essential individuals who should be prioritized and protected. This theory, known as liberation theology, gives a voice and identity to defenseless people. Because indigenous populations are poor and vulnerable, their prioritization during natural disasters aligns with the principles of liberation theology.

Liberation theology aligns with the field of integral human ecology because it advocates for the intrinsic worth of every human being, especially the poor. This stance supports the paper's purpose by emphasizing the need to recognize indigenous populations and their hardships in disaster settings. Another point of convergence between the two concepts is their focus on the interconnectedness of social and environmental issues, which must be addressed holistically. When liberation theology prioritizes the needs of the poor and marginalized indigenous communities, it also considers their ecological concerns. Recognizing the cultural and historical knowledge indigenous people hold, as well as their role in maintaining the ecosystem, means that failing to prioritize them during disasters could lead to their extinction, along with all their knowledge about protecting their environment from the harmful effects of climate change.

Consequently, this paper is divided into three parts. The first part introduces liberation theology, highlighting three key principles. These include God's memory, understanding the living conditions of the poor, and the responsibilities owed to the poor. The rest of the paper is structured in accordance with the principles. The second section examines the living conditions of indigenous populations through case studies of the Oromo and Sengwer people. It notes that the deprivations they face do not align with God's biblical teachings. Having assessed these conditions, the third section applies the principle of framing responsibilities owed to indigenous populations to suggest a path forward. This involves developing sustainable policies for pre- and disaster relief that address their social, economic, and political deprivations.

While this section does not detail specific guidelines, it emphasizes that effective and sustainable policies and solutions must be rooted in human rights principles.

17.2 Liberation Theology

Several scholars have enunciated on liberation theology and its principles. However, this paper will primarily rely on Gustavo Gutiérrez, whose teachings align with the paper's core argument.

Two main principles in liberation theology include "preferential love for the poor" and the "universality" of God's love (Gutierrez 2008, p. 32–33). The poor, downtrodden, and defenseless on earth constitute a crucial part of our population. Efforts should be made to protect and support them within society through "solidarity with the poor." This solidarity has been described as the best way to bring God's reign to earth, considering that God is for the poor and vulnerable (Gutierrez 2008, p. 29).

Conversely, God's love is universal for all humans made in His image and likeness, regardless of race, group, or gender. While these principles generally reflect God's message to all humanity, the challenge lies in applying them simultaneously in practice (Gutierrez 2008, p. 33). This is because, by stating that God's love is universal, it implies that there should be no favorites, as all are equal before Him, which negates the preference for the poor and the oppressed. However, Gutierrez points out that these principles are not contradictory but complementary and must work together (Gutierrez 2008, p. 33). In addressing the needs of all humans to uphold the universality principle, special attention must be given to the poor and those who are suffering, because not all humans on earth can be equal, and those who are suffering should not be left to their fate.

In disaster situations, when a population is affected, their needs and rights must be addressed; however, special attention should be given to the most vulnerable, especially the poor, to lessen their suffering. Evidence shows that indigenous populations are among the poorest due to their cultural lifestyles and pre-existing marginalization, making them the most impacted social group during natural disasters and deserving of prioritized aid. Gutierrez refers to indigenous people as a group that faces exclusion and marginalization, and that much of the interest in them lies in their natural resources, with little to no concern for the people (Gutierrez 2008, p. 32). His acknowledgment of indigenous populations as far back as when he wrote this piece proves that the exclusions indigenous populations face have been an existential issue facing them. It is important to note that he cites examples of these populations, such as the Arawak peoples of the Amazon, the Kuna of Panama, and the Mapuche peoples of Chile (Gutierrez 2008, p. 32). Although he does not mention either the Oromos and Sengwer, on which this paper relies, indigenous populations face similar threats of exclusion and marginalization, which is the reason for protecting them in disaster contexts. Thus, advocating for their prioritized protection and relying on liberation theology as an ethical approach are essential for ensuring this marginalized group's protection in various disaster situations.

One reason that justifies a focus on indigenous populations is that they live in poverty. The Pope's Encyclical states that an integrated approach to human ecology involves merging justice for the poor with environmental issues (Francis 2015, p. 35, para 49). To obtain justice for the poor involves describing poverty, understanding its causes, and taking steps to prevent it (Gutierrez, p. 31).

One pressing question is how, in disaster situations, one can balance addressing the needs of the entire affected population with the principles of universality and with prioritizing the most impacted disaster population. Consideration will be given to the human rights principles of equality and equity. Upholding equality in this context means that all disaster-affected individuals are cared for and that their diverse needs are met during disasters. This assertion aligns with the principle of "universality," which protects all victims and their rights without discrimination.

Conversely, when applying equity, it recognizes that society is generally unequal, and we must be aware of groups that are poor and vulnerable, making them deserving of priority protection. Indigenous populations are one such group due to their marginalization and deprivation. It aligns with the integrated approach of merging environmental and social crisis, where poverty is eliminated and dignity restored to excluded populations (Francis 2015, p. 104, para 139). This approach aligns with the "preferential love for the poor," and they deserve special protection in natural disaster scenarios (Gutierrez 2008, p. 33).

Indigenous populations are not the only groups affected by natural disasters; however, there is sufficient scholarship focused on the protection of other marginalized groups other than indigenous populations. Consequently, this paper takes a new approach by advocating for Indigenous communities affected by disasters. It uses liberation theology as an ethical foundation, offering preferential protection to socially marginalized groups, particularly the poorest, who are most impacted by disasters. Doing this affirms God's presence on earth (Gutierrez 2008, p. 29).

Life as given by God presents us with three interrelated relationships: one with God, with neighbors, and with the earth (Francis 2015, p. 48, para. 66). One aspect of the relationship of humans with God is being made in His image. Indigenous populations are created in God's image and therefore present a proper context to argue that just like other human populations, they are deserving of protection. Consequently, as other humans navigate their relationship with their neighbors, they must appreciate the conditions of these populations and frame policies that protect them.

In support of prioritizing the poor, Gutierrez points out that people living in poverty contradict God's plan for humanity (Keenan 2012, p. 200). Similarly, the deprivations and living conditions of indigenous populations before and after disaster aftermaths do not align with God's plan for humans.

To prevent them from suffering after disasters, three methods in liberation theology are highlighted to ensure the poor receive preferential treatment. They include reflecting God's memory, understanding the poor's identity and living conditions, and mapping out the responsibilities owed to the poor (Gutierrez 2007, p. 20).

The best way to reflect on God's memory is through biblical examples and doctrines that embody God's teachings. Gutierrez cites biblical passages, such as Deuteronomy 15:4, in which God emphasizes that no one should be poor (New

Catholic Bible 2024, p. 22). As members of society, we must understand that having poor people is part of life. However, we are called to open ourselves to the poor to ensure that their conditions of poverty and marginalization are eliminated from the face of the earth. The best way to demonstrate this is to support the poor and the most deprived (Gutierrez 2007, p. 22–23). Everyone is encouraged to return to God's original design for humans and actively work toward creating conditions that align with God's plan (Gutierrez 2007, p. 22). These include eliminating poverty in society and ensuring that the poor achieve equality. Therefore, social, economic, and political structures that maintain poverty and inequality should be dismantled to honor God's memory.

The second method involves focusing on understanding the identity and living conditions of the poor. Living in poverty contradicts the human life God gives. Poverty here is both physical and cultural, involving deprivation of social amenities and subjugation based on ethnic, racial, or gender affiliations, respectively (Gutierrez 2007, p. 27). Giving preferential treatment to the poor requires an understanding of their "sociological" background (Watson 2003, p. 519). To understand the identity and living conditions of the indigenous population, case studies will be highlighted that discuss these living conditions. Doing this aligns with their lived experiences, as their pre-disaster conditions are assessed to understand the political, economic, and social factors they face. In analyzing the living conditions of indigenous populations, the argument for prioritizing their protection is strengthened.

The third approach to liberation theology discusses responsibilities toward the poor. Gutierrez states that all people, including the poor, must stand up against their deprivation and living conditions (Gutierrez 2007, p. 30). One suggested way is first to follow Christ's example in his interactions with the poor and marginalized in society (Gutierrez 1973, p. 172). In emulating Christ's interactions with the poor, Matthew 25:35–36 serves as a key example, where Christ commands everyone to feed the hungry, give water to the thirsty, clothe the naked, and comfort the suffering (New Catholic Bible). Similarly, Matthew 25:40 states, "Amen, I say to you, whatever you did for one of the least of these brethren of mine, you did for me" (New Catholic Bible). Christ's teachings here remind us of our duty to the poor and highlight God's remembrance.

In the following sections of this paper, these three methods will be explained and linked to their support for the prioritization of indigenous populations.

17.3 God's Memory and the Living Realities of Indigenous Populations

God's memory does not align with having poor populations that are marginalized and excluded. Natural disasters create situations in which some populations are excluded from others, particularly in disaster response and protection. Although mainstream responses target vulnerable groups like women and children, indigenous populations are not generally included as a protected group deserving of protection. This is often because they already experience marginalization from the State and therefore will not be considered a priority in such situations.

Considering the issues caused by natural disasters, does it reflect God's memory for Indigenous populations to suffer intense hardships due to their socio-cultural living conditions? The answer is no, because God's plan for human life, even in disaster situations, is that no one experiences suffering. God's nature is to align with the poor, defenseless, and those who are deprived (Gutierrez 2007, p. 24). Since indigenous populations are consistently marginalized, they fall into the category of people who are deprived and defenseless; they are a marginalized population, making them among those with whom God's nature aligns.

The poor are impoverished due to societal causes, such as existing societal structures resulting from human choices and policies. Therefore, there is a collective obligation for everyone to unite and develop solutions that change the narrative (Gutierrez 2007, p. 25). These solutions must address historical issues like the social and political exclusions of indigenous peoples. With the growing impacts of climate change, mainly caused by decisions related to technological advancement, there have been significant harmful effects on Indigenous populations. Consequently, there is a duty to ensure that human-made policies address their disaster vulnerabilities. Anything contrary to this does not align with God's memory and concern for the poor and oppressed.

One concern is the tendency to see the poor as individuals unworthy of either priority protection or rights. Policy makers must remember that poor people have rights because they are humans made in God's image and likeness (Tamez 2007, p. 48). It does not honor God's memory to have them living in extreme poverty and facing multiple vulnerabilities during natural disasters. They experience marginalization and exclusion due to poor human choices, which undermine God's purpose and cause their suffering. During natural disasters, their rights are God's rights, and every time their rights are violated, God's rights are also violated. Therefore, their rights must be recognized and enforced.

In understanding and upholding their rights, there is debate over whether liberation theology is a "theology for the poor" or "of the poor." (Watson 2003, p. 519). When a theology is "for the poor," it is created to benefit poor people and carried out through external agents such as the government and NGOs that provide aid to the poor. In contrast, when a theology is "of the poor," it means a design or approach that originates from and belongs to the poor people, who are both the beneficiaries and participants. Such a design must reflect the characteristics of the poor. Therefore, the "theology of the poor" aligns with the principle of participation in liberation theology. Participation involves actively shaping and executing processes so that ownership is attributed to participants. As a result, participation ensures the poor are active contributors to processes designed for them, which ultimately helps preserve their human dignity. Watson further emphasizes that liberation theology is a "theology of the poor," comparing it to the biblical story of Jesus identifying with and walking alongside the poor (Watson 2003, p. 519).

Consequently, applying participation in natural disaster situations for Indigenous populations is a "liberation theology of Indigenous populations," evidencing that they are active participants in the rebuilding process, restoring their lives. Their multiple needs, as identified through case studies, reflect their economic, social, and cultural identities, which are used to frame adequate protection measures.

An important point to note is that this paper is the first of two that advocate for prioritizing indigenous populations during disasters. Establishing the ethical foundation provides moral justification for prioritizing indigenous populations and an ethical basis for their preferential protection. The second paper will thoroughly analyze the practical steps taken to protect indigenous populations affected by disasters legally.

17.4 Understanding the Identity and Living Conditions of the Poor/Indigenous Populations

The complexity of an ecological crisis requires solutions that examine the human lives of its victims (Francis 2015, p. 45). Similarly, there is a need to explore different ways of interpreting issues from other cultural contexts. The same is true for making a case for indigenous populations, which requires understanding their living conditions and how these issues intersect with their lived experiences in disaster contexts.

In Gutierrez's second approach to applying liberation theology, focusing on understanding the identity and living conditions of the poor, he notes that living in poverty contradicts the human life God gives. Poverty here is both physical and cultural, involving deprivation of social amenities and subjugation based on ethnic, racial, or gender affiliations, respectively (Gutierrez 2007, p. 27).

The overall goal here is to recognize that Indigenous populations face these issues due to social structures shaped by poor policy choices. This supports Gutierrez's idea that the poor are not poor in isolation but are poor because of societal structures driven by "human decisions" (Gutierrez 2007, p. 25). Thus, analyzing their living conditions through case studies justifies prioritizing and protecting them during disasters. The aim is to make human protection a top priority for disaster-affected indigenous populations, the primary objective of this research. Therefore, understanding their socio-cultural conditions through case studies is necessary to justify the study.

One question that may come up is why the living conditions of indigenous populations are analyzed before disasters rather than after. Research on their conditions post-disaster is limited; however, multiple studies have shown that indigenous populations are the most affected during disaster events (The Nation's Health 2022). Therefore, since this paper demonstrates that the starting point for indigenous populations is low economically, socially, and culturally, natural disasters tend to worsen these conditions, highlighting the need to prioritize their support.

The case studies used here include the Oromo people of Ethiopia and the Sengwer people of Kenya.

17.4.1 *Oromo Population in Ethiopia*

The Oromo people are mainly farmers in Ethiopia's Amhara and Oromiya regions and are the largest indigenous group in Ethiopia, comprising various subgroups. Despite their significant size, they are marginalized by the Amhara region due to cultural and living conditions. The Amhara have an organized state structure that governs them, unlike the Oromo way of life (Tulu/Roro 2022, p. 1, 3). They were isolated from the Oromiya population, where they were dominant, and now it is difficult for them to survive in the Amhara region because they cannot thrive in a non-cultural environment (Tulu/Roro 2022, p. 5).

The Ethiopian government implemented its development policy to promote political and economic progress. These projects, focused on the Lower Omo River Valley and Lake Turkana, have resulted in the displacement of indigenous populations (GoE 2011). The government developed the "Addis Ababa Master Plan" to boost economic growth (Tulu/Roro 2022, p. 5). Although they launched a program to relocate and settle indigenous communities, these populations lost access to their ancestral lands and rivers used for pastoral farming and cultivation, despite their unsuccessful protests for land rights (Human Rights Watch 2016). Mainstream media reported that the Master Plan increased the country's GDP at the expense of Oromo farmers' livelihoods, who mainly engage in pastoral farming, and has hindered their ability to continue practicing it (Southard 2017, p. 12).

It also blocked farmers from accessing lands needed to feed their livestock, especially ancestral lands. These traditional communal lands were restricted, preventing herders from reaching them. The restrictions on accessing these lands have caused multiple food and water insecurities among indigenous peoples (Oakland Institute 2019). Despite repeated protests, the government has ignored these issues and responded with violence. Over time, this has increased the Oromos' vulnerabilities, leaving them more exposed to drought and disease (Southard 2017, p. 12).

Furthermore, many people have fled Ethiopia for neighboring countries. Due to the displacement, numerous Indigenous populations have been forced to move to survive, making them susceptible to diseases, rising poverty, food shortages, and loss of livelihoods. The Oromos have shifted from independence to dependence, vulnerability, and exposure to various risks (Southard 2017, p. 12).

With the multiple vulnerabilities they encounter before disasters, disasters will worsen these vulnerabilities, increasing their poverty. These circumstances demonstrate a lack of alignment with God's memory.

17.4.2 *Sengwer Indigenous Community in Kenya*

The Sengwers are a marginalized community facing forced evictions from their ancestral lands, which significantly impact their livelihoods as hunters. These evictions have caused various consequences, including health problems for women and children and the loss of cultural knowledge, values, and identity, especially as a hunting community. The suppression of their cultural values originates from colonial times, when the colonial government aimed to undermine their cultural heritage. As early as 1918, the Sengwers were described as a "minority, unrecognized, marginalized, oppressed, and discriminated hunter-gatherer Indigenous group" (Mamati/Maseno 2021, p. 77, 1). This description has persisted to the present day. It has worsened due to the dispossession of land and displacement caused by the Kenyan government and EU/World Bank projects aimed at reducing the effects of climate change (Mamati/Maseno, 2021, p. 2).

The hunting community is known for collecting various animals, which they use for their survival and consumption. They are also deeply communal, paying respect to the earth, the supreme being, and their ancestors (Mamati 2018, p. 35). An example of their reverence for the earth is that when trees are cut for spiritual reasons, they must be worshiped because they are believed to house spirits. Therefore, they think their environment and everything within it should not be disturbed to avoid punishment from their supreme beings (Mamati/Maseno 2021, p. 38).

The Sengwers have faced exclusion by the government through its laws, initially by depriving them of access to their lands, which prevented them from practicing their religious beliefs and protecting their natural resources. The colonial government took steps to safeguard the forest areas (which were the community's home) by converting them into government-owned properties (Mamati/Maseno 2021, p. 44–45). This disrupted the communal ownership of hunting lands and the organized systems for managing natural resources that existed before this period. Furthermore, the government introduced cattle ranches and potato plantations as new livelihood options without respecting or preserving the Sengwer people's hunting traditions. These oppressions and deprivation of livelihoods continued even after the colonial period, with successive presidents in the post-colonial era extending the government's policies. Lands covered by government forests and reserves were reallocated to other communities, depriving the Sengwer of their livelihoods (Mamati/Maseno 2021, p. 4–7). It also forced them to interact with people from other communities despite differences in cultures, traditions, and ways of life. For example, the Keiyo and Marakwet people generally do not believe in natural resource conservation, which conflicts with the Sengwer's view that it is central to their values and practices (Mamati/Maseno 2021, p. 49).

As a result, promoting and protecting their cultural identity has become difficult, and this has also increased the population in these areas, putting pressure on natural resources and making them insufficient to meet needs. Different communities have had to clear traditional farmlands to build homes (Mamati/Maseno 2021, p. 47). The ongoing deprivation has caused conflicts between the Sengwer and the government, with the government often prevailing due to unequal power dynamics. The government's actions ignore the provisions of the Kenyan Constitution that protect minority rights to community lands. An example of such deprivation is the eviction of the Sengwer from the Embobut forest, which deprived them of their ancestral lands, sacred sites, and ability to practice their religious beliefs (Mamati/Maseno 2021, p. 47–48).

Based on the living conditions described above, the Oromos and the Sengwer share similar living conditions, despite being in different countries. Some of these similarities include health issues, hardships caused by land dispossession and increased displacement from their ancestral lands, and their inability to sustain themselves, which leads to greater poverty and higher vulnerability to disaster risks. This also puts them at a disadvantage, making them more severely affected during natural disasters.

Elizondo points out that when someone is born into an excluded social group or culture, they experience "existential or cultural poverty" (Elizondo 2007, p. 159). This kind of poverty occurs when different groups exist within a society, with one or more dominant groups exerting control over others, thereby stripping the oppressed of their dignity and self-respect (Elizondo 2007, p. 160). A key trait of the dominant group is to enforce its will and culture on others through force, a pattern that has persisted since the beginning of time, primarily through ongoing attacks on Indigenous populations (Elizondo 2007, p. 160).

Similarly, Indigenous populations face "cultural poverty" because they are marginalized mainly and subjugated by dominant societal groups. This marginalization causes multiple deprivations that leave them vulnerable to both poverty and extreme poverty before and after natural disasters, which opposes the idea that human life originates directly from God (Gutierrez 2007, p. 27).

When indigenous populations are excluded from post-disaster resilience efforts, they experience cultural poverty because it worsens their existing hardships and poor living conditions, given their lives of deprivation before the disasters. As a result, Elizondo advocates for an "inclusive society" that includes everyone without subjugating one group over another (Elizondo 2007, p. 167). Such a society reflects "Jesus' inclusive table fellowship," where all those marginalized and ignored by society are invited to share in fellowship with Jesus (Elizondo 2007, p. 167).

Such inclusive gatherings should also be promoted during disaster situations, where rebuilding efforts include all affected victims, with priority given to the poorest and most vulnerable, predominantly indigenous populations. This aligns with the paper's goal of prioritizing the protection of Indigenous peoples in disaster contexts, grounded in liberation theology.

17.5 Framing Responsibilities Owed to the Poor/ Indigenous Populations

In the words of Guiterrez, framing responsibilities must stem from emulating Christ's teachings about the poor and his interactions with them. Human life has three existential relationships: with God, with other human neighbors, and with the earth (environment) (Francis 2015, p. 48).

In preserving this relationship, all humans must ensure that, in their relationship with God and with their neighbors, they understand the biblical teachings that require unconditional love for every human as our neighbor and that we express this love in a manner we would express it to ourselves. This aligns with the fact that all humans share a common lineage, being made in God's image. Understanding that indigenous populations are our neighbors and made in God's image deserves protection just like every other disaster-affected population. To do this, we must acknowledge their living conditions to understand the responsibilities owed to them.

Thus, the question is how to replicate Christ's teachings in our responsibilities toward indigenous populations in disaster situations. Gutierrez offers a way by highlighting two main approaches: "solidarity with the poor" and "protest against the inhuman situation of poverty in which they live" (Gutierrez 2007, p. 30). Solidarity means supporting the poor by recognizing their suffering and the societal issues that contribute to it. For indigenous populations affected by disasters, these living conditions, as seen in the case studies, stem from land dispossessions, loss of livelihoods, loss of dignity, and shortages of food and water—all of which heighten their vulnerability to natural disasters and worsen their living situations.

The second approach, which focuses on protesting poverty, emphasizes developing solutions to eradicate poverty among society's vulnerable groups (Gutierrez 2007, p. 30). This includes designing sustainable policies to eliminate poverty within society. After identifying the difficulties faced by the Oromo and Sengwer, creating solutions involves framing and implementing policies that specifically address and prioritize the issues raised in their living conditions (Brown 1993, p. 7). Doing this strengthens their resilience in withstanding natural hazards and the losses associated with them.

Framing these policies requires evaluating living conditions from the perspective of indigenous populations to understand their deprivations and develop sustainable policies that prioritize and protect them after disasters. Effectively doing this requires Farmer's "accompaniment" principle, where one puts oneself in the shoes of the indigenous population to gain a clearer understanding of their living conditions and hardships (See Farmer 2011).

When developing policies that prioritize their protection, they must be tailored to human rights principles, including non-discrimination, self-determination, and the right to sustain livelihoods. Combining an understanding of socio-cultural and political contexts with human rights principles yields targeted disaster policies that not only meet the needs and concerns of indigenous communities but also sustainably enhance their resilience to disasters. This approach respects God's memory in disaster situations by prioritizing the most vulnerable and affected.

17.6 Conclusion

"No one can escape the task of avoiding the destruction of our environment, but we in this continent should be particularly attentive to the situation facing the weakest in humanity" (Gutierrez 2008, p. 35). This statement emphasizes the importance of paying attention to the most vulnerable members of the environment.

The overall goal of this paper was to demonstrate that the living conditions of indigenous populations require prioritization in natural disaster responses, using liberation theology as both a moral and ethical foundation to support this prioritization. Applying the principles and methods of liberation theology, this paper argues that the lived experiences of the Oromo and Sengwer peoples show that their conditions do not reflect God's memory, making them among the most affected by natural disasters. Furthermore, it highlights the inseparability of the environment and indigenous populations, noting that their deprivations and marginalizations are crucial in addressing ecological issues.

Having demonstrated that adopting a preferential option for indigenous populations in the context of natural disasters is essential to reducing suffering, the next point of inquiry is how to implement this preference through targeted policies.

References

Brown, R.M. (1993), *Liberation Theology: An Introductory Guide.* Kentucky: Westminster John Knox Press.

Farmer, P. (2011) 'Accompaniment as Policy', *Science of Implementation Policy.* At: https://siidata.org/paul-farmer-accompaniment-as-policy/ (30 March 2025).

Francis. (2015). *Laudato si': On Care for our Common Home.* Vatican Press.

Government of the Federal Democratic Republic of Ethiopia (GoE) (2011a) *Ethiopia's Climate-Resilient Green Economy: Green Economy Strategy.* Addis Ababa: GoE.

Gutierrez, G. (1973) *A Theology of Liberation: History, Politics, and Salvation.* Edited by S.C. Inda and J. Eagleson. Maryknoll, NY: Orbis Books.

Gutierrez, G. (2007) 'Memory and Prophesy', in Groody, D. (ed.) *The Option for the Poor in Christian Theology.* Notre Dame, IN: University of Notre Dame Press, p. 20.

Human Rights Watch (2016), *Such a Brutal Crackdown: Killings and Arrests in Response to Ethiopia's Oromo Protests.* At: https://www.hrw.org/report/2016/06/16/such-brutal-crackdown/killings-and-arrests-response-ethiopias-oromo-protests (24 February 2025).

Interviews conducted with Felix Horne of Human Rights Watch and Bonnie Holcomb of the Oromo Studies Association (June–August 2016), cited in Southard (2017) *supra* note 1 at p. 12.

Keenan, J. (2012), *A History of Catholic Moral Theology in the Twentieth Century: From Confessing Sins to Liberating Consciences.* New York: Continuum, p. 200.

King'asia Mamati, and Maseno, L. (2021), 'Environmental Consciousness amongst Indigenous Youth in Kenya: The Role of the Sengwer Religious Tradition', *Theological Studies*, 77(2).

King'asia Mamati (2018), 'An African Religious Worldview and the Conservation of Natural Environmental Resources: A Case Study of the Sengwer in Embobut Forest in Kenya', *FERNS*, 1(1), p. 35.

Mazdiyasni, O. and AghaKouchak, A. (2020), 'Natural Disasters Are Prejudiced Against Disadvantaged and Vulnerable Populations: The Lack of Publicly Available Health-Related Data Hinders Research at the Cusp of the Global Climate Crisis', *GeoHealth*, 14(1).

New Catholic Bible, Deuteronomy 15:4. At: https://www.biblegateway.com/passage/?search=Matthew%2025:35-40&version=NCB (24 March 2024).

Oakland Institute (2019) *How They Tricked Us: Living with the Gibe III Dam and Sugarcane Plantations in Southwest Ethiopia.* Available at: https://www.oaklandinstitute.org/report/how-they-tricked-us (25 October 2024).

Rowland, C. (ed.) (2008), *The Cambridge Companion to Liberation Theology.* 2nd edn. Cambridge: Cambridge University Press, p. 29.

Southard, N. (2017), *The Socio-Political and Economic Causes of Natural Disasters*. CMC Senior Theses, 1720. At: http://scholarship.claremont.edu/cmc_theses/1720 (25 October 2024).

The Nation's Health (2022) At: https://www.thenationshealth.org/content/52/7/5 (25 October 2024).

Tulu, A. and Roro, A.G. (2022) 'Regionalisation or Marginalisation? The Case of the Oromo Living in the Amhara Region of Ethiopia,' *Heliyon*, 10(11), p. 1.

United Nations Framework Convention on Climate Change (UNFCCC) (2016), *Technology and the UNFCCC: Building the Foundation for Sustainable Development*. Bonn: UNFCCC Secretariat.

United Nations High Commissioner for Refugees (UNHCR) (2023) *UNHCR Policy on Emergency Preparedness and Response Emergency Handbook*. At: https://emergency.unhcr.org/emergency-preparedness/emergency-policy/unhcr-policy-emergency-preparedness-and-response (16 November 2024).

Watson, N. (2003) 'Liberation Theology', in Houlden, L. (ed.) *Jesus: The Complete Guide*. London: Continuum, p. 519.

Chapter 18
Integral Ecology and Animals in the Anthropocene

Brett Beasley

Abstract An array of moral philosophers and theologians have lamented that in Pope Francis's integral ecology, there is no 'new' teaching about the moral status of non-human animals. They note that the growing use of animals for human purposes, especially due to the increasing demand for meat and other animal products and the resultant rise in intensive animal agriculture, is an urgent environmental concern that seems to demand a response. Thus, for these scholars, integral ecology's lack of specific focus on animals hampers its ability to respond to this problem and is thus a significant shortcoming. In this paper, I interpret and respond to integral ecology's lack of focus on animals. I argue that, despite the lack of specifics, integral ecology, through a combination of its Franciscan and liberationist themes, can nevertheless provide guidance on issues related to animal welfare and animal agriculture. I illustrate how this is the case by considering integral ecology's application to the ongoing "livestock revolution" occurring in many developing countries around the world.

Keywords More than human · Animals · Laudato Si · Husbandry

18.1 Introduction

"In certain areas of the world, too, it would be appropriate to consume less meat: this too can help save the environment" (Pope Francis, 2022). This piece of advice, delivered from Pope Francis to participants in the European Union Youth Conference in 2022, made headlines around the world. (See coverage in El Financiero 2022; CNN Español 2022; Yahoo News UK 2022; etc.) The fact that this statement triggered strong reactions is illuminating. It speaks not only to views about the impact of diet on the environment. The fact that so many were surprised at the former pontiff's

Brett Beasley, University of Notre Dame, bbeasle1@nd.edu.

B. Beasley (✉)
University of Notre Dame, Notre Dame, IN, USA
e-mail: bbeasle1@nd.edu

R. Marcantonio et al. (eds.), *Integral Ecology in the Anthropocene: Africa and Beyond*, The Anthropocene: Politik—Economics—Society—Science 49,
https://doi.org/10.1007/978-3-032-32125-1_18

remarks says much about the character of his writing in his earlier environmental writings, especially his encyclical letter, *Laudato si'*, which has been called the "Magna Carta of integral ecology" (Boff 2015). In the document, first promulgated in 2015, Francis offered almost no explicit advice about animal agriculture. As Wendell Berry has pointed out, "eating is an agricultural act." Nevertheless, not only does *Laudato si'* not specifically address questions related to *eating* animals, but it also offers no direct acknowledgement of the global rise in livestock production and resulting environmental crises (Berry 2010, p. 145). This is despite the fact that the document does draw attention to other specific environmental concerns, such as, for example, deforestation, loss of biodiversity, water scarcity—and problems that interrelate closely with animal agriculture and especially with the ongoing steep rise in intensive forms of animal agriculture known as "factory farming."

Clearly, when it comes to understanding and applying *Laudato si'* and the broader movement of integral ecology it represents, there appears to be a curious—perhaps even puzzling—"animal gap": A pope whose own papal name was drawn from the patron saint of animals, St. Francis of Assisi, and who wrote eloquently and extensively about 'creatures,' was nevertheless surprisingly silent about this most striking feature of human beings' relationship with animals today.

What does this silence mean? How should we respond to this "animal gap"? One can understand the response of Peter Singer, written less than two weeks after the death of Pope Francis. The well-known utilitarian philosopher and animal rights advocate summed up Francis's legacy as one of "a well-intentioned man who often said the right things but failed to use his position to push through the necessary reforms." Singer added, somewhat wistfully if also very pointedly, "What would have happened if Pope Francis had urged the Church's followers not to be complicit in this disordered use of animals, and to avoid, whenever possible, the products of factory farming? It's hard to know. Perhaps then *Laudato si'* would have really made a difference" (Singer 2025).

Is Singer correct? Does the work of integral ecology really add up to nothing more than a missed opportunity regarding animals and animal agriculture? In this chapter, I address the "animal gap" in integral ecology in its rhetorical, practical, and conceptual dimensions. First, we need to explain the "animal gap" on the level of rhetoric. We need to fully articulate, in other words, why explicit advice about animals and animal agriculture is absent. We then need to respond to the "animal gap" on a conceptual level: How can we responsibly extend what the document says to the question of animal agriculture? Finally, we need to respond to the "animal gap" in application. Given a concrete instance, what would integral ecology bring, and what, if anything, would be distinctive in its approach? I specifically raise the issue of what has been called the "livestock revolution" and attempt to articulate what the resources of integral ecology would suggest by way of a response to the ecological challenges it represents. While I grant that *Laudato si'* "lacks a direct treatment of certain ways in which nonhuman animals are used," I argue nevertheless that animals are the *linchpin* as well as the *emblem* of integral ecology's distinctive contribution to our understanding of "the Anthropocene" (Aseneta 2017).

In calling our treatment of animals a *linchpin*, I mean something eminently practical: Reconsidering dominant approaches to treating animals is one of the most directly actionable steps we can take in the manner that *Laudato si'* urges us to. Increases in industrial animal agriculture are leading to an alarming rise in greenhouse gas emissions, deforestation, biodiversity loss, habitat destruction, and the emergence of new threats to human health. Addressing these concerns requires resisting or reversing the current trajectory of animal use. Additionally, Pope Francis writes that we must not assume we can solve problems by extending human control over the natural world. Instead, we must resist the "technocratic paradigm" that has led us to perceive other creatures as "merely as potential 'resources to be exploited.'" Hence, my claim that our treatment of animals is *emblematic*. Animals, as the fellow creatures with which we share the most proximate and recognizable kinship, form an essential test case. Our treatment of animals reveals whether we have adopted an anthropocentrism of domination or whether we have adopted a 'situated' anthropocentrism of cultivation, one characterized by "responsible stewardship" and an acknowledgment that "human life is incomprehensible and unsustainable without other creatures." These features culminate in a powerful and distinct contribution to our understanding of and response to "the Anthropocene," which is defined by our growing awareness "that human activity is having a dominating presence on multiple aspects of the natural world and the functioning of the Earth system, and that this has consequences for how we view and interact with the natural world—and perceive our place in it" (Malhi 2017).

One of the most prominent features of the human domination of the Earth indicated by "the Anthropocene" is the shifts in Earth's biomass that are a direct cause of human activity, especially the global decline of wild animals. An analysis by Yinon M. Bar-On, Rob Phillips, and Ron Milo revealed that wild mammals account for just 4% of Earth's mammal biomass, compared to 62% for livestock. (Humans account for the remaining 34%.) Wild birds account for 29% of Earth's avian biomass compared to the 71% accounted for by poultry (see Fig. 18.1).

18.2 In Search of Animals in the Rhetoric of *Laudato si'* and *Laudate deum*

Singer was by no means the first to note the lack of specific attention to animal agriculture in *Laudato si'* and its follow-up exhortation *Laudate deum* (2023). In fact, this absence has sometimes provoked the consternation of ethicists and moral theologians interested in articulating a concern for the moral and legal status of nonhuman animals. Charles Camosy writes that in *Laudato si'*, "Shockingly, at least for those of us who work on animal ethics, there is nothing explicit in the encyclical about factory farming [...] it would be difficult to come up with a topic more in need of a critique coming from the encyclical's commitment to integral ecology. And yet, inexplicably, we find no direct mention of the practice" (Camosy 2016). Similarly, writing

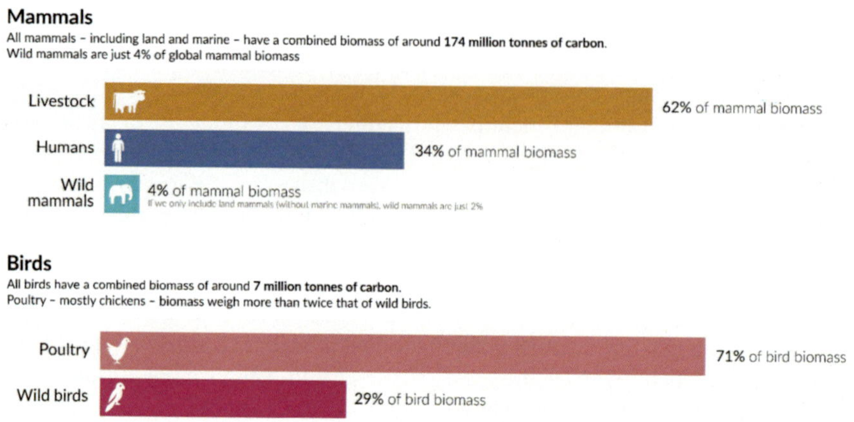

Fig. 18.1 Wild mammals and birds are just a fraction of humans and our livestock. Animals are compared in terms of biomass, measured in tonnes of carbon. *Source* Our World in Data, open access

on behalf of the organization Catholic Concern for Animals after the appearance of *Laudate deum,* Ruby Alemu noted that "industrial animal agriculture, or factory farming [...] is once again an item which has been missed off the agenda in Pope Francis' papal writings. We believe the lack of explicit engagement with industrial animal agriculture to be a serious missed opportunity" (Alemu 2023). In recognition of the need to move *Laudato si'* into action, which is a major theme of *Laudate deum* as well, a broad coalition of groups representing a wide array of Christian faith traditions undertook the commitment to develop "the resources for the Laudato Si' Action Platform regarding the place of animals in integral ecology" (Christians for Animals—International 2022). In 2025, in response to debate among stakeholders, the Laudato Si' Action Platform amended its goals to include a greater recognition of the role of animals in putting integral ecology into practice. For example, for the goal of fostering "Ecological Economics," the Laudato Si' Action Platform added text to indicate the importance of avoiding "exploitative animal agriculture," and to the goal of the "Adoption of Sustainable Lifestyles," it added text to indicate the importance of "opting for a more plant-based diet and making judicious choices about meat and dairy consumption" (Dicastery for Promoting Integral Human Development 2025).

The disappointment among Christian animal ethicists has not been only about putting integral ecology into practice. It also has to do with expectations about the trajectory of Christian ethics discourse, which, at the time *Laudato si'* appeared, had been experiencing a sudden groundswell of interest. In 2014, a year before *Laudato si'* appeared, John Berkman and Celia Deane-Drummond introduced the first-ever special issue of the *Journal of Moral Theology* devoted to the moral status of non-human animals, noting that "No longer excluding animals from view, there is a veritable flowering of interest in non-human animals among Catholic moral theologians" (Berkman/Deane-Drummond 2014). The hope among many was that

Pope Francis would add to the momentum. But this is not what *Laudato si'* provides. In fact, when we collate the mere 13 mentions of "animal" or "animals" in *Laudato si'*, we are likely to be disappointed or even confused in any attempt to draw out a specific message about non-human animals and their moral status. As we will see, the general (and deeply significant) tendency of the document is to prefer the term "creatures," a term that appears 80 times in several different forms. By contrast, the term "animal" appears 13 times. And even in these instances, several uses seem to break down the distinction between animals and other creatures rather than highlight it. In three of these instances, "animal" occurs in the phrase "animals and plants," or "plant and animal." Thus, even a resounding demand on behalf of animals lacks specificity. Pope Francis is concerned about biodiversity. He writes:

> There is the loss of biodiversity, worsened when "Highways, new plantations, the fencing-off of certain areas, the damming of water sources, and similar developments, crowd out natural habitats and, at times, break them up in such a way that animal populations can no longer migrate or roam freely. (Pope Francis 2015, §35)

Biodiversity is important because it is precisely in the variety of animals God created that he is glorified, quite apart from whether those animals are valuable to us for sustaining material life.

> It is not enough, however, to think of different species merely as potential "resources" to be exploited, while overlooking the fact that they have value in themselves. Each year sees the disappearance of thousands of plant and animal species which we will never know, which our children will never see, because they have been lost forever. The great majority become extinct for reasons related to human activity. Because of us, thousands of species will no longer give glory to God by their very existence, nor convey their message to us. We have no such right. (Pope Francis 2015, §33)

Here, one might still ask: Is the extinction of a lichen as important and worthy of our concern as that of a lemur? Do both give God equal glory? Do both have an equal message to convey to us?

This is hardly cleared up by the several passages that seem to more openly recognize that animals are crucial insofar as they support or contribute to the human good. Francis writes, for example:

> "changes in climate, to which animals and plants cannot adapt, lead them to migrate; this in turn affects the livelihood of the poor, who are then forced to leave their homes, with great uncertainty for their future and that of their children" (Pope Francis 2015, §25).

Or later, Francis recognizes that cruelty to animals and cruelty to people are interconnected: "We have only one heart, and the same wretchedness which leads us to mistreat an animal will not be long in showing itself in our relationships with other people" (Pope Francis 2015, §92).

In many ways, this serves to highlight the interconnectedness of all things that the document is continually drawing our attention back to, and yet this in itself can be frustrating because it does not identify the destructiveness of intensive forms of animal agriculture.

In other places, Pope Francis rehearses the traditional view that recognizes that although all species can glorify God, human beings, as more perfect, can legitimately use them for our purposes. He states that "Human power has limits" and that "it is contrary to human dignity to cause animals to suffer or die needlessly" (Pope Francis 2015, §130). Interventions have to "favor its development in its own line," to respect the createdness of the animal or plant. Similarly, the document's key thrusts, relating to "integral ecology" and "the technocratic paradigm" do not, for the most part, single out animals among the created things of the world for special care of a kind not already directed toward all of creation. In fact, there is an elevation of cultural life: "The imposition of a dominant lifestyle linked to a single form of production can be just as harmful as the altering of ecosystems" (Pope Francis 2015, §145).

Clearly, explaining what integral ecology has to say about animals will also entail explaining why it does not make its statements more direct or explicit. I argue that the best way to do this is to recognize that Francis's main effort in developing integral ecology is to articulate a perspective on the environment that is both Franciscan and liberationist. In doing so, Francis is doing the deep, preliminary work of revaluation. The main effort, as befits a papal encyclical letter, is to strike the keynote necessary to reorient a world deeply in need of transformation. Only after we have built a new foundation can we move forward effectively. Ultimately, in order to mount a critique of what he calls the "technocratic paradigm," Francis develops the Franciscan and liberationist perspectives in tandem. His task is not presenting a new argument about the moral status of animals so much as it is re-presenting a deep ethic of care for creation that is already implicit in the Christian story. This calls to mind John Milbank's claim that we should see "theology itself as a social science," one whose task is "not apologetic, nor even simply argument. Rather, it is to tell again the Christian *mythos*, pronounce again the Christian *logos*, and call again for Christian *praxis* in a manner that restores their freshness and originality. It must articulate Christian difference in such a fashion as to make it strange" (Milbank, 383). When it comes to integral ecology, Pope Francis's goal is especially to "tell again the Christian *mythos*"—and also to tell that *mythos* over and against the cluster of *mythoi* that define the "technocratic paradigm." In this way, the relative dearth of specific comments and advice related to animals and animal agriculture may be seen as something more and other than liability; it can be seen as an effort to focus on the re-narration and re-presentation of Christian truth.

18.3 Franciscan Themes

To say that *Laudato si'* is Franciscan may appear so obvious as to be unnecessary. Pope Francis bookends the encyclical by explicitly grounding its perspective within the vision of his namesake, St. Francis of Assisi. Pope Francis makes clear that by invoking the perspective and attitude of St. Francis, he is not engaging in "naive romanticism" (§11). Here, Francis makes a claim that is crucial for understanding not just his perspective on ecology but also the way his encyclical proceeds and how it aims to affect us. He writes:

> If we approach nature and the environment without this openness to awe and wonder, if we no longer speak the language of fraternity and beauty in our relationship with the world, our attitude will be that of masters, consumers, ruthless exploiters, unable to set limits on their immediate needs. By contrast, if we feel intimately united with all that exists, then sobriety and care will well up spontaneously. The poverty and austerity of Saint Francis were no mere veneer of asceticism, but something much more radical: a refusal to turn reality into an object simply to be used and controlled. (Pope Francis 2015, §11).

As the document unfolds, it becomes clear that Francis takes this point quite seriously. The main aim is to cultivate "openness to awe and wonder" and to reinvigorate "the language of fraternity and beauty in our relationship with the world." Without these, the necessary ecological changes we must make to heal our wounded world will be impossible. But with them, many of the right changes will become inevitable.

It is within this perspective that Francis draws our attention to 'creatures,' his preferred term for all living beings, both sentient and non-sentient. And many of the references to the natural world and what it contains tend toward the familiar and the elemental, owing to the inspiration of St. Francis of Assisi's "Canticle of the Sun." There is, in this Franciscan imagination, a tendency to focus on what is elemental: "the smallest of animals" as well as the sun, the moon, flowers, etc. Francis quotes Bonaventure: "from a reflection on the primary source of all things, filled with even more abundant piety, he would call creatures, no matter how small, by the name of 'brother' or 'sister.'" The focus of the document is in putting kinship ahead of specific dicta about lifestyles, diets, or blanket prohibitions about the eating or rights of animals, and it also offers kinship instead of a biocentrism and a kind of sentientism whereby we would care for animals insofar as they had the capacity to feel pain, for example. In effect, it is nearly the reverse of that; St. Francis invites us to feel kinship with what is fundamental: the sun, the moon, wind, water, and fire.

It is understandable why this Franciscan perspective could raise concerns, especially in the way it seems to elide distinctions between different types of beings. Gradations of intelligence and the distinctions between sentient and non-sentient beings or even organic and inorganic matter seem to be elided. Not only lichen and lemur, as we have noted, but even limestone would appear to belong to the one category of "creatures." However, the possibility of concrete guidance for our actions becomes clearer when we consider three points together:

1. First is the interconnectedness that the document continually draws our attention to. Its regular refrain is that "everything is connected" or "everything is interrelated." We cannot pursue the human good without reference to the good of the rest of creation.
2. At the same time, we are also called to pursue them for their own sake. Francis writes, "It is not enough, however, to think of different species merely as potential 'resources' to be exploited, while overlooking the fact that they have value in themselves" (§33). This is a point Francis reiterates in several different forms, later stating, "We must forcefully reject the notion that our being created in God's image and given dominion over the earth justifies absolute domination over other creatures" (§67).

3. Finally, Pope Francis writes that creation has "an intrinsic value" which is "independent of [its] usefulness. Each organism, as a creature of God, is good and admirable in itself" (§140).

What this line of thinking culminates in is Francis's distinctive form of anthropocentrism. On the one hand, he openly denies one extreme form of anthropocentrism, namely "When human beings place themselves at the centre [and] they give absolute priority to immediate convenience and all else becomes relative" (§122). What Francis introduces instead is what he will eventually describe in *Laudate deum as* "situated anthropocentrism" (Pope Francis 2023, §66). To adopt "situated anthropocentrism" is to recognize that "human life is incomprehensible and unsustainable without other creatures. For 'as part of the universe… all of us are linked by unseen bonds and together form a kind of universal family, a sublime communion which fills us with a sacred, affectionate and humble respect'" (§66). Concomitantly, we will "stop thinking, then, of human beings as autonomous, omnipotent and limitless, and begin to think of ourselves differently, in a humbler but more fruitful way" (§67).

18.4 Liberationist Themes

If the main thrust of the Franciscan theme within *Laudato si'* is to look at creation from the perspective of the most elemental, incorporating a spirit of wonder that resists instrumentalization, then the liberationist theme connects closely to and builds upon this idea. The influence of liberation theology, especially the works of Latin American liberation theologians Gustavo Gutiérrez and Leonardo Boff, on *Laudato si'* has long been a topic of discussion. In 2016, Daniel Castillo noted that the precise meaning of 'integral ecology' remains somewhat elusive. This lack of precision reflects not only the highly complex nature of the term, but also the pope's elliptical writing style" (Castillo 2016, p. 354). He argued that "the concept of integral ecology should be construed as a liberationist concept, one that calls for a paradigm shift away from the structural and ideological dimensions of the contemporary 'global system'" (Castillo 2016, p. 354). Castillo connects Pope Francis's concept of "integral ecology" to Gutiérrez's concept of "integral liberation," noting that "Whereas Gutiérrez's denunciation of developmentalism was accompanied by his call for a socialist order, Pope Francis's denunciation of the contemporary globalization project is marked by a call for dialogue." For Castillo, Pope Francis's more "circumspect" approach to liberation nevertheless remains a major challenge to the global order, specifically in its rejection of "sustainable development," which mirrors Gutiérrez's denunciation of "developmentalism" (Castillo 2016, p. 354). The influence of liberation theologian Leonardo Boff is widely acknowledged, even by Boff himself. Most notably, Pope Francis's framing of the problem of integral ecology as the problem of listening both to the "cry of the poor" and the "cry of the Earth" echoes Boff's *Cry of the Earth, Cry of the Poor*. What is crucial to note about the liberationist theme for our purposes is that it calls us not just to recognize how environmental issues intersect with the poor;

it also asks us to respond to creation as poor and/or a victim, especially a victim of our "violence." At the outset of the document, it draws attention to the "sickness" that, through human violence, can be found in the "soil, in the water, in the air and in all forms of life" with the result that "the earth herself, burdened and laid waste, is among the most abandoned and maltreated of our poor" (§2).

We should also recognize then that even as the document asks us to take "a frank look at the facts" (§61) it calls us to do so in a liberationist mode, with an eye "not to amass information or to satisfy curiosity, but rather to become painfully aware, to dare to turn what is happening to the world into our own personal suffering and thus to discover what each of us can do about it" (§19). In other words, *Laudato si'* draws upon its Franciscan emphasis on the most elemental aspects of creation to then also invite us to see what has been specifically cast out, victimized, neglected, harmed or done violence to in some way with the recognition that what allows violence to occur is often precisely our sense that this violence is hidden. It is that hiddenness that Pope Francis asks us to overcome, to see what we would rather not look at.

It is with this aim in mind that the document goes on to mention methane, fossil fuels, deforestation, water issues, and the like. Pope Francis could indeed have identified that the rise in meat consumption on the demand side and factory farming on the supply side exacerbating these issues. Nevertheless, we should acknowledge that there is a definite preference for smaller forms of agriculture, even if no specific condemnation of factory farming. Pope Francis lifts up "small-scale food production systems which feed the greater part of the world's peoples, using a modest amount of land and producing less waste" (§129).

The key, then, as Pope Francis would make perhaps even more clear in *Laudate deum*, is that we exercise our care for creation in a way that recognizes and responds to the fact that "the other creatures of this world have stopped being our companions along the way and have become instead our victims" (Pope Francis 2023, §15). It is in light of this focus on victims that we should interpret statements about self-sacrifice in Pope Francis's writings. In engaging with the world, we need to be attentive not just to what is small but what is easily neglected, ignored, marginalized, devalued—and it is through self-sacrifice that we discover and participate in an alternative kind of valuation of the world. What is necessary is a "new lifestyle" (§202). However, we do not achieve this new lifestyle by sheer resignation or even by accepting a specific new set of moral demands. Rather, what we need is "a culture and spirituality genuinely capable of setting limits and teaching clear-minded self-restraint" (§105).

It is worth asking, then, whether rising meat consumption and the dramatic intensification of animal agriculture in factory farming continue because we do not think animals have any special moral status or whether it is because we fail to fully live out the convictions most of us already have. Instead, there are stories about progress and advancement that we think make these forces inevitable. So, in this sense, we can say that although the problem of intensive animal agriculture is not singled out, it is the linchpin for enacting that critique, and it is a special emblem of the very kind of concern the document has in mind, and it is to that issue that we now turn.

18.5 Integral Ecology and the Livestock Revolution

Armed with the conceptual resources we have assembled from our consideration of the interrelated Franciscan and Liberationist themes in Pope Francis's integral ecology, we are now in a position to consider what practical difference these resources can make for decision-making, action, and advocacy on environmental issues. Here, I want to consider in particular the issue of the "livestock revolution" as a way of illustrating what integral ecology has to offer to conversations about animals in continued discussions of the Anthropocene.

The "Livestock Revolution," broadly speaking, names the fact that the demand for meat is rising and is expected to continue to rise throughout the developed world. When the term was first introduced, it identified the combined forces of urbanization, population growth, and rising incomes as the leading causes for the dramatic rise in demand: Meat consumption in developing countries was 10 kilos per person per year in 1964/66 and is projected to rise to near the global average of 37 kilos by 2030.

From the time it was introduced, the term "Livestock Revolution" has been used to name not just the increased demand for animal products but also the complex of problems that need to be 'solved' in order to make the revolution 'successful.' But resisting the 'revolution' has often been framed as unwise and, indeed, futile:

> Some want to halt the livestock revolution. But the ongoing nutritional transformation in developing countries, driven by income, population, and urban growth, leaves little room for policy to alter the widespread increase in demand for animal food products. (Delgado et al. 1999, p. 29)

In fact, the coiners of the term "Livestock Revolution" explicitly reject moral appeals as relevant to the problem. They write:

> In sum, it is unwise to think that the Livestock Revolution will somehow go away in response to moral suasion by well-meaning development partners. It is a structural phenomenon that is here to stay. How bad or how good it will be for the populations of developing countries is intricately bound up with how countries choose to approach the Livestock Revolution. Policies can significantly improve poverty alleviation, environmental sustainability, and public health, but only if new actions are undertaken. Failing to act risks throwing away one of the few dynamic economic trends that can be used to improve the lives of poor rural people in developing countries. (Delgado et al. 2001, p. 65)

Thus, according to these development economists, the primary way to respond to the livestock revolution is through top-down policy changes that steer the poor in rural areas into the types of work and situations in which they can benefit the most economically from the revolution. This is, in fact, the approach that has been adopted by the African Development Bank, which writes in its Livestock Investment Master Plan:

> The continent is seeing demand for animal source foods increase rapidly as incomes rise, driven by population growth, economic growth and urbanisation, which offers an unprecedented opportunity to reposition livestock as a business activity with the potential to significantly improve food and nutrition security, drive inclusive growth, create millions of new jobs along several value chains and support the continent's efforts to become a major player in global export markets. (African Development Bank 2021)

While the report recognizes increased risks, it calls for new systems and technologies to be developed to minimize or mitigate the concerns related to health, food security, and the environment.

While for many economists, the increasing demand for farmed animals is simply "market potential," the moral vision of *Laudato si'* suggests an alternative view. Much of the writing about the "Livestock Revolution" assumes that this trend is inevitable. It assumes that economic forces will drive developing nations throughout the developing world will follow a trajectory similar to that of the United States, where today, more than 90% of all agricultural animals are in concentrated animal feeding operations (CAFOs, or "factory farms") and where 80% of antibiotics are used on animals.

Integral ecology would not assume that "every increase in power means 'an increase of "progress" itself' and would recognize that "reality, goodness and truth" do not "automatically flow from technological and economic power as such" (Pope Francis 2015, §105). It would take seriously the idea that reasserting our values and developing conscience and simplicity of life is a legitimate response to the "Livestock Revolution." Indeed, from the perspective of integral ecology, and especially through its skepticism regarding the "technocratic paradigm," we can see that many of the promises of the "Livestock Revolution" have proven and will continue to prove empty. The 'problems' presented by the revolution are not ones we can 'solve' if we are to maintain a holistic view of true integral human development.

If we consider, for example, the problem of "food-feed competition," we encounter a permanent biological feature of our world that cannot be engineered out of natural systems. Intensive forms of animal agriculture, unlike traditional forms of animal agriculture, generate "food-feed competition" because they rely on feeding cereals to livestock that could be fed to humans. Thus, feedlot systems inherently consume much more human-edible protein than they produce. By contrast, traditional, smaller-scale forms of animal agriculture involve feeding animals waste products such as food scraps or allowing them to graze, where they can feed on grasses or insects not easily edible by humans. Thus, the technological solution of moving smaller producers to participate in and embrace the "Livestock Revolution" is likely to exacerbate food insecurity while blocking efforts to arrive at more moderate, more sustainable food systems, especially as CAFOs drive out smaller family farms.

Similarly, health concerns posed by intensive animal agriculture are not easily engineered out of systems through technology or policy-based solutions. One of the most serious health concerns arising from intensive animal agriculture is the spread of disease from zoonotic spillover, where a disease spreads from animals to humans. Factory farms involve the warehousing of large numbers of animals in conditions that make the spread of disease more likely and more rapid. While many producers attempt to control this problem with antibiotics, doing so can lead to the development of new, antibiotic-resistant pathogens, particularly when, as is often the case, antibiotics are administered 'subtherapeutically,' meaning before an infection has occurred. However, attempts to spread large numbers of animals over wider spaces lead to the destruction of natural habitats, making zoonotic spillover from wild animals more likely. Recognizing this double-bind, environmental scientists have pointed out that 'Animal agriculture creates a 'trap' of rising infectious diseases, and escaping it requires limiting meat consumption" (Hayek 2022).

The liberationist theme in integral ecology should guide us to regard with skepticism any merely technical solutions. It calls upon us to "pay renewed attention to reality and the limits it imposes" and to use as a guide "the message contained in the structures of nature itself' (Pope Francis 2015, §116 and §117). The deeper point is that we must recognize the misguided forms of anthropocentrism that have led us to disregard the message of nature in the first place. And ultimately, doing so involves more than just pointing out the impossibility of technological solutions; it also calls upon us to retell the human story and to tell it as a counternarrative against the misguided narrative of progress that condones the human domination of nature.

The liberationist theme in integral ecology can also help use recognize that the framing of the "Livestock Revolution" has always had technocratic and developmentalist underpinnings. After all, a diet centered around meat consumption is by no means 'natural,' however much we might associate it with cavemen or 'paleoman.' In his history *Red Meat Republic,* Joshua Specht characterizes the rise of the meat-centric diet as a product of nineteenth-century America. And while the story of the intensification of animal agriculture is sometimes told as a story of technological progress, involving extensive railroad networks paired with refrigerated train cars, Specht points out that "Centralized production was as much the consequence of institutional developments and social conflict as of technological change" (Specht 2019, p. 260). With this point in mind, we can more easily recognize that the "Livestock Revolution" is not just naturally occurring in the developing world; it is actively being exported to the developing world through not just the culture but also the companies of the developed world. Similarly, the liberationist theme in integral ecology would call for greater solidarity between the developed and the developing world. It would refuse to frame the "Livestock Revolution" as a problem for low- or middle-income countries meant to be solved by solutions designed by and imported from high-income countries. Rather, it would call upon high income countries to voluntarily partake in the necessary sacrifices and lead the way in limiting the demand for livestock and enact the "change in lifestyle [that] could bring healthy pressure to bear on those who wield political, economic and social power" (Pope Francis 2015, §206).

The Franciscan theme in integral ecology also suggests a number of insights for addressing the "Livestock Revolution." First, and most fundamentally, it redefines what the act of raising and caring for livestock ought to be. In place of animal management (treating animals as an asset from which to maximize productive outputs), the Franciscan vision of integral ecology suggests a renewed form of animal husbandry. This kind of husbandry, as defined by Wendell Berry, is "the name of a connection." It is "the art of keeping tied all the strands in the living network that sustains us" (Berry 2005).

What does it look like to put husbandry into practice? As Berry also notes, "In its original sense, it is the name of the work of a domestic man, a man who has accepted a bondage to the household." Whereas for much of the development discussions surrounding the "Livestock Revolution" communities, cultures, organizations, household structures, etc., are all negotiable and open to change to make way for progress, integral ecology recognizes that it is from within renewed and counter formative communities that true social and ecological change is grounded—even if those communities do not enact political change in the traditional sense. Pope Francis writes:

> Not everyone is called to engage directly in political life. Society is also enriched by a countless array of organizations which work to promote the common good and to defend the environment, whether natural or urban. [...] Around these community actions, relationships develop or are recovered and a new social fabric emerges. Thus, a community can break out of the indifference induced by consumerism. These actions cultivate a shared identity, with a story which can be remembered and handed on.

Thus, renewed agriculture and renewed relationships with the natural world will also be rooted in new types of institutions. This stems from the fact that Pope Francis's critique is fundamentally a fresh retelling of the Christian *mythos* against the *mythoi* of the technocratic paradigm: Because the critique is on the level of *mythos*, the transformation it calls for cannot be merely individual; but must be received, embodied, and passed on by communities. Depending upon the harms, challenges, and affordances in their context, we could imagine communities that embrace integral ecology by reducing their exposure to intensive animal agriculture by eating less meat or no meat at all. As we have already had occasion to mention, "eating is an agricultural act" and that means "Eaters [...] must understand that eating takes place inescapably in the world, that it is inescapably an agricultural act, and that how we eat determines, to a considerable extent, how the world is used (Berry 2010, p. 149). Integral ecology, like the larger body of Catholic Social Teaching that it adds to, prioritizes *participation* as a key criterion. The antidote to the consumerist approach to agriculture that has resulted in intensive forms of animal agriculture cannot be limited to changes in our habits as consumers but must, above all, be greater participation in agriculture (see Pope Francis 2015, §144).

On one level, supporting husbandry could take the form of supporting smaller, less specialized farms, mainly family farms, that do not define their relationship to animals in narrow, highly specialized, highly managerial terms. These would be farms, in other words, with deep knowledge of the animals present, farms that define their responsibility to animals in a way that emphasizes their intrinsic value and not only their exchange value. Supporting husbandry in this sense could entail supporting those practices that pre-date the Industrial Revolution and the rise of modern "agricultural science" and "animal science." These might include, for example, practices and rights of pastoralists, who are able to travel with animals and thus allow those animals to maintain their seasonal migration and journeys and preserve their habitats. (See *Laudato si'* §35 for a discussion on protecting the migration patterns of animals.) It may also include support for traditional agricultural practices that involve a positive interaction between humans and animals. Examples of such practices might include grazing on savannas and prairies, which, in combination with prescribed burning, enhances carbon sequestration in soil (Bai/Cotrufo 2022).

The kind of animal husbandry suggested by *Laudato si'* is, it is clear, is incompatible with a view of business that would conceptualize the profit maximization as the sole responsibility of a business. Nevertheless, it is not incompatible with profit and, in fact, could have as its aim "wealth creation" in a rich sense. This broader conception of wealth creation would be by nature long-term and would include "physical, financial, human, and social capital." This kind of wealth creation is not simply for enriching oneself but "encompasses private and public wealth" and would insist that wealth creation is not synonymous with possession or acquisition of wealth but is equated with "making something new and better" (Enderle 2009, p. 287).

In sum, this exploration of the "Livestock Revolution" has revealed that the Franciscan and liberationist themes of integral ecology do indeed have a distinctive contribution to make to our way of regarding animals in the Anthropocene. The liberationist theme of integral ecology insists that human beings can and should recognize the ways animals become our victims and should attempt to live without making animals our victims. The Franciscan theme of integral ecology offers a way of combining a non-instrumentalizing, appreciative view of animals with one that retains a qualified and "situated" anthropocentrism: humans retain the distinctive, central role of cultivating the flourishing of all creation. It would recognize that renewed ways of regarding animals are more than a matter of personal commitments. They also involve the cultivation of new forms of community that embody a different way of human social existence—one that resists the individualism of the technocratic paradigm and can then include, by extension, a renewed way of regarding animals.

References

Aseneta, A. A. R. (2017) 'Laudato Si' on Non-Human Animals', *Journal of Moral Theology*, 6(2), pp. 230-245.

Alemu, R. R. (2023) *CCA Response to Pope Francis' Apostolic Exhortation Laudate Deum.* At: https://catholic-animals.com/activities/cca-response-to-pope-francis-apostolic-exhortation-laudate-deum/ (1 September 2025).

Bai, Y. and Cotrufo, M.F. (2022) 'Grassland soil carbon sequestration: Current understanding, challenges, and solutions', *Science*, 377(6606), pp. 603-608. https://doi.org/10.1126/science.abo2380.

Berry, W. (2005) 'Renewing Husbandry', *Orion Magazine*, September/October. At: https://orionmagazine.org/article/renewing-husbandry/ (1 September 2025).

Berry, W. (2010) *What are people for?: Essays*. New York: Catapult.

Boff, L. (2015) 'The Magna Carta of integral ecology: Cry of the Earth, Cry of the poor', *Earth Charter*, 29 June. At: https://earthcharter.org/article-by-leonardo-boff-on-the-popes-encyclical/ (5 May 2025).

Castillo, D. P. (2016) 'Integral Ecology as a Liberationist Concept', *Theological Studies*, 77(2), pp. 353–376.

Christians for Animals - International (2022) *Integral ecology needs to include the animals to succeed*. At: https://laudatosiactionplatform.org/wp-content/uploads/2022/07/our-common-home-integral-ecology-needs-to-include-the-animals-to-succeed.pdf (20 May 2025).

CNN Español (2022) 'Papa Francisco dice que dejar de comer carne ayudaría a salvar el planeta', *CNN Español*, 13 July. At: https://cnnespanol.cnn.com/2022/07/13/papa-francisco-carne-renuncia-argentina-orix/ (5 May 2025).

Delgado, C., Rosegrant, M., Steinfeld, H., Ehui, S. and Courbois, C. (1999) *Livestock to 2020: the next food revolution.* IFPRI Food, Agriculture, and the Environment Discussion Paper 28. Washington, DC: IFPRI.

Delgado, C., Rosegrant, M., Steinfeld, H., Ehui, S. and Courbois, C. (2001) 'Livestock to 2020: the next food revolution', *Outlook on Agriculture*, 30(1), pp. 27-29.

Dicastery for Promoting Integral Human Development (2025) *Laudato Si' Goals.* At: https://laudatosiactionplatform.org/laudato-si-goals/ (1 September 2025).

El Financiero (2022) '¿Quieres salvar al mundo? El Papa Francisco tiene una simple sugerencia: deja de comer carne', *El Financiero*, 12 July. At: https://www.elfinanciero.com.mx/mundo/2022/07/12/quieres-salvar-al-mundo-el-papa-francisco-tiene-una-simple-sugerencia-deja-de-comer-carne/ (5 May 2025).

Enderle, G. (2009) 'A rich concept of wealth creation beyond profit maximization and adding value', *Journal of Business Ethics*, 84(S3), pp. 281-295.

Francis I (2015) *Laudato Si': On Care for Our Common Home.* At: https://www.vatican.va/content/francesco/en/encyclicals/documents/papa-francesco_20150524_enciclica-laudato-si.html (8 May 2025).

Francis I (2022) *Message of His Holiness Pope Francis to the Participants in the International Youth Conference* (Lisbon, 27-31 July 2022). At: https://www.vatican.va/content/francesco/en/messages/pont-messages/2022/documents/20220706-messaggio-conferenza-giovani.html (5 May 2025).

Francis I (2023) *Laudate Deum: Apostolic Exhortation to all people of good will on the climate crisis.* At: https://www.vatican.va/content/francesco/en/apost_exhortations/documents/20231004-laudate-deum.html (8 May 2025).

Hayek, M. N. (2022) 'The infectious disease trap of animal agriculture', *Science Advances*, 8(44), p. eadd6681. https://doi.org/10.1126/sciadv.add6681.

Malhi, Y. (2017) 'The concept of the Anthropocene', *Annual Review of Environment and Resources*, 42(1), pp. 77-104.

Rulli, M. C., D'Odorico, P., Deplano, C. T., Chiarelli, D. D. and Tuninetti, M. (2022) 'The water-energy-food-environment nexus of animal-to-plant-based diet transitions', *Science Advances*, 8(44), p. eabo0158. At: https://www.science.org/doi/10.1126/sciadv.abo0158 (5 May 2025).

Singer, P. (2025) 'Francis on Animals', *Project Syndicate*, 2 May. At: https://www.project-syndicate.org/commentary/pope-francis-reject-human-domination-but-no-practical-change-by-peter-singer-2025-05 (5 May 2025).

Specht, J. (2019) *Red Meat Republic: A Hoof-to-Table History of How Beef Changed America.* Princeton, NJ: Princeton University Press.

Yahoo News UK (2022) 'Pope tells young people eating less meat can help save the planet', *Yahoo News UK*, 11 July. At: https://uk.news.yahoo.com/pope-tells-young-people-eating-110310764.html (5 May 2025).

Chapter 19
"Be Welcoming": Ecology of Hospitality as a Core Practice of Integral Ecology

Sister Damien Marie Savino

Abstract This chapter begins by exploring the notion of an "ecology of hospitality," keeping these dimensions of integrality at the forefront. Next it addresses how a religious perspective and theological understanding enrich the concept. Then it reflects on the process of cultivating an ecology of hospitality to God, others, and the earth. Finally, the chapter concludes with some comment on how an ecology of hospitality can help heal the wounds inflicted by pervasive ecologies of neglect that dominate so many social-ecological landscapes today.

Keywords Hospitality · Care for creation · Interior development

19.1 Introduction

In April 2025, a new Vatican-commissioned statue was unveiled in St. Peter's Square, located in the colonnade near the "Palace of the Poor," a converted nineteenth century Vatican palace that offers housing, hot meals, showers, medical assistance, and counseling for the homeless. The sculpture, when viewed from the front, portrays a homeless person with a staff and large sack on his back, sitting on a bench. When viewed from the back, the sack is revealed to be the wings of an angel, whose hand beckons the viewer to sit down next to him on an empty space on the bench. The sculpture, entitled "Be Welcoming," invokes the Scripture passage from Hebrews 13:2: "Be welcoming to strangers; many have entertained angels unawares."

Another translation of this Scripture verse reads: "Do not neglect hospitality, for through it some have unknowingly entertained angels." Take special note of the word 'neglect' in this passage. When one neglects someone or something, one does not pay attention to that person or thing; one literally does not pick up, collect, or gather

Sister Damien Marie Savino, University of Notre Dame, ksavino@nd.edu.

S. D. M. Savino (✉)
University of Notre Dame, Notre Dame, Indiana, USA
e-mail: ksavino@nd.edu

R. Marcantonio et al. (eds.), *Integral Ecology in the Anthropocene: Africa and Beyond*, The Anthropocene: Politik—Economics—Society—Science 49,
https://doi.org/10.1007/978-3-032-32125-1_19

it (fr. *legere*), but rather leaves it by the wayside (Etymonline—Online Etymology Dictionary, no date). In other words, not only do neglectful actions deny the value or worth of people or things, they also, even more egregiously, express indifference to their existence; they are blind to the very *being* or reality of the person or thing. It is to this neglect of both value and being that hospitality is called to respond.

This chapter explores the contours of an "ecology of hospitality" as one antidote to the widespread "ecologies of neglect" that plague contemporary culture, especially in the context of the summons to integral ecology articulated by Pope Francis in his environmental encyclical *Laudato Si': On Care for our Common Home* (2015). In *Laudato Si'*, Francis lamented the neglectful dispositions that deafen us to the cry of the earth and the cry of the poor (para. 49) and emphasized the need for an integral ecology attentive to both (paras. 137–162). Integral ecology demands a level of care to which an ecology of hospitality could make a meaningful contribution, as will be explored in this chapter.

19.1.1 Background and Origin of the Concept

The notion of an ecology of hospitality was envisioned by the author during a field visit to Bethany Land Institute (BLI) in Uganda with a student group from the University of Notre Dame in June 2024. "You are most welcome!" was a common greeting received throughout the visit. The radical experience of welcoming experienced during the visit inspired the idea that cultivating an ecology of hospitality could help build the culture of care needed to advance the mission of integral ecology.

The Bethany Land Institute is an educational program in regenerative agriculture for the rural poor, situated on several hundred acres of land in the Luwero District of central Uganda. The program educates youth in integral ecology and economic entrepreneurship through three branches: Mary's School, Lazarus' Forest, and Martha's Market (Bethany Land Institute, no date). The core of BLI's work and mission is the Caretaker Program, a two-year residential program that trains young persons to be 'caretakers' through personal formation and hands-on instruction in farming, agroforestry, and agribusiness. In the spirit of *Laudato Si'* and its call to "care for our common home," the vocation to caretaking goes beyond stewardship (as valuable as this is) and demands a commitment to love and service through deep personal investment and extension of self to others and to the earth.

The warm welcome and generosity extended to the Notre Dame group by the BLI instructors and caretaker trainees, the beauty of the land, and the intensive efforts toward social-ecological restoration of the land and local communities all pointed to a kind of radical "ecology of hospitality" at the heart of BLI's caretaking mission. It seemed that hospitality was integral, not only to the practical living of the caretaking mission but also to its spiritual expression, since BLI is scripturally anchored in the charisms of Mary, Martha and Lazarus, with their fundamental expressions of hospitality to Christ. These 'entwined' spiritual and practical dimensions are essential to envisioning an ecology of hospitality in the spirit of integral ecology.

What is integral ecology? This concept was introduced by Pope Francis in *Laudato Si'* to refer to an ecology that "takes us to the heart of what it is to be human" and is integral in respecting its human and social dimensions alongside its scientific ones (paras. 11, 137). Three senses of integrality are especially relevant here. First, integral ecology recognizes the interconnections between natural and human communities and is attentive to the 'groaning' or neglect of both (paras. 138–139). Second, it draws upon the language and values not only of science but also of religion, because "no branch of the sciences and no form of wisdom can be left out, and that includes religion and the language particular to it" (paras. 62–63). Further, since the life and example of St. Francis of Assisi were identified by Pope Francis as his inspiration, Franciscan spirituality represents an essential aspect of that religious language (para. 10). Third, it is directed to reconciling the three fundamental relationships for which humans are responsible, as articulated in paragraph 66 of *Laudato Si'*—with God, with others and with the earth.

This chapter begins by exploring the notion of an "ecology of hospitality," keeping these dimensions of integrality at the forefront. Next it addresses how a religious perspective and theological understanding enrich the concept. Then it reflects on the process of cultivating an ecology of hospitality to God, others, and the earth. Finally, the chapter concludes with some comment on how an ecology of hospitality can help heal the wounds inflicted by pervasive ecologies of neglect that dominate so many social-ecological landscapes today.

19.2 "Ecology of Hospitality" Defined

The word 'ecology' refers to the study of the 'ecos,' which is Greek for 'home' (Etymonline—Online Etymology Dictionary, no date). Broadly, ecology is the study of our common home and the interconnected relationships within it. It has burgeoned as a scientific discipline since the last half of the twentieth century and now encompasses the study, from micro to planetary scales, of the complex interrelationships between living organisms, including humans, and their physical environment (The Ecological Society of America, no date).

The word 'hospitality' is derived from the Latin *hospes*, meaning 'guest,' 'host,' or 'stranger,' and is also etymologically related to the word, *hostis*, meaning 'enemy' (Etymonline—Online Etymology Dictionary, no date). While they at first appear to be contradictory, these apparently divergent connotations add an important nuance. Hospitality refers to welcoming gestures and practices extended to strangers and even to enemies. "The stranger, perceived as an enemy, can become a guest. Hospitality, then, is a way of overcoming the animosity and making the stranger or enemy a friend" (Hagstrom 2013, p. 4).

The conversion from *hostis* to *hospes*, from stranger or enemy to friend, is a result of more than just the provision of food and material goods by the host, as important as these welcoming gestures are. It is also the host's provision of space—both physical and spiritual—that serves to affirm the innate dignity of the guest and creates "the free space where such a conversion can take place;" the conversion flows from an open, receptive stance on the part of guest and host that allows both "to receive instead of to conquer" (Hagstrom 2013, pp. 4–5).

Hospitality is not a one-way street, but a reciprocal movement between guest and host that fosters the transformation of both. The host initiates and capacitates a give-and-take with the guest that opens time and space and makes room for, rather than passing by or neglecting, the other. In this respect it is a radical counter to contemporary attitudes that perceive strangers as threats or disregard them in a kind of throw-away mentality. In hospitable exchanges, guest and host see each other through new eyes and call forth the personhood of one another as mutual self-gift.

An ecology of hospitality thus refers to this ecosystem of mutual exchange, give-and-take, between guest and host in the many varied interactions that are possible in the landscapes of our common home. This includes not only the ecological relationships, but also, in a faith context, the "spiritual ecosystem," defined by the three prime relationships for which humans are responsible: with God, with others, and with the earth. Religion, then is a helpful dialogue partner on the path to an ecology of hospitality.

19.3 What Can Religion Contribute to an Ecology of Hospitality?

As Pope Francis articulated so clearly in *Laudato Si'*, religion has an important role to play in addressing complex ecological questions, because "solutions will not emerge from just one way of interpreting and transforming reality," and to arrive at them, "no form of wisdom can be left out, and that includes religion and the language particular to it" (2015, para. 63). Following this reasoning, religious language has a valuable role to play in articulating a robust ecology of hospitality and could make the difference in moving the needle from ecologies of neglect toward those of care. This section suggests some key theological underpinnings for an ecology of hospitality.

First, Trinitarian theology is important, because hospitality is a virtue that begins with and flows from the triune God, who *is* hospitality. In the Catholic intellectual tradition, the ultimate origin and source of human hospitality is the Divine hospitality of the Trinity. As a communion of persons, the Trinitarian God pours himself out totally, exemplifying the self-diffusiveness of persons, from Father, Son, to Holy Spirit, that is the template, if you will, of all human acts of hospitality.

The medieval Franciscan theologian, Saint Bonaventure, conceived of the Trinity as the "fountain fullness"—a torrent of goodness, love and truth originating in the trinitarian relations between Father, Son and Spirit that overflowed into the created world. (Bonaventure 2002, 123–131). As the source of creation, the dynamic, self-communicating, trinitarian God is imprinted on creation and reflected in the diversity and order of the created world. According to Bonaventure, because "the Trinity has left its mark on all creation . . . *each creature bears in itself a specifically Trinitarian structure,*" . . . which means that the challenge for Christians is to learn "to read reality in a Trinitarian key" (LS 239).

The Trinitarian God in essence summoned creation into existence through an act of divine hospitality—an extravagant act of generosity which is the model of, and motivation for, human hospitality. Divine hospitality is communicated in a particular way in and by the Word, the second person of the Trinity. This incarnate Word both announces and witnesses to the 'cost' of hospitality, revealed in Christ's outpouring of his life on the cross. The message is that hospitality requires self-gift, and in its ultimate expression, being willing to lay down one's life for the good of the other. It demands the theological virtue of charity and the desire to imitate the self-giving love of Christ. Meeting such a radical demand is not possible by human effort alone but only with the help of supernatural grace. This further reinforces the value of religious faith and practice as an undergirding for an ecology of hospitality.

Two additional theological points deserve mention here. First, hospitality is biblical, as evidenced by the many injunctions to hospitality in the Old Testament (Janzen 2002)—too many to be elucidated here. It is also central to the New Testament and is a defining feature of the life of Christ, who fed the five thousand with five loaves of bread and two fish (Mt 14:15-21; Mk 6: 35-44; Lk 9:11-17; Jn 6:1-14) and washed the feet of his disciples (Jn 13:1-15), to invoke two prominent Scriptural examples. In fact, hospitality is "a lens through which we can read and understand much of the gospel" (Pohl 1999, p. 8). In the gospels, Christ extended hospitality in a special way to sinners, the poor, the neglected, and those on the margins. In imitation of this, hospitality is a practice by which we can continue to welcome Jesus himself into our midst, to which the sculpture described at the opening of this chapter testifies. The act of welcome at the heart of hospitality is a concrete step toward welcoming Christ, the incarnate Son of God, into our lives.

Second, because religious faith teaches that creation itself is an act of divine hospitality, the distinction between the notion of 'creation' and that of 'nature' is an important one. While our technocratic culture has increasingly replaced the word 'creation' with 'nature,' they are not interchangeable terms. Words themselves represent tangible, meaningful concepts. The word 'creation' acknowledges the reality of a transcendent dimension expressed in the world, while 'nature' represents a narrower vision of reality, one confined to physical, material realities. This is not to deny the value of the concept of 'nature' in our contemporary scientific culture, only to assert that it should not be to the exclusion of the word, and concept, of 'creation.' Creation includes nature, but it does not stop there, as it points beyond itself to the reality of the transcendent Creator. It is a "both-and" word. This is an important distinction, for the call to hospitality is deepened and intensified when we embrace the world as 'creation' and not just 'nature.'

19.4 Cultivating an Ecology of Hospitality

How is an ecology of hospitality to be cultivated and with whom is it to be shared? As stated earlier, an ecology of hospitality is both practical and spiritual, and it is bounded by the three fundamental relationships for which humans are responsible: with God, with others, and with the earth (LS 66). In the spirit of integral ecology, it is to these prime relationships that we are to extend hospitality, in both practical and spiritual contexts.

What an interesting reversal that God, the source of hospitality, is now one to whom we are called to extend hospitality! While before he was the host, he now becomes the guest. We have come full circle: When we cultivate hospitality by being welcoming, we do so in imitation of Christ, next to whom we sit on the bench. Christ is both host and guest. When we join him on the bench, we attend to Christ, and are at the same time sent out to sit next to and attend to others. And we do so having been given the gift of grace, which is participation in the life of God (Catechism of the Catholic Church, para. 1997). In the fullness of an ecology of hospitality as proposed here, humans are given the gift and awesome responsibility to extend Christ's hospitality to others—to literally be his eyes, hands, feet and heart.

This exemplifies one of the fundamental tenets of the Catholic faith—the principle of sacramentality. Sacramentality is grounded in the teaching that when the Word, the second member of the Trinity, became incarnate and walked on the earth, all creation was changed; the Divine imprint was left there as a vestige or footprint. This has implications for our human relationships and the care that is to be extended to others in the human family; Christ works through us as we attend to them and build a culture of care. It also resonates with ecological understanding; consider that, in the context of the interconnectedness of the water cycle, for example, the water that flowed from the side of Christ at his crucifixion is circulating somewhere on earth even now. The reality and closeness of the incarnate God shine forth in the created world; this is the sacramental lens through which Catholic Christians see all creation. More will be said about this shortly in relation to the call to care for the earth.

In the Catholic tradition, hospitality is extended to God through embodied worship, the communal sharing of ordinary elements of water, bread, wine, and oil, as they are consecrated by an ordained priest who "stands in" for Christ (*in persona Christi*). In the consecratory action by the priest, the *sacramental* elements of creation (bread and water) become *sacraments*, Christ's body and blood, which is a profound way in which Christ continues to be physically and spiritually present among his people. In the reception of the sacraments as understood in Catholic theology, the relationship is mutual and intimate, because, through Christ's sharing of his body and blood, the sacraments physically and spiritually confer grace upon the persons receiving them. The process also elevates creation and cannot occur apart from it. In all of this we are presented with an apex experience of Divine hospitality and its interconnected spiritual and physical dimensions.

Hospitality is regenerative, as the divine graces received through the sacramental, liturgical life are meant to be extended to others outside of the worship space, in whom the presence of God is also found. The traditional host-guest relationship is 'flipped' as God becomes the guest, seen in the face of the materially or spiritually poor and needy who are served by acts of kindness and hospitality. In and through them, it is God who is welcomed and cared for as we extend God's original hospitality. Human acts of hospitality can be offered to assuage not only the physical, but also the psychological or spiritual wounds of others, even if we are simply present and do no more than "sit on the bench" with them. In so doing, we help remediate toxic ecologies of neglect that drive persons to pass by others, hardly aware of their existence, and then leave the scene.

What has been said thus far focuses primarily on expressions of hospitality to God and others in this entwined physical-spiritual ecosystem. More needs to be said about the responsibility to express hospitality toward the third prime relationship for which humans are responsible—the earth. What does an ecology of hospitality entail in relation to caring for the created world?

An enlightening experience occurred one morning in Uganda when a group of students were visited by bees at the table where they were having a pancake breakfast. Disconcerted by the presence of the bees, the students departed, and a waiter came over, concerned. When informed that the students had left because of the bees, he responded immediately: "The bees only want breakfast." He then removed the remaining pancakes, poured syrup over the plate and left the plate on the table. The bees promptly returned, joined by many others, to eat their breakfast peacefully. What a striking image of extending an ecology of hospitality to creatures!

As with persons, we express hospitality to creation by making room "on the bench" (or at the table) for creation, by making time for creation and giving it space, so to speak. At the same time, as with hospitality to God and others, there is a mutuality in the relationship between humans and the earth because creation opens time and space for us. When we make time and space to be out in creation, it rejuvenates us, calling forth from within us attitudes of awe and gratitude. The created world extends the gift of time, space, and beauty to us as we are refreshed and revitalized by it.

Much research has shown that being in nature is essential for human health and well-being. Nature deficit disorder, as identified by Richard Louv in his book *Last Child Left in the Woods* (2005), refers to the disconnectedness of the "wired generation" from nature and the negative impacts this alienation is having on human development, including such manifestations as rises in obesity, attention disorders, and depression. His work emphasizes the spatial component of connectedness and the benefits of humans being physically connected to the outdoors.

More recent research also explores the "contributions of nature to human sense of time" (Correia 2024). As modern societies struggle with increasing time scarcity amidst the rapid pace of life, especially in urban areas, this research focuses on nature's contributions to a healthy sense of time. For example, "there is an emerging body of research indicating that time feels longer when spent in nature compared to the same time spent in urban environments" (Correia, 360). Spending quality time in nature also seems to decrease impulsive behaviors and feelings of detachment, while restoring the capacity for attentiveness (Correia, 361–362). The study concludes that experiences in nature "can shape our sense of time and help restore a healthier and more balanced relationship with time" that ultimately will result in improved human health and well-being, especially in the context of the increasing stress produced by time poverty in rapidly paced urban life (Correia, 362).

These studies demonstrate the value of spending time in natural spaces, with a focus on the physical value in terms of improved human health and well-being. Spending time in nature can also have spiritual value, as spending time outdoors opens a window for spiritual contemplation. This reinforces the entwined spiritual and practical dimensions of an authentic ecology of hospitality, particularly in relation to our relationship with the earth. Further, when this relationship is understood to be with 'creation' and not just 'nature,' the potential exists for establishing an even deeper connection, because the presence of the Creator can be experienced in and through his creation.

The conviction that the incarnate, Creator God communicates through the created world gave birth to the idea that creation can be read as a book, an analogy that was developed by the Patristic fathers in the first centuries after Christ. Augustine even referred to the book of creation as "a first revelation" before that of the Scriptures (Augustine 1991). Almost a millennium later, the analogy was picked up by Bonaventure, who taught that every creature is "a word of God" and that God created everything "not to increase his glory, but to show it forth and to communicate it" (Catechism of the Catholic Church 2000, para. 293).

Almost a millennium later, in the face of contemporary ecological challenges, contemporary popes revived the analogy. (The analogy also has scientific roots, with Galileo and other scientists viewing their science as one way of reading the book of creation.) John Paul II, for example, called creation "a gospel which speaks to us of God;" through this gospel, we are meant to "discover a transcendent presence" that "must lead us also to rediscover our kinship with the earth, to which we have been linked since our own creation" (John Paul II 2000, para. 5). Benedict XVI wrote, "*Nature expresses a design of love and truth*. . . . [I]t is a wondrous work of the Creator containing a 'grammar' which sets forth ends and criteria for its wise use, not its reckless exploitation" (Benedict XVI 2009, para. 48). Pope Francis wrote in *Laudato Si'*: "God has written a precious book, 'whose letters are the multitude of created things present in the universe'. . . This contemplation of creation allows us to discover in each thing a teaching which God wishes to hand on to us" (para. 85).

As a book, the created world speaks a language that humans have a responsibility to learn (Savino 2026). The created world is constantly communicating the revelation of God the Creator, so that creation is itself an expression of God's hospitality. In response, humans are called to reciprocate and be hospitable—by listening, being attentive, and taking good care of a created world that gifts humanity with not just material provisions, but also beauty, peace and the capacity to glimpse and comprehend the communications of an expressive creator.

Note that the notion of hospitality also applies to the nature of a place, such that when places are hospitable, they "offer a pleasant or sustaining environment" (Merriam Webster 2025). If creation is a book and an essentially communicative reality, a core element of the human responsibility for practicing an ecology of hospitality toward the earth is to learn to read the book of creation and obey the grammar written there. If humans do not act in accord with that grammar, nature itself will rebel and will no longer be a hospitable space.

Is this not the reality confronting contemporary society in the face of prevailing ecologies of neglect? We live in a wounded world in which both the non-human and the human creation are groaning. Not all this groaning is the result of human transgressions, but humans do bear responsibility for a significant portion of it. The final section of this chapter reflects on the value of an ecology of hospitality for healing this woundedness.

19.5 Concluding Thoughts

From the beginning of his pontificate, Pope Francis spoke about the church as a "field hospital" (Spadaro 2013), and it is interesting that the word hospitality is also related to the etymology of the word 'hospital.' Hospitality thus includes the connotation of healing as well as welcoming. The role of humans in the created world, especially as called for by integral ecology, is to be a healing force for both natural and human ecology. Putting an ecology of hospitality into practice could look like a "field hospital"—one, in other words, that is rather dirty and messy (or one might say, polluted) but in which the doctors, nurses and assistants are attuned to the cry of the earth and the cry of the poor and are doing all they can to heal them both.

As the best doctors and nurses are attentive to their patients, attentiveness is a fundamental requirement of an ecology of hospitality—attentiveness to the delicate, complex balances in the created world and to the needs of species other than us, along with attentiveness to other human persons, especially those on the peripheries. Further, attentiveness can help move persons out of surficial self-preoccupation and into a deeper inner life that is more attuned to, and touched by, the cry of the earth and the cry of the poor. Such interior development is essential for persons to find the motivation to see and anticipate the needs of others and of the created world. Without it, humanity will not be able to meet the demands of contemporary socio-ecological crises with meaningful, sustainable and enduring solutions. Indeed, the best thinking and most advanced technologies in the world will not be successful (and have not been, to date) at healing the wide-ranging socio-ecological 'dis-ease' plaguing the world; what is needed is a commensurate inner development of human persons—an inner ecology, in other words, capable of confronting and reconciling disordered external ecologies.

The work we are called to do in the field hospital of the world thus has to do with reconciling disordered internal ecologies, from the inside out and the ground up, as the path to healing diseased or polluted external ecologies. "Heal the wounds, heal the wounds," as Pope Francis responded when asked what the church needs most at this time (Spadaro 2013):

The thing the church needs most today is the ability to heal wounds and to warm the hearts of the faithful; it needs nearness, proximity. I see the church as a field hospital after battle. It is useless to ask a seriously injured person if he has high cholesterol and about the level of his blood sugars! You have to heal his wounds. Then we can talk about everything else. Heal the wounds, heal the wounds. . . And you have to start from the ground up.

An ecology of hospitality involves listening to the groaning of the poor and of the earth and working to heal wounded natural and human ecologies. Doing this will require attentiveness to, and healing of, distorted internal human ecologies. It will require a reorientation and healing of the three prime relationships for which humans are responsible—with God, others, and the created world.

When the Pope indicates that this healing must happen from the ground up, he seems to imply that it should happen locally, in small communities and small ways –by being together, taking time to work the land or walk together and be with one another in the pressures of daily life; by growing and sharing food together, as in the regenerative agriculture movement and the work being pioneered by the Bethany Land Institute; by sharing together in liturgies and the sacramental life; in praying for the grace to forgive and heal superficial and self-absorbed attitudes of consumerism and to reconcile the tensions between natural and human ecologies.

As noted earlier, the practical and spiritual dimensions of an ecology of hospitality are interconnected and interdependent. St. Francis of Assisi is excellent witness of these entwined dimensions and how they played out in a life of holiness through a strong witness of hospitality to God, others, and the created world. In fact, St. Francis has been aptly called an "Icon of the Hospitable Imagination" (Keating 2017). One simple gesture that represents this is his hospitality to worms, which he demonstrated by picking them up on the roadside so they would not be stepped on (Keating 2017). He did this, not only to be kind to worms, but also because worms made him think of Christ on the cross, harkening back to the Scriptures that referred to Christ as a "worm and no man." These entwined spiritual and practical manifestations of hospitality are central to Franciscan spirituality, as Francis sought to live his incarnate spirituality and literally configure himself to Christ. His prayer was answered when he received the stigmata—another entwined physical-spiritual manifestation of hospitality in his openness to receiving the actual wounds of Christ. While those seeking to practice integral ecology today may not actually receive the stigmata, they will be stigmatized and configured to Christ in other ways as they "get dirty" in the field hospital of the world.

An ecology of hospitality is one that is not afraid to "get dirty," to work hard in the field hospital of the church and the world, in degraded human and natural ecosystems, instead of staying clean and on the sidelines. It is one that is not afraid to pick up worms, analogically speaking, whether they be neglected people or polluted earth. It is not afraid to sit next to the stranger on the bench, or to allow the bees to dine. The Japanese practice a kind of pottery art called kintsugi, which pieces together and repairs a broken pot with a lacquer of gold. In the end, the finished pot is stunningly more beautiful than the original. The potential of an ecology of hospitality as proposed here is to participate in an analogous mystery of beauty through woundedness, one that extends into wounded communities and places and repairs them with the "gold" of an ecology of hospitality shared with God, others, and the earth.

References

Augustine of Hippo (1991). "Sermon 68, 6," in Rotelle, J.E., OSA, ed. and Hill, E. O.P., trans. *Sermons 51-94, The Works of Saint Augustine III: A Translation for the 21st Century.* Brooklyn, NY: New City Press, 226.

Benedict XVI (2009). *Caritas in Veritate*: Encyclical Letter on Integral Human Development in Charity and in Truth. Vatican.va.

Bethany Land Institute (BLI) (no date). *Forming Leaders in Integral Ecology for Rural Transformation in Uganda.* At: https://www.bethanylandinstitute.org/. (April 11, 2025).

Bonaventure. (2002) "Itinerarium Mentis in Deum" in Boehner, Philotheus and Hayes, Zachary, O.F.M, eds. *Works of St. Bonaventure.* Volume II. St. Bonaventure, NY: Franciscan Institute Publications, pp. 123–131.

Catechism of the Catholic Church (2000). San Francisco: Ignatius Press.

Correia, R. A. (2024) "Acknowledging and understanding the contributions of nature to human sense of time," *People and Nature* 6: 358–366.

Ecological Society of America (no date). At: https://esa.org/about/what-does-ecology-have-to-do-with-me/. (May 1, 2025).

Etymonline – Online Etymology Dictionary (no date). At: https://www.etymonline.com. (April 14, 2025).

Francis (2015). *Laudato Si: On Care for our Common Home.* Vatican Press.

Hagstrom, Aurelie A. (2013) 'The Role of Charism and Hospitality in the Academy,' *Integritas* 1(1), pp. 1–14.

iBreviary (no author, no date). Intercessions for the Celebration of the Passion of Our Lord. At: https://www.ibreviary.com (May 1, 2025).

Janzen, W. (2002). 'Biblical theology of hospitality,' *Vision* 6(1), 1–14.

John Paul II (2000). General Audience, January 26, 2000. Vatican.va.

Keating, J (2017). "St. Francis of Assisi: Icon of the Hospitable Imagination." *Church Life Journal* [online]. At: https://churchlifejournal.nd.edu/articles/st-francis-of-assisi-icon-of-the-hospitable-imagination/ (December 13, 2024).

Louv, R (2005). *Last Child Left in the Woods: Saving Our Children from Nature-Deficit Disorder.* Chapel Hill: Algonquin Press.

Merriam-Webster Dictionary (2025). At https://www.merriam-webster.com/dictionary (April 29, 2025).

Pohl, C. (1999) *Making room: Recovering hospitality as a Christian tradition.* Grand Rapids: Eerdmans.

Savino, D.M.S., FSE (2026). *Learning the Language of Creation: Catholic Social Teaching and Integral Ecology.* Collegeville, MN: Liturgical Press.

Spadaro, A., S.J. (2013). *A Big Heart Open to God: An interview with Pope Francis.* At: https://www.americamagazine.org/faith/2013/09/30/big-heart-open-god-interview-pope-francis (March 22, 2025).

Chapter 20
Disturbing the Silence: Integral Ecology as a Response to AI and the Technocratic Mindset

Heather M. Foucault-Camm

Abstract In designing a machine that can do the sorts of things our minds can do but *faster*, it is not surprising that the rise of digital technologies have gripped our imaginations. More accurate medical diagnoses, the prediction of weather events with greater precision, unlocking one's phone – it is hard to escape technology's impact on daily life. Hand in glove with these advancements, however, are serious and significant questions about our place in the world. Is our ability to think, to discern, and to create replicable in a 'thing'? Are we organic machines? When considered within a technocratic mindset, the tool becomes exalted above its maker; who we *are* is subsumed by what we can make as we respond 'yes' to these questions. Integral ecology, by contrast, provides the hermeneutic for distinguishing between what constitutes true progress and what threatens our ultimate good. By drawing attention to our being created in three-fold relationship to God, each other and creation, integral ecology provides the framework we need to rightly orient any tool to its user. It will be the goal of these comments, therefore, to consider the landscape of current developments in digital *tools* alongside the guiding influence of integral ecology. By consequence, a robust theology of progress will be presented in order to counter the tool-first culture that is eclipsing daily life. With hope, we will look to the principles of *Laudato si* as we explore the rise of technology and the technocratic mindset, seeking to build up a future where the God-given fruits of our ingenuity are in service to all.

Keywords Technology · Artificial intelligence · Imagination · Personhood

H. M. Foucault-Camm (✉)
University of Notre Dame, Notre Dame, United States
e-mail: hfoucaul@nd.edu

R. Marcantonio et al. (eds.), *Integral Ecology in the Anthropocene: Africa and Beyond*, The Anthropocene: Politik—Economics—Society—Science 49,
https://doi.org/10.1007/978-3-032-32125-1_20

20.1 Introduction

Hello darkness, my old friend
I've come to talk to you again
Because a vision softly creeping
Left its seeds while I was sleeping
And the vision that was planted in my brain
Still remains
Within the sound of silence
(Simon et al, 1964)

This opening stanza from *The Sound of Silence*, a song written by Paul Simon in 1963, points to a reality of concern sixty years ago that has become even more radically intrusive today. What is softly creeping is the persistent and pervasive vision that to be fully human in our current time equates to what we can make with increasing sophistication. There are few tools in our possession that exemplify this more than what can be grouped under the general classification of *digital technology*. These fruits of our ingenuity are *designed* to superimpose themselves on what we would almost, universally, classify as modes of action that are uniquely *human*—observation, judgement, synthesis. In the apostolic exhortation *Laudate Deum*, Pope Francis reminds us that we are immersed in a paradigm whose foundations stem from the, 'notion of a human being with no limits, whose abilities and possibilities can be infinitely expanded thanks to technology' (Francis 2023, n. 21). What is beautiful, what is good and what is true, especially where the concept of who a person *is* are concerned, becomes reduced to efficiency and control, profit and advantage.

While moments of technological progress in humankind's past have elicited similar concerns, there is something unique about this 21st century moment. When Romano Guardini commented that 'never before had humanity had such power over itself, yet nothing that ensures it is used wisely' in 1956, he could have been easily commenting on our current situation of supercomputers and social media. (Guardini, 1998, p. 174) As humanity continues to push towards a horizon where what we construct could actually prove sufficiently disruptive to our flourishing, what distinguishes technologies of the past from what is present now is not the *degree* of effect, but the *kind*. To say that technologies contained under the classification Artificial Intelligence (AI) are radical exemplifications of our preoccupation with technological advancement is to hold that, like no other tool before it, AI has the potential to revolutionize all aspects of our existence, especially self-concept. What we value, what is true, how we interact with each other and with the created world – all of this has begun, and will continue to be absorbed into, the domain of AI. While there is no denying that the goods afforded by this technology will save lives, these advantages cannot be disentangled from the risks of the atrophy of authentic human relationships and the distortion of what is actually 'true' in reality.

The question remains, then, that if Artificial Intelligence represents a mixed set of instances (i.e. risk and benefit), how should we interact with it? Even if one were to hold the extreme position that technology that could potentially disrupt our ability to discern what is 'real' should be avoided, this remains a stance that will quickly become untenable (by picking up one's iPhone and engaging facial recognition we make use of AI). Therefore, and where our interaction with any technology in the world, especially artificial intelligence, is concerned, we would do well to re-present for ourselves and each other what is truly meant by what it means to be *human*. By cultivating that we are more than just output based on input, that we are gifted with free agency, created to give and receive love by Love itself, we position ourselves to resist the reductive and superficial anthropology that is so easy to succumb to with AI.

It will be argued in this paper, therefore, that in the midst of the existential threat that a technocratic imagination radically exemplified by Artificial Intelligence poses, a recovery of the Christian concept of personhood is key to our learning to co-exist with the ever-advancing fruits of our ingenuity. Drawing, principally, on the thought of Pope Francis and commentaries on his recent writings, this paper will elucidate in theological language for both the Academy and the Church writ large, what makes the human person different, in kind, from a computer. This, it will be asserted, is the difference that the disrupting influence of a Christian anthropology can make—it asserts that humans are just that, radically different. It is precisely because this concept of personhood *necessarily* orders our gaze beyond ourselves that our tools can take on their true purpose—not as ends in and of themselves but as an assistance to us in the pursuit of ours, communion with each other, with creation and with our Creator.

20.2 An Imagination Confined by Progress

> 'Neath the halo of a street lamp
> I turned my collar to the cold and damp
> When my eyes were stabbed by the flash of a neon light
> That split the night
> And touched the sound of silence
>
> (Simon et al., 1964)

Before delving headlong into the technical aspects of progress in an age of technology, before dissecting what is *really* meant by Artificial Intelligence—ponder, for a moment, the scene being painted by this stanza. There is an element of violence one can sense as one considers as a visual metaphor the 'flash of neon light' that managed to 'split the night' and stabs the eyes. The key to unlocking the deeper meaning in this stanza is the choice of the word 'neon'—neon is a contemporary invention, it is synthetic illumination, neon is not soothing like a candle but is artificial, oftentimes intrusive. As a representation of the pervasiveness of technology, there are very few descriptions that could be superior to this one. Like the neon lights that are all around us, computers, cellphones, artificial intelligences—all of this is designed to impose itself on the individual. Moreover, and despite the unprecedented ability to communicate, there is an epidemic of loneliness. Holding onto this reality, one is almost cornered into asking 'Is this *progress*?'.

20.2.1 Thoughts on Progress

There can be no argument made, from a Christian perspective, for this being progress—without, as least, some qualification. In drawing upon the Catholic Church's social tradition, progress appears to be most closely related to the principle of the Common Good. This principle is described in *Gaudium et spes* as:

> The sum of those conditions of social life which allow social groups and their individual members relatively thorough and ready access to their own fulfillment [which] today takes on an increasingly universal complexion and consequently involves rights and duties with respect to the whole human race. (Paul VI 1965, n. 26)

There is nothing antithetical to technology in this statement. However, reading further into the document there is a clue as to what constitutes false progress.

Indeed today's progress in science and technology can foster a certain exclusive emphasis on observable data, and an agnosticism about everything else. For the methods of investigation which these sciences use can be wrongly considered as the supreme rule of seeking the whole truth. By virtue of their methods these sciences cannot penetrate to the intimate notion of things. Indeed the danger is present that man, confiding too much in the discoveries of today, may think that he is sufficient unto himself and no longer seek the higher things (Paul VI, 1965, n. 57).

False progress, then, is when persons are subordinate to things. Simply put, when the tool is more valued than the *person* who made and uses it.

20.2.2 The Technocratic Paradigm

This distorted ordering of persons to tools has an iron grip on the human imagination. We have become ever more focused on the increasing sophistication of what we can make as we attempt to control and confine. (Francis 2010, n. 106, 224) This so-called *technocratic paradigm* is a reality where:

Every increase in power means an increase of 'progress' itself, an advance in "security, usefulness, welfare and vigor; …an assimilation of new values into the stream of culture", as if reality, goodness and truth automatically flow from technological and economic power as such. (Francis 2010, n. 105)

This paradigm offers an easy solution to a world that can lack the appearance of order and which needs radical improvement. But the technocratic paradigm is a Trojan horse that transforms a duty of care mediated by scared responsibility into a mode of operating where what is most valued is what can be ultimately controlled and consumed. Human relationships do not matter in this paradigm, the world matters only in so far as it can be exploited.

And yet, many who are users of cellular phones, of computers, of GPS devices, many of these individuals would struggle to see how these devices are, themselves, so singularly problematic. Their confusion would be well-founded as ‘technology’ can have positive benefits and lead to tangible improvements in quality of life. Perhaps what is *actually* problematic is that the technocratic paradigm has become an epistemology? (Francis 2010, n. 107; Deane-Drummond 2023, pp. 102–104) Slowly at first, but at increasing speeds with every new advancement, we are persuaded to believe that the key to our fulfillment lies in efficiency, productivity, and control. Nothing, not international relationships, nor economies, nor research—all facilitated by persons immersed in this paradigm—are immune to the effects. Such an epistemology as the technocratic paradigm is fixated on the fantasy of limitless possibility as re-enforced by ever-advancing technology, with the ironic result being, ultimately, the limiting of our development into who we were actually *created* to be. But there is an irony here—that this epistemology's telos of limitless possibility does not make us ‘more’ but actually ‘less’; reduced to simply a cog in the techno wheel we are a shadow of who we are created to be.

20.2.3 Conclusion

The question being asked in this section—how should human persons relate to the fruits of their ingenuity—is of particular interest in the 21st century but, ultimately, represents a consideration that transcends computers and electronics. For as long as the human family has been looking to the natural world and seriously considering their place in it there has been a need to invent, to problem-solve, and to improve. What is unique about this current moment in history is that never before has such technological power been in the hands of relatively so few individuals—the power to split atoms, the power to alter our genetic code, the power to distort what is true on a vast scale. Moreover, the nature of this power, as mediated by technology, continues to advance at a pace that exceeds our ability to carefully discern the fullness of its effects. We stand, repeatedly, blinded by the ‘neon light that splits the night’, the radical artificialness that seeks to persuade the person that they are no different from the objects around them—to be used, controlled, consumed, discarded. This is what progress has become and Artificial Intelligence, the next great thing we have created, not only radically expands the boundaries of the technocratic paradigm but radically challenges who we *think* we truly are.

20.3 The Future Rolling Toward Us: AI as the Technocratic Paradigm

The radical exemplification of the technocratic paradigm that Artificial Intelligence (AI) is can be traced to AI's very function—its telos is not only to *mimic* the sorts of actions that human minds can do but surpass our own natural constraints. AI has a grip on our imagination because there is something addictive about a future without limitation—as possibly, one day, we may even cheat death by 'downloading' ourselves into a system for indefinite access. But is this the reality that we *want*, is this the meaning that our lives are created to discover and to question? There is also something rather stifling about this reality of limitlessness where we can also send a text at the speed of light but not connect or converse, where people, 'talk without speaking, hear without listening'. Artificial Intelligence becomes the radical exemplification of this danger because our interaction with it is novel—it has the *appearance* of creativity, it has the *appearance* of moral reasoning—and thus begets the very natural question, with whom or with what are we interacting when we use it.

Therefore, and in order to better understand what our response to AI should be, it seems prudent at this juncture to discuss what AI is. Simply, the objective of this technology is to make computers do the sorts of things that human minds can do (Frankish, Ramsey, 2014; Boden, 2018, p. 1). But, the following caveat should be added—'through human-engineered systems'—because AI is a human-generated, controlled, and deployed technology. The AI does not *think*, it does not possess *intuition*—it executes a program. The more data the AI has access to, whether one considers machine learning or large language models, the more accurate the output that the user receives will be.

What masquerades as 'creative' thought for artificial intelligence is simply an increasingly well-aligned 'if X, then Y' calculus—one that is entirely dependent on human input. As persons, by contrast, we are *free* to observe, to consider, to evaluate, to synthesize. Generative Artificial Intelligences are utterly dependent—they will not get better, they will not execute their pre-determined set of rules, or algorithms, without persistent interaction with human users. Moreover, and where the consideration of 'newness' is concerned, there is a profound distinction to be made, which will be offered in brief for the sake of time. There are very few persons who would not accept the probability that if an infinite number of monkeys were permitted to type on a computer that the works of Shakespeare would, eventually, appear. However, no one would equate this with the singular genius of Shakespeare, himself, penning *King Lear*. The same must be said where artificial intelligence and human creative thought are concerned – there is a difference in *kind* between the two.

Making the effort to understand how AI works is fundamentally important for broader considerations of the effect and impact of this technology. For, when we ask 'how does an AI learn?' or 'why is ChatGPT's answer so good?', there is an implicit comparison to our human capacities. From a completely secular perspective, the human person stands completely unable to make the distinction that we intuitively know exists. What, then, prevents our self-concept from being reduced to the level of another machine in the Leviathan that is the technocratic paradigm? To return to the song, the silence that grows like a cancer remains our slow surrendering to a way of knowing that is *entirely* constituted by things. There is no value but the immediate, anything can be thrown away and forgotten when its time has come.

20.4 The Difference the Christian Version of Personhood Makes

'Fools' said I, 'You do not know
Silence like a cancer grows
Hear my words that I might teach you
Take my arms that I might reach you'
But my words like silent raindrops fell
And echoed in the wells of silence
(Simon et al. 1964)

What, then, forms an element of a response to a world that is spiraling at an increasing rate into an epistemology that worships at the altar of consumption, optimization, and control? First and foremost it is not a return to the Stone Age. Indeed, what must be re-presented is a vision of progress that is not in competition with 'words that might teach and arms that might reach'—in short, a vision that re-orients the digital tool back in service to (and not as a replacement for) the user. Pope Francis counsels, to this effect, that:

We have the freedom needed to limit and direct technology; we can put it at the service of another type of progress, one which is healthier, more human, more social, more integral. Liberation from the dominant technocratic paradigm does in fact happen…when technology is directed primarily to resolving people's concrete problems, truly helping them live with more dignity and less suffering (Francis 2015, n. 112).

More human, dignity, less suffering—none of these are inherently incompatible with 'mainstream' digital technologies (i.e. understood to be but not limited to: cellular phones, laptop computers, artificial intelligences, etc…). Provided—what? It is key, and is thus in need of urgent re-presentation, that like no other time before human persons need to reclaim what is theirs by right, by divine creation—freedom. The operative technocracy corrupts this by persuading the individual that they are limitless, constrained only by a set of out of external constraints that are imposed upon them and that to choose and to choose *well* are identical. We have freedom *for* something—to either limit oneself by selecting blindly or to attain what perfection we can do on this side of the eschaton by living according to the Gospel.

And to live in this way is to seek out what was present in Eden before the Fall and which is remade in the ministry of Our Lord on earth and reaches a climax on the Cross and in the Resurrection—relationship. But with whom or what? We, human persons, are, 'grounded in three fundamental and closely intertwined relationships: with God, with our neighbor and with the earth itself' (Francis 2015, n. 66). Indeed, these three closely-intertwined associations were ruptured by sin but recent advances in technology have driven a wedge into them. We have expanded sin's fracture in living the modern life of disconnection and the Tradition, as Pope Francis reminds, calls us to something more—an integrated view of who we are. Such a vision takes up the development of the person and communities (integral human development) and contextualizes it within the fullness of creation and in relation to the Creator. Integral ecology, 'calls for openness to categories which transcend the language of mathematics and biology, and take us to the heart of what it is to be human', is 'inseparable from the common good' and the recovery of harmony (Francis 2015, n. 11, 156–158).

This recovery cannot be contrived or invented, is not searchable on the Web nor can it be represented in its essence by Artificial Intelligence, it must be found and uncovered in a new attitude—an attitude of the heart. To be open to Love, the language that gives life and which draws together, is what is requisite to be fully who we are and helps us to re-call that we are not sophisticated wind-up machines set in motion at birth. To be open to Love necessarily opens us to the value and necessity of a world and those persons and things in it as being linked in some way—in an integral way. On its own, what will not get us there is technology. Why? Because, Pope Francis explains, what has become the normative way of relating does not:

Demand the slow and gradual cultivation of friendships, stable interaction or the building of a consensus that matures over time, have the appearance of sociability…they do not really build community; instead, they tend to disguise and expand the very individualism that finds expression in xenophobia and in contempt for the vulnerable. Digital connectivity is not enough to build bridges. It is not capable of uniting humanity (Francis 2020, n. 43).

How can we unite humanity? How can we, if even for a moment, disturb the silence of the technocratic paradigm? It beings with no less than a recovery of who we are—as beings capable of acts that are free and as beings who are created to be in community. Both of these traits take time, require discernment and carry with them the real potential for joy and growth but also sadness and failure. And this is precisely where the false solution of the technocracy starts to disrupt our proper end as persons—offering a solution that appears to avoid any failure and in a lightning-fast timeframe. We are not becoming more, as our 'friends' on Facebook grow, we are lesser.

Therefore, a bold step must be taken, not to shun the fruits of our ingenuity but to resist becoming reduced to them. We are to remember that:

No algorithm will ever be able to capture, for example, the nostalgia that all of us feel, whatever our age, and wherever we live, when we recall how we first used a fork to seal the edges of the pies that we helped our mothers or grandmothers to make at home…Along with the fork, I could also mention thousands of other little things that are a precious part of everyone's life. All these little things, ordinary in themselves yet extraordinary for us, can never be captured by algorithms…all of these live on as precious memories "kept" deep in our heart (Francis 2024, n. 20).

Why can algorithms not *fully* capture this nostalgia? They are not free to make the pie or not but are obligated to present a composite of data points—a form of predestined concoction. In taking these words of Pope Francis further still, beyond the domain of nourishment, there is another more significant conclusion to be made—the flourishing of the *person* making those same pies cannot be reduced to algorithms either. Rather, who she is, fully, remains a tapestry knit together over a lifetime—and which takes into account *every interaction* she has.

While the digital technologies that drive our current technocratic paradigm promote a reductive epistemology, it is the broader vision of integral ecology which promotes the necessary knitting together of the interactions/relationships that constitute the person. Such a way of seeing, 'calls for openness to categories which transcend the language of mathematics and biology, and take us to the heart of what it is to be human' (Francis 2015, n. 11). In short, we are taken to love. Irreducible, in its fullness, to an algorithm and much more than an emotion—love takes us to the very foundation of what it means to be human. Love, theologically understood, wills the good of the other for their sake in a seeking out of the other. Love, theologically understood, does not seek to dominate, consume, and control but is a transforming influence for care and stewardship. Love is the, 'innermost core of reality' (Francis 2024, n. 16). To see the person in her entirety, a tapestry of interactions with God, with others, and with creation—integrated by Love—is to get to the heart of what it means to be human.

And, we mention again, relationship is inextricable from what it means to be human. The difference that an instantiation in love makes is that relationships become irreducible to an interaction without a *telos*. Void of love, any relationship can take on a 'what for' element that becomes instantly manipulatable to serve the ends not of the person but of a system, a device, a state. Rather, relationships instantiated by love permit the person to take a God's eye view of the world—an integral view—and makes possible a way of being that is focused on care. We see this as having been first presented to man in the Garden, where God, 'took the man, and put him into the garden of Eden to dress it and to keep it' (Genesis 2:15). Simply, we were created to be *stewards*—caretakers—of the world.

How we live out this vocation is inextricably linked to the gift of our uniquely *human* intelligence being 'directed toward the responsible use of reason and technical abilities' in this sacred role (Dicastery for Culture and Education et al, 2025, n. 1). This can be difficult to see in relation to digital technology, however. Typically, a consideration of creation takes up elements within nature—consider a forest, a mountain range or the divine simplicity of one's home vegetable garden. What our 21st century moment suggests, what the Tradition primes us for, is a taking up of digital technologies into this larger concept of creation. Just as we are called to be good stewards of the Earth, we are equally called to be good stewards of technology. Such a vision is one that takes up the pieces, the information we receive from our technology and the interactions that we have, in light of the larger whole. The great challenge that we face is to not simply seek out information—increasingly easy as our competency in the digital world improves but to cultivate wisdom (information and interaction in light of the whole). It is:

From this perspective of wisdom, [that] believers will be able to act as moral agents capable of using this technology to promote an authentic vision of the human person and society. This should be done with the understanding that technological progress is part of God's plan for creation—an activity that we are called to order toward the Paschal Mystery of Jesus Christ, in the continual search for the True and the Good.

This is what resisting the technocratic paradigm means, irrespective of the next tool that might be swiftly approaching. We are compelled to ask the question with each and every use whether we are seeking the greatest good for the greatest number (the operative mode of the paradigm, what artificial intelligences are primed to do) or are we seeking *through* that use what is good, what is true and what is beautiful?

In the end, as we return to the call 'to take arms so that we may be reached', we all know that oftentimes this offer is rejected as 'my words like silent raindrops fell' as 'the people bow and pray to the neon god they made'. If we are so immersed in the technocratic paradigm, if AI is becoming increasingly pervasive, what difference can one or even a handful of people even make? At this juncture it would appear as if the central claims of this work have been dismantled so close to its conclusion but one would be mistaken in this regard. Rather, it needs to be point out that while AI is a new problem, what it means to be Christian in the world is as old as the Church is. Speaking at a time when Christians were scattered, persecuted, and seriously considering the value of Christ's message, St Peter's advice to our earliest sisters and brothers in Christ remains equally valuable to our discussion of AI and the technocratic paradigm. St. Peter advises:

> In your hearts honor Christ, the lord as holy, always being prepared to make a defense to anyone who asks you for a reason for the hope that is in you; yet do it with gentleness and respect (1 Peter 3:15).

This is our mission in the world and what sustains us in the midst of the technocratic paradigm; hope that the fullness of who we are lies outside of what we can comprehend. That we work toward a world where others can see, in our conduct, glimpses of this reality, this is core to our Christian vocation. That we equate and take the outstretched arms of the crucified Christ that he 'might reach us' in all of that vulnerability and as an icon of the most perfect love—this, this is what being human *means* and this can never be reduced to the abilities of even the most sophisticated AI.

20.5 Conclusion

And the sign said, 'The words of the prophets
Are written on the subway walls
And tenement halls
And whispered in the sounds of silence'
(Simon et al., 1964)

On what it might take to be a prophet in the age of AI, it has been the objective of this work to present elements of a response. The point has been frequently made that as human persons we are in possession of something beyond the mere ability to weigh possible outcomes based on datasets—we are embodied beings, created in His image and likeness, with a vocation to relationship and the ability to freely choose. We are not linked through fiberoptic cables, we do not have a deterministic limit to our communication; ours is a journey freely taken toward communion with each other, with creation, and with our Creator.

However, as digital technologies continue to advance, inclusive of AI and whatever else may ask the next existential question, this same freedom will continue to be challenged as we seek to relate ourselves to our tools, to the world, and to each other. It is not inevitable that the technocratic paradigm will triumph and that the myth of the limitless autonomous person wise through access to endless data is the epitome of progress. Integral ecology affords us another path—one that re-presents original harmony, one where we are called to care as opposed to consume, one that directs us toward Someone and not some thing. While it is precisely this epistemology that *can* facilitate the fruits of human ingenuity to 'become sources of new cultural progress for humanity and not a threat to our deepest riches', victory will never be complete. (Francis 2015, n. 47) For a tension will always be present where the question of 'true' progress is concerned—a tension between the technocratic and the integral, between the temporal and the transcendent.

In response, we must seek out moments of choice *for* who we integrally are. To be open to the value of integral ecology will gift us with the certitude that these moments, lived by individuals in seemingly disconnected places, can become the beginning of a new way of conceiving of progress in the digital world. It remains incredibly ironic that the very freedom which the technocratic paradigm claims to liberate will, in the end, be its downfall. But who will take that first step? Who will disturb the silence?

References

Boden, Margaret A. *Artificial Intelligence: A Very Short Introduction.* Oxford, Oxford University Press, 2018, p. 1.

Deane-Drummond, C. (2023). Pope Francis's Integral Ecology Paradigm: An Exploration of Its Theological Foundations and Ethical Implications. *Irish Theological Quarterly*, 88(2), pp.102–104.

Dicastery for Culture and Education and Dicastery for the Doctrine of the Faith (2025). *Antiqua et nova. Note on the Relationship Between Artificial Intelligence and Human Intelligence.* At: https://www.vatican.va/roman_curia/congregations/cfaith/documents/rc_ddf_doc_20250128_antiqua-et-nova_en.html., n. 1, 117.

Pope Francis (2015). Laudato Si': On Care for Our Common Home. [online] The Holy See. Available at: https://www.vatican.va/content/francesco/en/encyclicals/documents/papa-francesco_20150524_enciclica-laudato-si.html., n. 11, 47, 66, 105, 106, 107, 112, 156-158, 224.

Pope Francis (2020). *Fratelli Tutti.* The Holy See. At: https://www.vatican.va/content/francesco/en/encyclicals/documents/papa-francesco_20201003_enciclica-fratelli-tutti.html., n. 43.

Pope Francis (2023). *'Laudate Deum': Apostolic Exhortation to all people of good will on the climate crisis* [online] The Holy See. At: https://www.vatican.va/content/francesco/en/apost_exhortations/documents/20231004-laudate-deum.html, n. 21.

Pope Francis (2024). *Dilexit Nos.* [online] The Holy See. At: https://www.vatican.va/content/francesco/en/encyclicals/documents/20241024-enciclica-dilexit-nos.html., n. 16, 20.

Frankish, Keith, and William M Ramsey. *The Cambridge Handbook of Artificial Intelligence.* Cambridge, United Kingdom, Cambridge University Press, 2014, Introduction.

Romano Guardini (1998). *The End of the Modern World.* Wilmington, Delaware: Isi Books, pp.174.

Pope Paul VI (1965). *Gaudium Et Spes.* [online] The Holy See. At: https://www.vatican.va/archive/hist_councils/ii_vatican_council/documents/vat-ii_const_19651207_gaudium-et-spes_en.html., n. 26, 57

Simon, P. and Garfunkel, A. (1964). *The Sound of Silence.* Sony/ATV Music Publishing LLC, Universal Music Publishing Group.

Chapter 21
Sacredness as Conservationist Environmental Ethics: A Qualitative Study of Indigenous African Practices

Obinna Ikejimba and Ikechukwu Alamezie

Abstract Environmental degradation has prompted significant responses from ethicists, primarily through conservationist efforts that focus on the environment's benefits to humanity. While these efforts highlight the necessity of preserving the environment for current and future generations, they often fail to incorporate diverse cultural values and spiritual beliefs. This study proposes integrating an alternative approach to environmental ethics that emphasises the sacredness of the environment, particularly drawing from African cosmological traditions. These traditions argue that natural forms should be conserved not only for their utility but because they are sacred, valuable in themselves and crucial for cosmic balance. Recognising the environment as sacred extends beyond the logic of anthropocentrism and introduces a moral and spiritual dimension to conservation, potentially galvanising broader global support for preservation. This study employs a qualitative research design using focus group discussions to explore the environmental practices of thirteen indigenous African communities. The goal is to uncover the natural forms that these societies protect due to their sacred status and understand the cultural and spiritual beliefs that drive their conservation practices. The research aims to demonstrate that sacred environmental ethics can complement and enrich existing conservationist approaches, offering a more holistic understanding of human-environment relationships. By integrating the sacred perspective, conservation strategies and policies could resonate more deeply with diverse cultural values and spiritual beliefs, fostering a more inclusive and effective global environmental movement. Ultimately, this research seeks to contribute to environmental ethics by presenting a compelling case for the sacredness of nature as a powerful motivator for stewardship. It aims to show that acknowledging the sacredness of the environment can unite global efforts to preserve the natural world for future generations.

O. Ikejimba (✉) · I. Alamezie
Pan-Atlantic University, Lagos, Nigeria
e-mail: oikejimba@pau.edu.ng

I. Alamezie
e-mail: ialamezie@pau.edu.ng

R. Marcantonio et al. (eds.), *Integral Ecology in the Anthropocene: Africa and Beyond*,
The Anthropocene: Politik—Economics—Society—Science 49,
https://doi.org/10.1007/978-3-032-32125-1_21

Keywords Conservation · Ethics · Indigenous · Anthropocentrism · Philosophical analysis

21.1 Introduction

The global environmental crisis is not merely an ecological or economic issue; it is fundamentally an ethical challenge. Climate change, deforestation, biodiversity loss, and pollution have escalated to alarming levels, threatening the stability of ecosystems and the well-being of both human and non-human life (Laurent 2021; Sharma and Saha 2021). In response, environmental ethicists have developed diverse moral frameworks to guide and justify conservation efforts. Western environmental ethics typically identifies four major value models that define who or what deserves moral consideration: anthropocentrism, sentientism, biocentrism, and ecocentrism (Rolston 2003; Palmer et al. 2014).

Anthropocentrism confines moral concern to humans, valuing nature primarily for its utility to human well-being. Sentientism broadens this concern to sentient animals capable of experiencing pleasure and suffering (Singer 2009). Biocentrism recognises the intrinsic value of all living organisms (Taylor 2011), while ecocentrism assigns moral worth to ecological collectives such as species, ecosystems, and the biosphere. Among these frameworks, anthropocentric conservationism has been particularly influential in shaping global environmental policy, including initiatives such as the Paris Climate Accord and the United Nations Sustainable Development Goals. Anthropocentrism promotes the protection of nature primarily based on its benefits to humanity, such as climate regulation, food security, and economic stability. While this approach has mobilised significant policy action, it remains ethically limited. By instrumentalising nature, it risks overlooking the intrinsic worth of non-human entities (Sandler, 2017). Moreover, it may alienate communities whose environmental ethics are grounded not in utilitarian reasoning but in spiritual or cosmological worldviews.

Many Indigenous and non-Western traditions perceive nature as sacred, with moral obligations that transcend human benefit (Posey, 1999; Kimmerer, 2013). Conflicts in conservation often arise not from a lack of concern but from divergent foundational beliefs about what constitutes or is deserving of moral consideration. For instance, tensions emerge between conservationists who advocate for the intrinsic value of all natural forms and those who prioritise human interests. These disagreements reflect deeper normative assumptions about what counts morally in conservation decisions. As values such as duty, care, and animal welfare increasingly influence conservation science (Dubois et al., 2017; Díaz et al., 2018), navigating value-based conflicts has become more urgent. This is because practical decisions often involve stakeholders with differing ethical commitments. Hence, explicitly engaging these underlying frameworks is essential to developing inclusive and context-sensitive conservation strategies.

In this context, the article argues that sacred environmental ethics, rooted in African cosmological traditions, offer a compelling and underexplored complement to dominant Western models. These traditions conceive of nature as embedded in a spiritual and moral order, where humans, animals, and natural entities are interdependent and mutually accountable (Mbiti, 1990; Ramose, 1999). Sacred rivers, forests, and mountains are protected not for their utility, but because their desecration would violate spiritual laws and disrupt cosmic harmony. This worldview challenges anthropocentric assumptions and advances a relational ontology that affirms the intrinsic value of nature.

Beyond its philosophical depth, sacred environmental ethics hold practical relevance for conservation. Across many African communities, sacred natural sites have been preserved for generations through spiritual beliefs and cultural obligations, often in the absence of formal legal protections (Sheridan and Nyamweru, 2008). These practices illustrate how conservation can be effectively sustained by moral worldviews that frame environmental stewardship as a sacred duty. Wangari Maathai's (2010) reflections on environmental healing link sacred values with practical activism. Drawing from Kikuyu traditions that revere trees and land as ancestral trust, she illustrates how sacred cosmologies can inspire grassroots restoration efforts, from reforestation to water protection. Maathai's integration of spiritual motivation with tangible conservation outcomes exemplifies how African sacred ethics can inform policy by embedding environmental protection within the moral fabric of community life.

Recognising such perspectives within global environmental discourse could enhance conservation outcomes by aligning strategies with culturally resonant ethical commitments. This sacred orientation is not confined to African traditions. Jonathan and Angela Scott's (2021) two-volume *Sacred Nature* documents how communities worldwide revere wildlife and landscapes as more than ecological assets. Their work demonstrates that sacred designations, whether in East African savannahs or Asian forests, often anchor conservation practices in moral commitments resistant to short-term political or economic pressures. Placing African sacred ethics in this wider frame reveals both its local specificity and its consonance with global movements to recover a sense of reverence for the natural world.

Armstrong (2022) advances a complementary thesis, contending that the ecological crisis stems in part from a spiritual alienation from the natural world. She calls for a reawakening to the sacred as a lived reality, not an abstract doctrine. This appeal resonates strongly with the African cosmological emphasis on reciprocity, ritual, and respect, suggesting that ecological restoration requires not only technological solutions but also moral transformation. The convergence between Armstrong's global spiritual ethic and African sacred environmental ethics strengthens the argument that recovering sacredness can serve as a unifying moral foundation for conservation.

This study employs a qualitative methodology, using focus group discussions with thirteen Indigenous African communities to explore how cosmological beliefs shape environmental practices. It investigates which natural forms are considered sacred, the spiritual rationales for their protection, and the broader implications for conservation. By centring Indigenous voices, the research aims to illuminate the ethical foundations of sacred environmental practices and assess their potential to enrich mainstream conservation paradigms.

Furthermore, sacred environmental ethics contributes meaningfully to broader debates in environmental philosophy. While critiques of anthropocentrism are central to movements such as deep ecology (Naess, 2017) and ecofeminism (Plumwood, 2002), these frameworks often underrepresent spiritual and cosmological perspectives. By bringing African traditions into dialogue with mainstream environmental ethics, this article affirms that spiritual worldviews are not peripheral but central to ethical engagement with nature. Rather than displacing scientific conservation, sacred ethics offers a deeper moral motivation rooted in reverence and responsibility. In engaging these perspectives, this article also brings African sacred environmental ethics into dialogue with Francis's (2015) *Laudato Si'*, whose call for an "integral ecology" offers a global moral framework resonant with African cosmological traditions.

The study unfolds in five sections. Section two involves a literature review, critically examines anthropocentric conservationism and its limitations. Section three outlines the research methodology and data collection process. Section four offers a discussion and philosophical analysis of findings. Section five discusses the policy implications of integrating these insights into conservation strategies, and the conclusion reflects on the broader relevance of this approach for fostering a more inclusive and morally grounded global environmental movement.

21.2 Literature Review

The global environmental crisis has catalysed urgent reflection on the ethical foundations that underpin conservation efforts worldwide. While dominant frameworks have traditionally drawn from Western anthropocentric paradigms, increasing scholarly attention has turned toward alternative moral worldviews, including sacred environmental ethics rooted in many African cosmological traditions. Mainstream conservation ethics have historically been grounded in anthropocentrism, a framework that centres human needs and values in environmental decision-making. This model gained traction during the post-industrial revolution era, where environmental degradation began to visibly threaten human health and economic stability (Norton, 2017). More recent articulations argue for conservation in terms of sustainability, intergenerational justice, and economic resilience. Conservation under this paradigm is thus justified primarily by its benefits to human beings, both present and future. For example, ecosystem services frameworks, now integral to global policy discourse, assess the value of nature in terms of tangible and intangible benefits to people, such as clean air, portable water, and recreational opportunities (Díaz et al., 2018).

However, this anthropocentric framing has been widely critiqued for its limited moral scope. Sandler (2017) and Kopnina et al. (2021) argue that valuing nature only in relation to its utility for humans fosters a utilitarian calculus that excludes non-human entities from moral concern. This not only weakens the ethical justification for biodiversity preservation but also reduces environmental stewardship to a strategic, transactional practice. Furthermore, such instrumental reasoning may fail to motivate long-term conservation behaviours, especially in communities whose environmental ethics are grounded in spiritual and relational worldviews (Raymond et al., 2021).

The inadequacy of anthropocentrism becomes especially apparent in cross-cultural contexts. Scholars have criticised mainstream conservation models for marginalising Indigenous and local knowledge systems that resist purely instrumental or economic framings of the environment (Aikenhead and Michell, 2011; Pascual et al., 2021). In African contexts, where cosmological beliefs often shape environmental interactions, this exclusion undermines the legitimacy and efficacy of conservation efforts (Adams and Mulligan, 2012). Calls for more inclusive frameworks emphasise the need to recognise ontologies, such as African sacred environmental ethics, which position humans, nature, and the spiritual realm in an interdependent moral order (Chu Ilo, 2022; Ramose, 1999).

Recent studies have attempted to explore these sacred environmental perspectives empirically. For example, Moyo and Dube (2014) investigated traditional ecological knowledge in Zimbabwean communities and found that many sacred groves have been preserved for generations due to spiritual taboos and religious rituals. The study used ethnographic interviews and participant observation to understand how local beliefs shaped land-use behaviour. The researchers concluded that spiritual reverence for nature often provided stronger conservation outcomes than externally imposed legal frameworks. Similarly, Shiyuka (2023) study among the Kalenjin people uncovered cosmological beliefs that designated certain hills and forests as abodes of ancestral spirits. The sacred status of these landscapes had, in many cases, effectively curtailed deforestation and commercial exploitation.

These empirical findings support the theoretical claim that sacredness, as an ethical category, offers a distinct mode of conservation motivation, one that is not reducible to cost-benefit analyses. However, they also reveal certain methodological limitations in the current literature. Many existing studies focus on isolated sacred sites or ritual practices without thoroughly unpacking the underlying cosmological logic that sustains them. Moreover, they often lack comparative analysis across different communities, thereby limiting generalisability. As highlighted by Sinthumule and Mashau (2024), the diversity of African cosmologies calls for more expansive and systematic research to map patterns and variations in sacred environmental ethics.

The conceptual divergence between anthropocentric ethics and African sacred cosmology is profound. While the former is grounded in Enlightenment humanism, which separates humans from nature and emphasises rational mastery over the environment (Descola, 2013), the latter embodies a relational ontology. In African thought, the self is constituted through relationships not only with other people but also with ancestors, spirits, and natural entities (Ramose, 1999). This metaphysical outlook assigns moral significance to nature based on its role in sustaining the web of cosmic interdependence. From this perspective, environmental ethics is not an optional social virtue but a sacred duty. As Angwafo et al. (2016) explain, in many African traditions, disrespecting nature is tantamount to disrupting ancestral authority and spiritual order. This moral imperative goes beyond legal or regulatory compliance and taps into a deeply held sense of spiritual accountability. Such motivations may offer more resilient foundations for conservation, particularly in regions where formal enforcement is weak or inconsistent.

Despite their promise, sacred environmental ethics face significant challenges. Scholars caution that sacred designations can be contested, eroded, or politically co-opted (Chidester, 2018). Generational value shifts, urbanisation, and religious conversion, such as those documented in Mozambique, can weaken the cultural force of sacred taboos. Additionally, sacred status may be invoked selectively or used to reinforce social hierarchies, raising questions about inclusivity and resilience (Tatay and Merino, 2023). These limitations highlight the need for a dynamic, context-sensitive understanding of sacred ethics.

Nonetheless, the enduring presence of sacred environmental values across diverse African communities suggests their relevance cannot be dismissed. Rather, what is needed is a more integrative approach that situates sacred ethics within broader socio-political and ecological contexts. Scholars like Molefe (2021) have proposed developing a pluralistic ethical framework that combines scientific, legal, and spiritual motivations for conservation. Such an approach could bridge the epistemic gap between Western and Indigenous worldviews, fostering more inclusive and ethically grounded conservation strategies.

The present study seeks to advance this discourse by conducting a qualitative inquiry into how sacred cosmological beliefs shape environmental ethics across thirteen Indigenous African communities. Unlike previous studies that focus on specific sacred sites or practices, this research aims to uncover the underlying ethical logics that inform what is considered sacred and how that status influences conservation behaviour. By foregrounding Indigenous voices and their moral reasoning, the study addresses a crucial gap in the literature and contributes to the philosophical and practical enrichment of conservation ethics.

The philosophical divergence between Western and African ethical systems is particularly salient when discussing the conceptual foundations of conservation. Western ethics often adopts a dualistic metaphysics separating the human from the non-human, the sacred from the secular, and the subject from the object (Callicott, 2020). Such a worldview underpins the dominant anthropocentric and utilitarian approaches, where moral worth is conferred based on utility, sentience, or ecological function. In contrast, African cosmological perspectives tend to reflect a non-dualistic metaphysics in which humans and the natural world exist in a state of ontological interdependence. This relational worldview posits that spiritual, moral, and ecological well-being are inseparably linked, a view that has profound implications for conservation ethics (Ramose 1999).

This ontological difference has been explored in recent studies that examine how African metaphysics frames nature not as a passive resource but as an active participant in moral life. A study by Banda et al. (2024), across Malawian communities, found that local environmental stewardship was grounded in a belief in ancestral spirits and spiritual taboos, rather than in scientific or policy rationales. Their work highlights how moral responsibility toward sacred groves and rivers was embedded in a wider cosmology of cosmic harmony and intergenerational accountability. Such findings underscore that conservation premised solely on human benefit fails to resonate with communities whose environmental ethics are inherently spiritual.

These critiques reveal the complexity of translating sacred ethics into universal conservation strategies. Unlike ecocentric theories such as deep ecology or land ethics, which seek broad ontological shifts in valuing ecosystems, African sacred environmental ethics operates through culturally situated epistemologies. It is precisely this localised, lived experience of sacredness that gives it moral weight in its context, but also raises questions about how such values interface with global, pluralistic environmental governance (Angwafo et al., 2016). Some empirical studies continue to explore this tension between localisation and generalisation. A multi-country study conducted by Tatay and Merino (2023), which included sites in Tanzania, Ghana, and Cameroon, utilised focus groups and participant observation to assess how sacred natural sites influence conservation. The researchers concluded that while sacred values enhanced community compliance and ecological protection, they also faced challenges of intergenerational transmission and conflicting land-use priorities. Crucially, the study found that where sacredness was supported by living traditions and community rituals, conservation outcomes were notably better than where such traditions had eroded due to urbanisation or religious change. This underscores the fragility but also the potential of sacred frameworks as culturally embedded conservation ethics.

These observations highlight the need for a nuanced ethical framework that accommodates pluralism without reducing all values to anthropocentric terms. Western frameworks, particularly anthropocentric conservationism, have been instrumental in shaping global environmental policy, yet they remain ethically thin when applied in culturally diverse contexts. As argued by Sandler (2017), utilitarian metrics often fail to capture moral motivations that derive from reverence, relationality, and spiritual obligation. Recent calls for "value pluralism" in conservation ethics (Chan et al., 2021) echo this concern, advocating for a broader moral vocabulary that includes spiritual, aesthetic, and relational values.

Sacred environmental ethics contributes to this pluralistic discourse by offering an indigenous, spiritually grounded account of moral responsibility to nature. While it does not offer a single universal framework, it invites deeper engagement with culturally situated worldviews and moral grammars. Moreover, it challenges the instrumentalist assumptions of dominant frameworks by shifting the ethical focus from outcomes to relationships between humans, nature, ancestors, and the cosmos. As Ndagala and Mosha argue in their qualitative study of sacred mountain sites in northern Tanzania, "protection is not derived from law, but from love, love for the spirits, for the land, and future generations" (2022, p. 233). Their study, based on interviews and local oral traditions, affirms that sacredness is not an abstract label but a lived moral experience. This recognition of moral and cosmological relationality provides philosophical justification for reconsidering dominant conservation frameworks. Where anthropocentrism values nature primarily for human survival or utility, sacred environmental ethics values nature for its participation in a moral and spiritual order. This shift is not merely conceptual but ethical: it invites practitioners and policymakers to consider conservation as a sacred duty rather than a strategic choice.

However, the literature reveals an important gap. While several studies explore how sacred values influence conservation behaviour (Tatay and Merino 2023; Banda et al. 2024), few provide a detailed philosophical analysis of sacredness as an ethical system in its own right. Furthermore, existing studies tend to focus on isolated case studies, often without examining how widespread such beliefs are across different communities or how they interface with policy and scientific paradigms. This study seeks to address that gap by examining sacred environmental ethics across thirteen Indigenous African communities, focusing not merely on the existence of sacred sites but on the ethical beliefs and cosmological narratives that sustain them. Through focus group discussions, the study aims to uncover the normative foundations of sacred conservation, contributing both empirical depth and philosophical insight to the broader discourse. From the foregoing, sacred environmental ethics is neither a replacement for scientific conservation nor a romantic holdover from the past. Rather, it is a living, dynamic moral system that demands serious philosophical and practical engagement. In a world facing urgent ecological crises and ethical disorientation, engaging with such traditions may offer new moral languages and practices for sustaining the planet not just as a habitat, but as a sacred home, our common home.

21.3 Methodology

This study adopted an exploratory qualitative research design to examine how sacred beliefs influence environmental conservation practices among Indigenous African communities. The design was appropriate given the study's dual objectives: to identify natural elements that are protected due to their sacred status, and to understand the cultural and spiritual beliefs underpinning these conservation practices. Thirteen Indigenous communities across sub-Saharan Africa, and specifically Nigeria, Kenya, Uganda and Zambia, were purposively selected for their strong ancestral ties to their present locations and their continued practice of traditional ecological knowledge. These criteria ensured that the communities held longstanding relationships with their local environments and maintained cosmological worldviews through which sacredness and environmental ethics are articulated.

Participants were recruited using a snowball sampling approach. Initial participants identified others within their communities who met the inclusion criteria: individuals aged fifty years and above who had lived continuously in the community for at least twenty years. This approach ensured that participants possessed deep-rooted knowledge of the spiritual and ecological traditions under investigation. A minimum of four participants were selected from each community, yielding a total of at least thirty-six participants. The study employed focus group discussions (FGDs) as the primary method of data collection. These were conducted online using platforms such as Zoom, Google Meet, and WhatsApp, depending on accessibility and participant preference. A semi-structured discussion guide was used to facilitate open dialogue.

Participants were asked to identify specific natural features, such as rivers, trees, hills, or animals, that their communities regard as sacred, and to explain the associated beliefs, rituals, and moral obligations for their protection. These discussions allowed participants to validate one another's responses and to clarify shared understandings.

To address the first research objective, participants were prompted to describe particular natural forms considered sacred and outline any practices, prohibitions, or rituals used to protect them. For the second objective, discussions focused on exploring the underlying belief systems, including ancestral reverence, spiritual laws, and cosmological principles, that inform and justify these practices. All FGDs were recorded and transcribed verbatim. Thematic analysis was employed to interpret the data, combining deductive coding based on key ethical concepts (such as sacredness, relationality, and intrinsic value) with inductive coding to allow novel themes to emerge from participant narratives. This process ensured that both the content of sacred conservation practices and the moral worldviews sustaining them were systematically examined.

21.4 Finding and Philosophical Analysis

This section presents and interprets findings from focus group discussions (FGDs) conducted across thirteen Indigenous African communities in Nigeria, Kenya, Uganda, and Zambia. The analysis addresses the two central research objectives: to identify natural forms considered sacred and protected in these communities, and to understand the cultural and spiritual beliefs underpinning their conservation practices. By weaving empirical data with philosophical reflection, particularly through the lens of sacred environmental ethics and relational ontology (Mbiti 1990; Ramose 1999; Angwafo et al. 2016), the study reveals a pan-African perspective that challenges anthropocentric conservation paradigms. This perspective finds profound resonance with Francis's (2015) *Laudato Si*, which envisions nature as a sacred gift within an integral ecology, offering a framework to enrich global environmental discourse. The seamless integration of these findings and their dialogue with *Laudato Si'* illuminates a spiritually grounded ethic that unites diverse worldviews in addressing the ecological crisis.

The thirteen communities include Akpoha (Ebonyi State, Nigeria), Ekeh-Olemgbecho (Benue State, Nigeria), Ihima (Kogi State, Nigeria), Ijebu-Omu (Ogun State, Nigeria), Isi-Ngwu (Enugu State, Nigeria), Uromi (Edo State, Nigeria), Udoh (Akwa Ibom State, Nigeria), Ndiakunwanta Arondizuogu (Ideato North, Imo State, Nigeria), Umuhu Village (Owelle-Court, Agwu, Enugu State, Nigeria), Kikuyu (Central Highlands, Kenya), Mijikenda (Mombasa-Kilifi Counties, Kenya), Kangulu mina (Central Region, Uganda), and Kamphemba Ward (Chama District, Zambia). These communities cut across diverse ecological settings, from highlands and riverine areas to coastal forests and savannahs, with livelihoods including farming, fishing, forest gathering, and ritual practices. Elders aged 50 and above, recruited via snowball sampling, participated in semi-structured virtual FGDs, providing rich narratives on sacred entities, protective mechanisms, and cosmologies. These narratives form the foundation for understanding how sacredness shapes environmental ethics across these communities.

The study finds a wide array of natural forms, rivers, trees, forests, rocks, mountains, animals, and wildlife habitats that are sacred across all communities, valued beyond their usefulness to human material needs now or in the future, but for their spiritual essence as abodes of ancestors, deities, or cosmic forces. In Akpoha, rivers like *Enyum Amachi* and *Egbara* house sacred crocodiles symbolising divine protection and fertility, while Ekeh-Olemgbecho reveres an Iroko tree and stream for rainmaking and ancestral communion. Ihima protects the *Ori-Ihima* rock, cave, and rivers, believed to wield supernatural agency over weather, and Ijebu-Omu holds the *Eiye* bird and woven mats as sacred, tied to town-founding myths. Isi-Ngwu considers Ngwu trees and rivers (*Ose*, *Ekulu*) deity abodes, with harvesting bans, while Uromi's *Okha* tree serves as a spiritual altar, and Udoh's *Enaa* River protects aquatic life as a divine residence. Ndiakunwanta Arondizuogu's *Nga-Eze* stream and Umuhu's *Amalokpo* stream, with its sacred chimpanzees signalling death, embody divine manifestations. In Kikuyu, the *Mugumo* tree, *Chania* and *Gura* rivers, and Mount Kenya are ancestral gifts symbolising divine mystery, with goats and sheep treated respectfully. Mijikenda's *Kaya* forests are ancestor dwellings for rituals, and Kangulu Mina's *Kalagala* stream, *Mabira* Forest groves, sacred rock, and wildlife (bats, crocodiles) house spirits like the god, *Dungu*. Kamphemba Ward's *Luangwa* River and North *Luangwa* National Park wildlife (elephants, lions, leopards, birds) are sacred for fertility and protection, linked to ancestral and nature spirits. Water bodies in 10 communities, trees/forests in eight, animals/wildlife in six, and rocks/mountains in four are believed to have value that transcends ecological utility, embodying intrinsic value rooted in metaphysical interdependence, aligning with yet extending beyond biocentric and ecocentric ethics.

The study found that the protection of these sacred forms relies on indigenous systems of taboos, rituals, and communal sanctions, which function as ethical regulators independent of formal legal frameworks. Taboos prohibit actions such as building, defecating, or farming near rivers (Kikuyu, Kamphemba Ward), cutting sacred trees (Isi-Ngwu, Kangulu mina), entering restricted forests or caves (Mijikenda, Ihima), and fishing or hunting (Akpoha, Ekeh-Olemgbecho, Kamphemba Ward).

The study also found that rituals were used to reinforce reverence, including sacrifices under *Mugumo* trees (Kikuyu), black-feathered hen offerings in *Mabira* forest (Kangulu mina) for fertility, elder petitions in Kaya forests (Mijikenda), rainmaking at Iroko trees (Ekeh-Olemgbecho), and ceremonies honouring *Luangwa* River spirits (Kamphemba Ward). Violations of pro-environmental practices trigger metaphysical sanctions such as death, illness, drought, curses, or bad luck, as seen in divine strikes in Kikuyu, community-wide consequences in Kamphemba Ward, or disappearance in Kangulu mina. These sanctions were enforced by custodians like elders, priests, or cult groups (e.g., *Obobayin*-Ihima in Ihima, Kaya elders in Mijikenda). These mechanisms, often outperforming formal conservation as seen in UNESCO-protected Kaya forests or Mabira's sacred groves, reflect a deontological ethic of absolute duty rooted in spiritual reciprocity, distinct from utilitarian calculations.

These protective practices are underpinned by cosmological beliefs that reveal a relational ontology, where humans, nature, and the spiritual realm form an interdependent moral web (Ramose 1999; Chu Ilo 2022). The study shows that sacred forms are animate participants in cosmic narratives. Examples, ancestors dwell in Kaya forests (Mijikenda) or manifest as rocks (Kangulu mina); deities control weather via *Ori-Ihima* rock (Ihima) or embody mystery in Mount Kenya (Kikuyu); Spirits ensure fertility through the *Luangwa* River (Kamphemba Ward) or protect via crocodiles (Akpoha). Desecration disrupts harmony, inviting retribution, as with *Umuhu's* chimpanzees signalling death or *Enaa* River's divine enforcement in Udoh.

Three structures or patterns emerge from the study. (1) Sacredness is determined by spiritual significance, not ecological utility, and this challenges anthropocentrism; (2) conservation is understood as a ritualised duty, a habit that fosters emotional bonds and resilience; and (3) moral authority rests with local custodians, promoting inclusivity over centralised policies. This shows that despite variations, such as wildlife focus in Kamphemba Ward versus tree emphasis in Kikuyu, unity lies in a non-dualistic metaphysics that resists Western subject-object divides, offering a holistic view of human-nature relations (Callicott 2020; Descola 2013).

This relational worldview finds a profound echo in *Laudato Si'*, which frames nature as a "sacred gift from the Creator" where "everything is interconnected" (Francis 2015, paras. 84, 240). The encyclical's vision of creation as a "web of relationships" (Francis 2015, para. 139) mirrors the communities' ontology, where rivers like *Nga-Eze* in Ndiakunwanta, trees like *Okha* in Uromi, and animals like the *Eiye* in Ijebu-Omu are moral agents within a cosmic order. The sacredness of Mount Kenya or Kaya forests exemplifies *Laudato Si'*'s call to find "mystical meaning" in nature (Francis 2015, para. 233), offering tangible practices of contemplation that hear creation's "paradoxical and silent voice" (Francis 2015, para. 85). Similarly, the communities' stewardship through taboos like pollution bans in Umuhu or rituals like sacrifices in Kangulu mina embodies the encyclical's call for humans to be "protectors of God's handiwork" with "gratitude and gratuitousness" (Francis 2015, paras. 217, 227). The concept of "ecological conversion" (Francis 2015, para. 217), a moral transformation toward care, parallels ritualised renewal in practices like rainmaking in Ekeh-Olemgbecho or *Mugumo* tree replacements in Kikuyu, which restore cosmic balance.

This alignment deepens with *Laudato Si'*'s explicit valuation of indigenous worldviews, noting that for such communities, "land is not a commodity but rather a gift from God and from their ancestors who rest there, a sacred space" (Francis 2015, para. 146). This resonates with practices in Akpoha, Mijikenda, and Kamphemba Ward, where sacred sites are ancestral trusts fostering accountability, akin to the encyclical's vision of a "universal family" (Francis 2015, para. 89). The communities' elder-led governance, as in Ihima or Mijikenda, models *Laudato Si'*'s call for dialogue with indigenous peoples as "principal partners" (Francis 2015, para. 146), embodying "prophetic and contemplative lifestyles" free from consumerism (Francis 2015, para. 222). Challenges like urbanisation (e.g., encroachments in Kikuyu) or religious shifts (e.g., Christianity in Ijebu-Omu) are addressed by the encyclical's advocacy for education in "environmental responsibility" (Francis 2015, para. 211), suggesting a hybrid approach that blends spiritual renewal with modern activism to counter cultural dilution.

Despite such challenges, including generational shifts (e.g., family-dependent adherence in Kangulu mina) and cultural erosion (e.g., Kamphemba Ward), these communities demonstrate resilience, as seen in UNESCO-protected Kaya forests and persistent rituals in Kangulu mina. Adaptability, such as replacing dried *Mugumo* trees in Kikuyu, aligns with Francis's (2015) *Laudato Si'*, which calls for dynamic, inclusive strategies. Resonating with deep ecology (Naess 2017) and ecofeminism (Plumwood 2002), this framework positions African cosmologies as vital contributors to global ecological discourse, offering a spiritually grounded, pluralistic ethic that unites diverse worldviews to address the planetary crisis through reverence, reciprocity, and renewal.

21.5 Conclusion

This study aimed to investigate how sacredness serves as an environmental ethical framework within Indigenous African communities. The research focused on two primary objectives: (1) to identify the natural forms that these communities protect due to their sacred status, and (2) to understand the cultural and spiritual beliefs underpinning such conservation practices. Adopting an exploratory qualitative design, the study engaged thirteen Indigenous communities across Nigeria, Kenya, Uganda and Zambia through virtual focus group discussions involving participants aged 50 and above with longstanding ties to their local contexts.

The findings revealed that sacredness is attributed to a diverse range of natural forms, including rivers, trees, rocks, animals, and culturally significant artefacts. These elements are not merely symbolically important but are perceived as spiritual entities deserving of reverence and protection. Conservation practices are governed by a complex system of taboos, rituals, and ancestral obligations that are embedded within community cosmologies. These beliefs and practices reflect a relational ontology in which humans, nature, and the spiritual realm are interconnected. Unlike mainstream anthropocentric ethics, which justify conservation through human utility or ecological function, sacred environmental ethics are motivated by moral obligations to non-human entities perceived as spiritually significant. These practices have persisted despite the influence of Christianity/Islam, the absence of formal legislation, demonstrating the regulatory power of these traditional belief systems.

Theoretically, the study advances discourse in environmental ethics by introducing sacredness not simply as a cultural practice, but as an autonomous ethical framework. This framework challenges dominant Western paradigms that often marginalise Indigenous ontologies and instead affirm the legitimacy of moral systems grounded in cosmology, ancestral reverence, and spiritual accountability. Practically, the findings urge a reconsideration of conservation strategies that rely exclusively on legal, economic, or scientific instruments. Integrating sacred values into conservation policy could enhance cultural legitimacy, strengthen community participation, and improve ecological outcomes. Recognising traditional custodianship and embedding rituals and taboos into local governance structures may also provide culturally sustainable pathways for environmental protection.

This study fills a significant gap in the literature by examining sacred environmental ethics through a multi-community, comparative lens. Unlike previous research that often isolates sacred sites or specific rituals, this work explores the broader cosmological narratives and ethical reasoning behind such practices. It contributes empirical and philosophical depth to debates on conservation pluralism and the role of Indigenous worldviews in environmental governance. However, several limitations plague the study. First, all data were collected virtually due to logistical constraints, which limited the researcher's ability to observe embodied expressions of belief and community interactions with sacred natural forms. Second, while the study references an African context, the communities studied were all located in Nigeria. As such, the findings are not representative of the full diversity of sacred environmental ethics across the African continent. Additionally, language translation from local dialects to English may have resulted in the loss of nuanced cultural meanings. The lived practices documented here echo Armstrong (2022) call for moral reawakening, Maathai (2010) fusion of sacred values with grassroots activism, and the Scotts (2021) global documentation of sacred conservation traditions. These parallels suggest that African sacred environmental ethics offer moral resources capable of enriching both local governance and global conservation discourse.

Future research should include communities across multiple African countries to better understand regional and cultural variations in sacred environmental ethics. Comparative studies between Indigenous African, Asian, or Latin American traditions could offer valuable cross-cultural insights. Further work could also explore the intergenerational transmission of sacred values and how these beliefs are adapting or resisting change under pressures from modernity, religious shifts, or environmental decline. Environmental ethicists and conservation practitioners are encouraged to collaborate with traditional custodians, not only in protecting sacred sites but in acknowledging and preserving the ethical systems that sustain them. While the paper highlights a powerful convergence of indigenous wisdom and ecological spirituality that can inspire more holistic and culturally grounded conservation initiatives, efforts should be made to integrate sacred environmental values into formal conservation planning while ensuring that such incorporation is respectful, non-extractive, and community-led.

In summary, in a world fragmented by ecological crises and moral disconnection, sacred environmental ethics found in many Indigenous African communities is like the vision in Francis's (2015) *Laudato Si'*. It offers more than an alternative framework; it animates the cosmos, reminding us that the mechanisms of sustainability surpass technical challenges to a more sacred covenant, one that binds humans, ancestors, and the natural world in a web of reciprocal care. As Francis (2015, para. 1) implores, a sentiment echoed by the study's participants, the Earth is not 'ours' to exploit but 'ours' to cherish as a common home. In this understanding, conservation efforts acquire a spirit, a holy vocation towards nature-based solutions, addressing ecological needs from our heart, a heart that cares.

References

Adams, W.M. and Mulligan, M. (2012) *Decolonizing nature: strategies for conservation in a post-colonial era.* London: Routledge.

Aikenhead, G.S. and Michell, H. (2011) *Bridging cultures: Indigenous and scientific ways of knowing nature.* Toronto: Pearson.

Angwafo, E.T., Billa, S.F. and Fotang, C. (2016) 'Contribution of traditional institutions to the sustainable management of sacred forests: Case study of Mankon sacred forests northwest region, Cameroon', *Journal of Biodiversity and Environmental Sciences*, 9(4), pp. 1–19.

Armstrong, K. (2022) *Sacred nature: restoring our ancient bond with the natural world.* London: Vintage.

Banda, L.O.L., Banda, C.V., Banda, J.T. and Singini, T. (2024) 'Preserving cultural heritage: A community-centric approach to safeguarding the Khulubvi Traditional Temple Malawi', *Heliyon*, 10(18).

Callicott, J.B. (2020) 'From civil rights to nature's rights', *Journal of World Philosophies*, 5(1), pp. 183–187.

Chidester, D. (2018) 'World religions in the world', *Journal for the Study of Religion*, 31(1), pp. 41–53.

Chu Ilo, S. (2022) *African ecological ethics and spirituality for cosmic flourishing: an African commentary on Laudato Si'.* Eugene, OR: Wipf and Stock Publishers.

Descola, P. (2013) *The ecology of others.* Chicago, IL: Prickly Paradigm Press.

Díaz, S., Pascual, U., Stenseke, M., Martín-López, B., Watson, R.T., Molnár, Z. and Shirayama, Y. (2018) 'Assessing nature's contributions to people', *Science*, 359(6373), pp. 270–272.

Dubois, S., Fenwick, N., Ryan, E.A., Baker, L., Baker, S.E., Beausoleil, N.J. and Fraser, D. (2017) 'International consensus principles for ethical wildlife control', *Conservation Biology*, 31(4), pp. 753–760.

Francis, Pope (2015) *Laudato Si': on care for our common home.* Brooklyn, NY: Melville House Publishing.

Kimmerer, R.W. (2013) *Braiding sweetgrass: Indigenous wisdom, scientific knowledge and the teachings of plants.* Minneapolis, MN: Milkweed Editions.

Kopnina, H., Spannring, R., Hawke, S., Robertson, C.D., Thomasberger, A., Maloney, M., Morini, M., Lynn, W., Zakari Muhammad, N., Santiago-Ávila, F.J., Begovic, H. and Baranowski, M. (2021) 'Ecodemocracy in practice: exploration of debates on limits and possibilities of addressing environmental challenges within democratic systems', *Visions for Sustainability*, 15, pp. 9–23.

Laurent, É. (2021) 'Human well-being and the biosphere: connecting the dots by connecting the circles', *The Tocqueville Review*, 42(1), pp. 121–130.

Maathai, W. (2010) *Replenishing the earth: spiritual values for healing ourselves and the world.* New York: Doubleday Religion.

Mbiti, J.S. (1990) *African religions and philosophy.* 2nd edn. Oxford: Heinemann.

Molefe, M. (2021) 'African ethics and agent-centred duties', in Murove, M.F. (ed.) *Essays on contemporary issues in African philosophy.* Cham: Springer International Publishing, pp. 107–124.

Moyo, P. and Dube, T. (2014) 'Edging towards a tipping point? An appraisal of the evolution of livelihoods under climate change in semi-arid Matobo, Zimbabwe', *International Journal of Development and Sustainability*, 3(6), pp. 1340–1353.

Naess, A. (2017) 'The shallow and the deep, long-range ecology movement: a summary', in DesJardins, J.R. and Pojman, L.P. (eds.) *The ethics of the environment*. London: Routledge, pp. 115–120.

Norton, B.G. (2017) 'Environmental ethics and weak anthropocentrism', in DesJardins, J.R. and Pojman, L.P. (eds.) *The ethics of the environment*. London: Routledge, pp. 333–350.

Palmer, C., McShane, K. and Sandler, R. (2014) 'Environmental ethics', *Annual Review of Environment and Resources*, 39, pp. 419–442.

Pascual, U., Adams, W.M., Díaz, S., Lele, S., Mace, G.M. and Turnhout, E. (2021) 'Biodiversity and the challenge of pluralism', *Nature Sustainability*, 4(7), pp. 567–572.

Plumwood, V. (2002) *Feminism and the mastery of nature*. London: Routledge.

Posey, D.A. (1999) *Cultural and spiritual values of biodiversity*. London: United Nations Environment Programme / Intermediate Technology Publications.

Ramose, M.B. (1999) *African philosophy through Ubuntu*. Harare: Mond Books.

Raymond, C., Matthews, T., Horton, R.M., Fischer, E.M., Fueglistaler, S., Ivanovich, C. and Zhang, Y. (2021) 'On the controlling factors for globally extreme humid heat', *Geophysical Research Letters*, 48(23), e2021GL096082.

Rolston, H. III (2003) 'Environmental ethics', in Bunnin, N. and Tsui-James, E.P. (eds.) *The Blackwell companion to philosophy*. Oxford: Blackwell Publishing, pp. 517–530.

Sandler, R. (2017) *Environmental ethics: theory in practice*. Oxford: Oxford University Press.

Scott, J. and Scott, A. (2021) *Sacred nature 2: reconnecting people to our planet*. Cape Town: HPH Publishing.

Sharma, P. and Saha, D. (2021) 'Augmenting ecological balance and health security through human activity', *SSRN*. Available at: https://ssrn.com/abstract=3778837 (Accessed: 7 June 2025).

Sheridan, M.J. and Nyamweru, C. (eds.) (2008) *African sacred groves: ecological dynamics and social change*. Oxford: James Currey; Athens, OH: Ohio University Press; Pretoria: UNISA Press.

Shiyuka, E.K. (2023) *Religious experience of the Kalenjin of Kerio-Valley cultural complex, Kenya, 1800–1965*. PhD thesis. Kenyatta University. Available at: https://ir-library.ku.ac.ke/handle/123456789/27365 (Accessed: 3 May 2025).

Singer, P. (2009) *Animal liberation: the definitive classic of the animal movement*. New York: HarperCollins Publishers.

Sinthumule, N.I. and Mashau, M.E. (2024) 'Sacred forests as repositories of local biodiversity in Africa: a systematic review', *Journal for the Study of Religion, Nature and Culture*.

Tatay, J. and Merino, A. (2023) 'What is sacred in sacred natural sites? A literature review from a conservation lens', *Ecology and Society*, 28(1).

Taylor, P.W. (2011) *Respect for nature: a theory of environmental ethics*. Princeton, NJ: Princeton University Press.

Chapter 22
The Sacred Earth in Igbo Cosmology: Lessons for Environmental Ethics and Policy in Africa

Jeremiah Anene Nwankwegu

Abstract This paper explores the environmental ethics embedded in Igbo cosmology—particularly the sacred conception of *Ala/Ali* (Earth deity)—and examines how these indigenous beliefs and practices can inform contemporary environmental policy in Africa. The central question guiding this study is: *What lessons can Igbo ecological traditions offer for developing integrative and sustainable environmental policies rooted in African cultural realities?* The hypothesis is that traditional Igbo environmental ethics, expressed through sacred sites and ecological taboos, embody a holistic ecological consciousness that aligns with the global concept of *Integral Human Ecology* as articulated in Pope Francis's *Laudato Si'*. Using qualitative and ethnographic methods, fieldwork was conducted in two south-eastern Nigerian communities—Izhi and Ikwo—through interviews with elders, shrine custodians, and community leaders, as well as direct observations of sacred forests, ponds, shrines, and ritual practices. This was supplemented by a critical review of relevant literature on indigenous African environmental knowledge. Findings reveal that sacred ecological sites and taboos are not merely spiritual practices but function as effective ecological governance systems. However, these are increasingly eroded by urbanization, religious transformation, and state—led development. The study concludes that reviving and integrating indigenous knowledge into environmental education, policy, and land—use planning can provide culturally grounded pathways for sustainable development in Africa. The paper is significant in that it connects Igbo ecological ethics to *Integral Human Ecology* and, in so—doing, demonstrates that environmental care and social justice are inseparable, and that indigenous cosmologies offer essential contributions to both African and global ecological futures.

Keywords Cosmology · Environmental ethics · Spiritual foundation · Sacred sites · Taboo

---

J. A. Nwankwegu (✉)
Ebonyi State University, Abakaliki, Nigeria
e-mail: Jeremiah.nwankwegu@ebsu.edu.ng

R. Marcantonio et al. (eds.), *Integral Ecology in the Anthropocene: Africa and Beyond*, The Anthropocene: Politik—Economics—Society—Science 49,
https://doi.org/10.1007/978-3-032-32125-1_22

22.1 Introduction

The accelerating environmental crisis in Africa and around the globe calls for urgent ethical and sustainable responses. In many indigenous African societies, such responses are not entirely new; they are embedded in longstanding cultural and spiritual worldviews that govern human relationships with the land and non—human entities. Among the Igbo of south-eastern Nigeria, the earth—revered as *Ala* (also *Ani*, *Ana*, or *Ali*)—is not merely a physical entity but a sacred, moral authority that regulates social and ecological order. The Igbo cosmology, which centres on *Ala* as the goddess of justice, fertility, and heritage, promotes practices that conserve biodiversity, enforce moral behaviour, and ensure intergenerational equity. Sacred forests, ancestral lands, community ponds, and forbidden paths are maintained not through formal environmental laws but through communal taboos and spiritual sanctions rooted in reverence for *Ala*.

However, modern pressures such as urbanization, industrial development, religious transformations, and globalization have led to the erosion of these traditional ecological systems. Sacred sites are being cleared for housing, infrastructure, and religious buildings; ancestral groves are sold to private developers; and cultural taboos are increasingly dismissed as superstitious. These disruptions have not only compromised ecological balance but have also weakened communal morality and the traditional custodianship structures that ensured environmental stewardship. The result is a rising trend of deforestation, erosion, pollution, and biodiversity loss across Igbo communities and beyond.

This study is motivated by the urgent need to revalorize the environmental ethics embedded in Igbo cosmology and to explore their relevance to contemporary environmental discourse and policymaking. In a global context where Western, technocratic, and anthropocentric approaches dominate environmental management, indigenous knowledge systems—such as the Igbo ecological worldview—offer a vital alternative grounded in spirituality, relationality, and communal responsibility. These indigenous practices resonate with global frameworks such as Pope Francis's *Laudato Si'*, which advocates for "integral ecology"—the recognition that environmental, social, and spiritual well—being are deeply interconnected.

The overarching aim of this study is to investigate how the sacred conception of the earth in Igbo cosmology, as reflected in the practices of Izhi and Ikwo communities in Ebonyi State, can inform environmental ethics and contribute to policy design in Africa. The specific objectives are:

1. To identify and analyse the typologies and cultural significance of sacred sites and ecological taboos in Izhi and Ikwo.
2. To examine how traditional Igbo ecological practices align with principles of conservation, social equity, and moral governance.
3. To explore the disruptions brought about by modernity and their impact on indigenous environmental systems.

4. To evaluate how traditional ecological knowledge can be revived, adapted, and integrated into contemporary environmental education, land-use planning, and policy frameworks.

In addressing these objectives, the study contributes to the broader project of decolonizing environmental knowledge by foregrounding African worldviews in ecological debates. It also seeks to demonstrate that traditional environmental ethics are not relics of the past, but evolving systems that, if properly understood and supported, can offer effective responses to current environmental challenges. Ultimately, the study advocates for a hybrid governance model in which indigenous custodianship, legal protection of sacred sites, and cultural environmental education are recognized and institutionalized.

22.2 Literature Review

The sacredness of the Earth in Igbo cosmology offers a compelling ethical framework for understanding and addressing environmental degradation. In this worldview, *Ala/ Ali,* the Earth goddess, is the spiritual custodian of morality, justice, fertility, and ecological order. As several scholars have affirmed, this cosmological perspective provides valuable insights for integrating indigenous knowledge into contemporary environmental discourse.

The work of Madu (1996) underscores the deeply relational and moral nature of the Igbo worldview, in which the spiritual and physical realms are interconnected. Environmental care is not viewed as a separate or technical concern but as a moral obligation governed by divine principles. *Ala* serves as the embodiment of these principles, enforcing taboos (*nso ala*) to preserve both moral and ecological order.

Similarly, Onwu (2002) emphasizes that traditional African religions possess an inherently ecological consciousness. For the Igbo, nature is not a passive resource but a living, spiritual presence, and violations against it are seen as both ethical and cosmological disruptions. In this sense, Igbo environmental ethics aligns with the core values of *Integral Human Ecology*, which Pope Francis describes as a holistic framework that links environmental care, moral responsibility, and cultural identity.

Expanding this framework further, Tosam (2019) argues that African environmental ethics, when properly understood and revived, can contribute to sustainable development and climate resilience. Drawing on Kom traditions in Cameroon, Tosam illustrates how precolonial African societies used taboos, values, and moral teachings to promote environmental sustainability. These indigenous systems extended moral consideration beyond humans to include animals, plants, and spiritual forces. However, he notes that colonial impositions, the erosion of traditional institutions, and anthropocentric modern ideologies have displaced this holistic ethic, leading to unsustainable exploitation and weak governance structures (Tosam, 2019).

This diagnosis is echoed in Igbo contexts by Nwobodo (2021), who explores how taboos once maintained socio-political harmony and ecological balance in

Igbo society. These taboos, rooted in respect for *Ala*, served as moral guardrails against acts that could harm the environment or fracture communal peace. However, Nwobodo laments that globalization, Christianity, and Western education have drastically reduced the observance of these taboos, threatening the moral and ecological fabric of the society. He calls for a reinvigoration of traditional practices—not as rigid dogma but as culturally resonant tools for restoring moral, ecological, and political equilibrium.

The spiritual foundation of ecological ethics in Igbo cosmology is further elaborated by **Agama and** Onyeakazi (2022), who provide a philosophical examination of taboo as it relates to spirituality. They contend that taboo (*nso ala*) is more than a moral code; it is a spiritual link between the visible world and the realm of the ancestors and deities. In their analysis, *Ala* is not merely symbolic; she is an active moral agent whose desecration has both spiritual and physical consequences for individuals and communities. The authors note that failure to observe these taboos can lead to illness, death, famine, or communal misfortune, thereby reinforcing the role of ancestral authority in ecological and moral governance (Agama & Onyeakazi, 2022).

These scholars advance the argument that traditional ecological systems in Igbo society were not just socially constructed but spiritually enforced. The obligation to preserve sacred sites, avoid desecration of the earth, and maintain ancestral compounds stemmed from a cosmology where human life, natural resources, and spiritual forces coexisted in a delicate moral balance. In this sense, *nso ala* operates as both a cultural code and an environmental protection mechanism.

This body of literature reinforces the need to recognize indigenous ecological knowledge as a vital component of environmental policy in Africa. Shomanah and Zvingowanisei (2024)**,** for instance, argue that sacred sites across Africa serve as informal conservation zones, maintaining biodiversity and ecological integrity. These areas are protected not by state legislation but by spiritual taboos and cultural reverence, making them cost-effective and community-owned conservation tools.

Adam (2012) similarly advocates for the legal recognition of sacred natural sites in Kenya, proposing that indigenous custodianship can be incorporated into national land-use frameworks. This resonates with the Igbo practice of community-based resource management, where elders, shrine priests, and traditional authorities play key roles in ensuring sustainability.

Deopa and Rinaldo (2023)**,** working in the context of Benin, reinforce the idea that African traditional religions embody ecological values capable of mitigating the environmental crisis. They argue that the dismissal of these traditions in favour of Western models has weakened community engagement and spiritual accountability.

Moreover, Fabiano, Schulz, and Branas (2021) highlight the kinship ontology in African thought systems, where humans are understood not as dominators but as stewards and relatives of nature. This relational ethic parallels Pope Francis's Laudato Si', which critiques technocratic and consumerist approaches and calls for an "integral ecology" that links the environment, the poor, and cultural traditions.

Finally, Anyokwu (2008) offers a cultural dimension to this discourse through his study of *Mbari* houses—ritual art forms that embody environmental respect and divine homage. These structures reflect a worldview in which art, religion, and ecology converge, affirming the centrality of culture in environmental ethics.

22.3 Methodology

This study adopts a qualitative and ethnographic approach, relying on both primary and secondary sources. Fieldwork was conducted in two Igbo communities (Izhi and Ikwo) in Ebonyi State, south-eastern Nigeria. Data were collected through oral interviews, direct observations, and participatory discussions with elders, shrine custodians, traditional leaders, and other knowledgeable community members.

The resource persons from Izhi include Chief Hon. Patrick Orogbo, Chief Ogbonna Nwiboko, Mr. Williams Nwofe, Mr. Nwizzi Onyibe, Dr. Chukwuemeka Nwamini, Barr. Paul Nwankwegu, and Mr. Ifeanyi Nwophoke. In Ikwo, the consultants included Chief Dr. Emmanuel Obona, Dr. Chigbo Nwode, Dr. Uche Jacob, and Prof. Anthony Itumo.

Through these engagements, the study documented sacred ecological sites, catalogued ecological taboos and their social enforcement mechanisms, and analysed how these traditional systems interact with or resist modern environmental pressures. The data were thematically analysed to identify patterns of environmental philosophy, ritual regulation, and cultural logic that inform indigenous ecological governance.

22.4 Typologies and Cultural Significance of Sacred Sites and Ecological Taboos in Izhi and Ikwo

Among the Igbo communities of Izhi and Ikwo in south-eastern Nigeria, the environment is not merely a physical space for habitation and agriculture; it is a sacred domain governed by spiritual, moral, and social codes. These codes manifest in the form of sacred sites and ecological taboos that regulate human interaction with the land, forests, waters, and ancestral compounds. These culturally sanctioned structures represent an indigenous environmental ethics system that has, for generations, maintained ecological harmony, communal order, and spiritual wellbeing.

Sacred sites in Izhi and Ikwo fall into distinct but interrelated typologies, each serving specific ecological and moral purposes. One of the most culturally significant sacred sites is the Ọchịedja/Ịkịrikpọ, commonly referred to as the *evil forest*. These forests are forbidden zones designated for the burial of individuals who committed abominable acts (*nso Ala*), such as incest, ritual killing, or witchcraft, or those who died mysteriously or under morally troubling conditions. They serve both a spiritual and ecological function: spiritually, they remove moral pollution from the community; ecologically, they serve as preserved forest ecosystems where human activities such as logging, farming, or building are strictly prohibited. Fear of spiritual reprisal ensures their continued conservation.

Another typology is the Eze Unuphu, or ancestral compounds, which are treated as sacred because they are the original dwelling places of family patriarchs. These spaces are typically not for sale or reallocation and are excluded from commercial development. Their continued reverence helps ensure land preservation and restricts land speculation and environmental degradation within ancestral boundaries.

Agwa, or shrine groves, are sites of worship dedicated to local deities and spirits. Located within forests or near rivers, they are maintained as sacred by prohibitions on defecation, urination, logging, and farming. Each shrine has unique taboos, but the common denominator is environmental protection grounded in reverence for spiritual beings believed to dwell there. A closely related type of site is Enya Alị (Eyes of the Earth), which are naturally occurring spiritual locations known for mysterious events or unexplainable manifestations. These are not human—made shrines but naturally sacred spots, often surrounded by protected groves and bushland.

Additionally, sacred streams and ponds (*ngele* and *okpuru*) are regarded as homes to spiritual beings or reincarnated ancestors. These water bodies are ritually protected from pollution and overexploitation. In some cases, fishing is forbidden, while in others, it is allowed only during annual communal expeditions. The belief in the presence of ancestral spirits within certain fish species adds another layer of protection to aquatic biodiversity. For instance, some communities distinguish between "men's ponds" and "women's ponds," each with its own sacred rules and rituals, reinforcing the communal responsibility for environmental management.

Beyond sacred spaces, ecological taboos serve as everyday regulatory mechanisms. These include prohibitions against defecating on anthills, sacred paths, or palm logs; cutting bamboo or sacred trees without permission; building on ancestral boundaries; or violating rotational farming arrangements. Violations are believed to incur severe consequences such as infertility, death, madness, or persistent misfortune. These taboos act as moral compasses and practical conservation strategies. For example, rotational land use (*ala egụ)* allows fallow periods that promote soil fertility, while also regulating human access to grazing and hunting zones.

Thus, the sacred sites and taboos in Izhi and Ikwo are not merely spiritual conventions but embedded systems of ecological regulation, with legal, moral, and social consequences that discourage unsustainable behaviour and promote long—term environmental stewardship.

22.5 Traditional Igbo Ecological Practices and Their Alignment with Conservation, Social Equity, and Moral Governance

Traditional ecological practices in Igbo communities such as Izhi and Ikwo reflect a worldview that places human beings in a reciprocal and sacred relationship with nature. Unlike Western anthropocentric paradigms that often see nature as a resource to be exploited, the Igbo ecological model is ecocentric—it values nature for its intrinsic worth and spiritual essence.

These practices align with modern principles of conservation in several ways. First, through the establishment and maintenance of sacred groves, evil forests, and protected water bodies, the communities effectively preserve biodiversity and prevent overexploitation of natural resources. These preserved spaces act as carbon sinks, breeding grounds for wildlife, and reservoirs of medicinal plants. The restriction on cutting trees in sacred forests or defecating near ponds ensures minimal human interference and pollution, echoing principles in contemporary conservation biology.

Second, traditional land rotation systems enforce ecological sustainability. These systems designate which portions of land are to be cultivated during specific years, allowing other areas to lie fallow, replenish, and serve alternative purposes such as grazing or hunting. This not only sustains soil fertility and local food systems but also encourages social equity by ensuring communal access to resources. Everyone in the village is subject to the same rotational rules, and violators, regardless of status, face communal sanctions. Such a system discourages elite capture of land and supports equitable distribution of environmental benefits.

Additionally, Igbo ecological ethics are deeply tied to moral governance. The land (*Ala*) is not a neutral space but a spiritual force that punishes moral transgressions. Taboos against land desecration, whether by incest, unjust land acquisition, or ecological violations, are spiritually enforced through a belief in divine retribution. Offenders may suffer illness, infertility, or social ostracism. This form of moral enforcement promotes responsible environmental behaviour, often more effectively than external regulation. Importantly, the process of appeasing *Ala* when transgressions occur involves public confession, community adjudication, and ritual restitution—practices that reinforce collective accountability and moral order.

Moreover, traditional rules governing the management of shared natural resources, such as communal fishing expeditions, serve not only conservation goals but also promote inclusivity and participatory governance. In these events, resources are shared, rules are agreed upon by consensus, and penalties for violation are enforced communally. Women and men have distinct roles, especially in relation to gendered ponds or forests, reflecting a culturally embedded logic of shared stewardship that respects social roles without diminishing ecological responsibility.

In summary, the environmental practices of Izhi and Ikwo communities represent a coherent system that aligns with core principles of conservation, equity, and governance. Their effectiveness lies in their holistic integration of the spiritual, social, and ecological dimensions of human life. These practices offer valuable models for modern environmental policy that seeks to be ethical, inclusive, and sustainable.

22.6 Exploring the Disruptions Brought About by Modernity and Their Impact on Indigenous Environmental Systems

The environmental knowledge systems of indigenous Igbo communities, especially in Izhi and Ikwo, have historically ensured environmental stability through culturally embedded practices. However, these systems are increasingly disrupted by the forces of modernity—urbanization, globalization, religious transformation, and state-centric development policies. These disruptions threaten both the tangible ecosystems that these traditions helped preserve and the intangible knowledge systems that underpinned their management.

One of the most visible disruptions is *urbanization*, which has led to the rapid clearance of sacred groves, evil forests, and ancestral lands for the construction of housing estates, markets, schools, and places of worship. Formerly taboo spaces like *ọchịedja* are now being reclaimed and commodified as building plots. The profaning of sacred spaces not only compromises biodiversity but also erodes community cohesion and the moral authority once associated with the land. In Izhi and Ikwo, it was observed during fieldwork that community leaders lament the diminishing respect for spiritual and cultural boundaries, as real estate speculation pressures families to sell land that was once considered untouchable.

Religious transformation, particularly the spread of Pentecostal Christianity, has also played a role in undermining indigenous ecological ethics. Many Christian converts reject traditional practices and sacred sites as idolatrous or demonic. Shrines (*agwa*), sacred streams, and ancestral compounds are abandoned, vandalized, or destroyed. Some churches have been deliberately constructed on former sacred groves to symbolize the triumph of new faiths over "paganism." In this process, traditional ecological knowledge is delegitimized and custodians are stigmatized as primitive or spiritually unclean.

Western education and state policies have further contributed to this epistemic disruption. School curricula and government environmental policies often ignore or marginalize indigenous knowledge systems, focusing instead on Euro—American scientific paradigms. The disconnection between formal education and cultural knowledge has resulted in younger generations being alienated from the sacred geography of their communities. Field interviews in Izhi, for example, revealed that many youths are unaware of the significance of ancestral sites or the consequences of violating ecological taboos.

In addition, state-led land reforms and individual land tenure systems have diluted the communal land stewardship model that undergirded traditional environmental governance. Under traditional arrangements, land was collectively owned and used under the guidance of elders and deities, with clear moral obligations and social checks. Modern legal systems, however, allow for private land acquisition and transfer, disrupting communal control and creating conditions for unsustainable land use, environmental degradation, and conflict.

Finally, modern consumerism and extractive economic models are displacing the indigenous ethic of restraint. The Igbo concept of *mma na ndu* (beauty in moderation and life) that once guided agricultural and consumption practices is giving way to profit—driven extraction of timber, sand, and minerals, often without community consent. The encroachment of multinational agribusiness and resource-extraction industries poses serious threats to the natural and spiritual landscapes of communities like Izhi and Ikwo.

These disruptions reflect a broader shift from holistic, moral—ecological frameworks to fragmented, technocratic systems that prioritize economic growth over cultural and environmental sustainability. The loss of traditional custodianship, taboos, and spiritual accountability has weakened indigenous environmental resilience and created a moral vacuum in land governance.

22.7 Reviving, Adapting, and Integrating Traditional Ecological Knowledge into Contemporary Environmental Policy

Given the challenges outlined above, there is an urgent need to revive and adapt traditional Igbo ecological knowledge and integrate it into contemporary environmental education, land-use planning, and policy frameworks. This revival must be intentional, community-driven, and culturally sensitive, recognizing both the intrinsic value of indigenous systems and their practical relevance to modern sustainability challenges.

First, revival begins with documentation and recognition. Sacred ecological sites in Izhi and Ikwo, such as evil forests, ancestral groves, and ritual water bodies, should be identified, mapped, and catalogued as cultural and ecological heritage sites. Legal frameworks at the local, state, and national levels must recognize the spiritual and conservation value of these spaces. The designation of such areas as "Community Ecological Reserves" or "Cultural Conservation Zones" would help protect them from destruction by development or religious encroachment.

Second, traditional custodians and elders should be formally included in land-use planning and natural resource governance. Their ecological knowledge, transmitted orally through stories, rituals, and social practices, can complement scientific approaches to biodiversity management. Hybrid governance structures can be created where traditional authorities co-manage resources alongside state agencies and NGOs, ensuring local ownership and cultural continuity.

Third, environmental education must be decolonized to include indigenous knowledge systems. School curricula in south-eastern Nigeria should be revised to teach students about *Ala*, ecological taboos, rotational farming, and the role of shrines and sacred forests in conservation. Storytelling, field trips to sacred sites, and the use of indigenous languages in instruction can help bridge generational gaps in ecological awareness. This would not only promote environmental consciousness but also restore cultural pride.

Fourth, community-based conservation initiatives can be strengthened through cultural festivals and rituals that renew collective identity and ecological values. For example, seasonal agricultural festivals like *ọbọ iphe, Ojiji, Ajị* can be used as platforms to promote messages on reforestation, waste reduction, and water preservation, all tied to the cosmological role of *Ali*. Engaging youth through cultural clubs, ecological theatre, and social media campaigns in indigenous languages can make these practices more accessible and relevant.

Fifth, policy integration should focus on moral and spiritual dimensions of environmental governance. Policymakers must move beyond technocratic frameworks and embrace ethical worldviews that see land not just as property, but as sacred. Drawing from the Igbo belief that violating the earth has spiritual consequences, environmental regulations can be enriched with moral narratives that emphasize long-term stewardship, communal accountability, and intergenerational justice.

Lastly, research partnerships between universities, cultural institutions, and local communities are essential for sustaining this integration. Collaborative research can validate the efficacy of traditional practices (e.g., fallow systems, ecological taboos, forest zoning) and provide data for policy innovation. These partnerships should ensure that knowledge production is reciprocal, with benefits flowing back to the communities whose traditions are being studied and preserved.

In sum, the revival and integration of traditional Igbo ecological knowledge offer a path toward "integral ecology" as envisioned in *Laudato Si'*—an approach that binds care for the earth with care for culture, ethics, and the poor. By placing *Ala* at the centre of environmental policy, Nigeria and Africa at large can develop culturally grounded responses to climate change, land degradation, and biodiversity loss.

22.8 Conclusion and Recommendations

This study has examined how the Igbo cosmological view of the earth as sacred (*Ala*) manifests in environmental ethics through the preservation of sacred sites and enforcement of ecological taboos. The typologies identified in Izhi and Ikwo (evil forests, ancestral lands, sacred ponds, and shrines) demonstrate a coherent indigenous framework of environmental management grounded in moral governance, communal stewardship, and spiritual accountability. These practices align with the principles of *Integral Human Ecology*, which calls for the interconnection of environmental conservation, human dignity, cultural integrity, and social justice.

Key findings indicate that these traditional ecological systems have historically ensured biodiversity protection, sustainable resource use, and equitable land distribution. However, these systems are increasingly threatened by modern pressures such as urban expansion, religious conversions, state—led land commodification, and cultural erosion. The weakening of indigenous environmental ethics has led to the degradation of sacred sites, the collapse of communal land systems, and the loss of ecological knowledge among the youth.

Reviving and adapting traditional practices is therefore not a nostalgic return to the past, but a necessary step toward building ecologically sustainable and culturally relevant futures. To achieve this, several recommendations emerge:

1. Policy Integration: Governments and environmental agencies should formally recognize sacred sites as protected ecological heritage zones. Legal frameworks must accommodate indigenous concepts of land stewardship.
2. Participatory Governance: Traditional custodians should be included in land—use planning, natural resource governance, and environmental impact assessments.
3. Educational Reform: School curricula should be decolonized to include indigenous ecological knowledge, taught in local languages, and grounded in cultural values.
4. Community Engagement: Cultural festivals, rituals, and ecological storytelling should be revitalized as tools for environmental education and intergenerational transmission of knowledge.
5. Research and Documentation: There should be collaborative efforts between academia and local communities to document, analyse, and archive traditional ecological knowledge for use in policy and education.

In conclusion, the Igbo model of environmental ethics offers more than cultural insight. It provides a living system of *Integral Human Ecology* that recognizes the earth as sacred, human life as morally bound to nature, and ecological justice as inseparable from cultural and spiritual values. Harnessing this wisdom within contemporary governance structures is essential for Africa's response to climate change and environmental degradation.

References

Adam, A.H. (2012). *Recognising sacred natural sites and territories in Kenya.* Gaia Foundation. At: https://gaiafoundation.org/post-library/report-recognising-sacre-natural-sites-and-territories-in-kenya/ [13 May 2025].

Agama, C.S. and Onyeakazi, J.C. (2022). The concept of taboo in Igbo spirituality: A philosophical examination. In: I.A. Kanu, ed. *African eco – philosophy: Cosmology, consciousness and the environment*. Nigeria: Association for the Promotion of African Studies (APAS), pp.137–140.

Anyokwu, C. (2008). Mbari and the Igbo concept of art in Chinua Achebe's *Things Fall Apart. Journal of Language and Literary Studies*, 16(3).

Deopa, N. and Rinaldo, D. (2023). Sacred ecology: The environmental impact of African traditional religions. *arXiv preprint*. At: https://arxiv.org/abs/2401.13673 [17 May 2025].

Fabiano, G., Schulz, M. and Branas, C. (2021). On religion and the environment: An analysis of the interface of African environmental ethics and the worshipping traditions of African Indigenous

Churches. *African Thought: A Journal of Afro—centric Knowledge*, 1(3). Available at: https://journals.co.za/doi/10.10520/ejc-atjack_v1_nse3_a13 [17 May 2025].

Madu, J.E., 1996. *Religion and the unity of the Nigerian nation.* Nsukka: Fulladu Publishing Company.

Nwobodo, R.E.E. (2021). A dialogic of taboo in traditional Igbo society and the 21st century mentality. *Trinitarian International Journal of Arts and Humanities*, 1(2), pp.100–110.

Onwu, C. (2002). *African traditional religion: A conceptual and philosophical analysis.* Enugu: Snaap Press Ltd.

Shomanah, L. and Zvingowanisei, C. (2024). Ecological sustainability for "life on land": Wellspring of indigenous knowledge. *Religions*, 16(3), p.311. At: https://www.mdpi.com/2077–1444/16/3/311 [25 May 2025].

Tosam, M.J. (2019). African environmental ethics and sustainable development. *Open Journal of Philosophy*, 9(3), pp.172–192. Available at: http://www.scirp.org/journal/ojpp [25 May 2025].

Chapter 23
Sacred Limits: Sovereignty, Development, and the Eco-Theopolitics of Land

Geleta Tesfaye Berisso

Abstract Is land just a commodity? This question sits at the heart of the discussion around modern politics, economics, and ethics. Across the globe, land has increasingly been treated as an object to be surveyed, securitized, traded, and extracted; and of course, its value reduced to the price it can fetch on markets or the revenues it can yield to sovereign states. I diagnose the commodification of land as a theological-political phenomenon that arises from a modern paradigm that stripped land of sacred meaning and relational value, thereby enthroning the human sovereign, whether state, investor, or technocrat, in place of divine or ecological order. In response, I propose an alternative vision grounded in eco-theopolitics is proposed that reclaims land as sacred trust rather than mere commodity of the state. Juxtaposing Indigenous Oromo concepts of *safuu*, an Oromo word for *a restrained moral limit*, and *nagaa*, that is peace, or right relation, with Catholic social teaching on Integral Ecology, I contend that peace and integral development require honoring the sacred limits on human sovereignty over nature.

Keywords Sovereignty · Land · Ownership · Conflict · Development

23.1 Introduction: Is Land Just a Commodity?

Is land just a commodity? This question sits at the heart of the discussion around modern politics, economics, and ethics. Across the globe, land has increasingly been treated as an object to be surveyed, securitized, traded, and extracted; and of course, its value reduced to the price it can fetch on markets or the revenues it can yield to sovereign states. Scholars illuminate how modern sovereignty, from the imperial era to neoliberal globalization, operates as if land were a passive canvas upon which the sovereign will inscribes its order. Development theorists like Arturo

G. T. Berisso (✉)
University of Notre Dame, Notre Dame, United States
e-mail: gberisso@nd.edu

R. Marcantonio et al. (eds.), *Integral Ecology in the Anthropocene: Africa and Beyond*, The Anthropocene: Politik—Economics—Society—Science 49,
https://doi.org/10.1007/978-3-032-32125-1_23

Escobar (1995) and James Ferguson (1994) have shown how land in the Global South was reframed as "idle" or "underutilized" until absorbed into circuits of capital. In Ethiopia, as instance, successive regimes, spanning from the imperial, socialist, and neoliberal, each justified the reordering of land through different framings spanning from civilizing, collectivizing, modernizing; yet: all converged on the reduction of land to a commodity and its peoples to obstacles of progress. The consequences have been devastating: dispossession, ecological degradation, cultural erasure, and, in times of drought, spirals of famine and conflict.

Standing on the phenomenon in Ethiopia's pastoral lowlands, particularly among the Guji and Borana Oromo recurrent droughts in recent years, I diagnose the commodification of land as a theological-political phenomenon that arises from a modern paradigm that stripped land of sacred meaning and relational value, thereby enthroning the human sovereign, whether state, investor, or technocrat, in place of divine or ecological order. In response, I propose an alternative vision grounded in eco-theopolitics is proposed that reclaims land as sacred trust rather than mere commodity of the state. Juxtaposing Indigenous Oromo concepts of *safuu*, an Oromo word for *a restrained moral limit*, and *nagaa*, that is peace, or right relation, with Catholic social teaching on Integral Ecology, I contend that peace and integral development require honoring the sacred limits on human sovereignty over nature. To advance this thesis, the article proceeds in four steps. First, I frame modern sovereignty as a theological-political formation that desacralizes land. Second, I trace Ethiopia's historical trajectory of land governance across imperial, socialist, and neoliberal regimes as exemplars of commodification. Third, I analyze how these processes constitute structural violence in the pastoralist regions of Guji and Borana, deepened by recent droughts and ecological crises. Finally, I articulate an alternative eco-theopolitical paradigm that integrates Indigenous and Catholic visions, contending that only by reclaiming land's sacredness can Ethiopia, and the wider world, achieve just peace and ecological sustainability.

23.2 Sovereignty, Political Theology, and the Question of Land

Modern concepts of state sovereignty have an entrenched theological underpinning. Carl Schmitt's own Political Theology (1922) defined sovereignty as the power to declare the state of exception: "Sovereign is he who decides on the exception". In other words, the sovereign stands as a kind of god of the political realm, with absolute authority to suspend ordinary law in the name of order. This idea casts a long shadow: it implies that land and life within a state's territory ultimately fall under the totalizing will of the sovereign. Theologically, one might say that modern sovereignty relocated the locus of ultimate authority from God to man, particularly to man as ruler of a territorial state. The land, in this framework, turns into a mere Nomos (law or order) that the sovereign carves out by act of power; land is the passive canvas on which sovereign power inscribes its will.

From a political-theological perspective, this move is a shift in the imagined relationship between humans, land, and the divine. Now, it is important to note that pre-modern and many Indigenous cosmologies regarded land as imbued with sacred significance: as creation, as gift, as home of ancestral spirits, or as under the sovereignty of a deity. By contrast, the modern Westphalian order, and of course, its globalized successor, largely treats land as property of the state or individuals, alienated from any intrinsic sacred status. In secular modernity, land is enclosed in what Jürgen Moltmann calls the "godless" realm of human dominion (Moltmann 1985). William Cavanaugh (2024) further critiqued this anthropocentric sovereignty, arguing that it amounts to a form of idolatry: an unchecked human lordship that recognizes no higher moral limits. Nevertheless, as philosophers Lovász and Pető (2025) observe, Schmitt's own framework, ironically, remained anthropocentric: he grounded political order in decisive human authority, with no role for non-human or divine agency. Schmitt's political theology "remains within the bounds of anthropocentrism," assuming that "only human subjects can be the subjects of politics" (Lovász and Pető 2025).

Recent scholarship seeks to move beyond this limitation. Political theologians influenced by ecological and post-humanist thought argue that we must expand Schmitt's critique of modernity and envision a theopolitics not confined to humans alone. Lovász and Pető (2025) propose this: a "posthumanist elaboration of political theology" that disempowers the human species and empowers nature (Lovász and Pető 2025). They suggest that sovereignty itself need not be abolished but must be reconceived in ecocentric terms (Lovász and Pető 2025). In other words, rather than a man-god ruling over land-as-property, we imagine sovereignty as vested in the whole creation or in the divine as present in the land. This is a return, in a new key, to the idea of sacred limits. Within this framework, the sovereignty of the state or the market over land is not absolute; it is bounded by a higher order, the inviolability of creation, or the rights of the earth. Political theology thus gives us language to see that treating land as a pure commodity is a choice grounded in a specific "theological" narrative: a narrative of human dominion unrestrained by any sacred obligation to the land.

Land is, in fact, the very site where theological and political ordering meet. Who owns the land? Who claims the right to extract its wealth, to define its boundaries, to allocate its use? These are political-juridical questions, but they are also moral and cosmological ones. Schmitt's theory informs us that sovereignty is about the power to decide, and decisions about land use, whether to build, to mine, to conserve, to allocate to some people and not others, are decisions that reveal what we venerate and what we deem dispensable. Modern secular sovereignty tends to venerate abstract notions of development or security at the expense of local communities and ecosystems. It has, effectively, de-sacralized land, making it a nearly profane object subject to technocratic management. But as we will see, alternative theological visions invert this picture: they sacralize the land and de-center human power.

To frame the problem concisely: within the dominant modern paradigm, land is treated as if it were "merely one more commodity": fungible, alienable, without inherent worth except what value human markets assign. This assumption underlies global capitalism's approach to everything from real estate speculation to resource extraction. Yet, this view stands in contrast to voices in both the Global North and South that insist land is not just a commodity. Pope Francis (2015), for example, elevates this very point in his encyclical Laudato Si'. In a section on "cultural ecology," Francis writes that for indigenous communities, "land is not a commodity but rather a gift from God and from their ancestors who rest there, a sacred space with which they need to interact if they are to maintain their identity and values" (Francis 2015, §146). This statement underscores that land has an irreducible sacred and cultural dimension; it is a gift, not a good to be bought and sold. Similarly, theologian Matthew Fox observes that the "heart of the current worldview which treats earth as an object ripe for extractive capitalism" is a loss of the sacredness of the Earth (Fox 2018). Recovering a sense of the Earth's sacredness, Fox contends, is essential to counteract the dominant exploitative mindset. Thus, how we answer the question "Is land just a commodity?" ultimately depends on what we consider sacred. If profit and state power are sacred, land will be sacrificed to those gods; if God, creation, or relational harmony is sacred, then land must be treated as holy ground, with limits on our control.

23.3 Development and the Commodification of Land

Modern sovereignty's commodifying stance toward land found full expression in the era of development economics and neoliberal globalization. The mid-20th century ushered in what Arturo Escobar (1995) calls the development discourse, a regime of knowledge and practice that framed "the so-called Third World" as a terrain to be transformed and uplifted by Western models (Escobar 1995). In his Encountering Development, he showed that development institutions, from the World Bank to state planners, constructed poverty and land use in the Global South as technical problems to be solved by external expertise (Escobar 1995, 45–52). This technocratic approach went hand in hand with commodification. Land was to be surveyed, mapped, zoned, and ultimately integrated into circuits of capital. As Asebe et al. (2019) observes, in the development imagination, land in Asia, Africa, and Latin America was often seen as "underutilized" or "idle" unless it was producing cash crops or minerals for the world market. Communal lands, pastoral rangelands, forest commons, all these were treated as wastelands awaiting improvement. The logic of development thereby extended the capitalist commodification of land, which it established long in Europe, to the farthest reaches of the post-colonial world.

Anthropologist James Ferguson further underscored how development interventions depoliticize land and resources. In The Anti-Politics Machine (1994), Ferguson examined development projects in Lesotho and found that they consistently failed to address; or even acknowledge, the political and class dynamics underpinning poverty and land tenure. Instead, development agencies presented their work as purely technical, apolitical efforts to "improve" livelihoods. The effect, Ferguson argued, was to expand bureaucratic state power into rural areas while deflecting attention from questions of who owns the land and who benefits. Significantly, Ferguson showed that development discourse renders political issues technical. For instance, treating landlessness or land degradation as a technical problem solvable by expert intervention, rather than as a result of historical dispossession or inequality (Ferguson 1994, 68–69). This is what he memorably called the "anti-politics machine" at work.

This development paradigm is not just a thing of the past; it has evolved into contemporary neoliberal approaches that explicitly treat land as a global asset class. Since the 1990s and especially after the 2008 financial crisis, we have witnessed a worldwide "land rush", large-scale land acquisitions by corporations and foreign investors, particularly in Africa and Asia. Ethiopia offers a vivid example: beginning in the mid-2000s, the Ethiopian government leased out millions of hectares of fertile land to investors from China, India, Saudi Arabia, and other countries, aiming to spur commercial agriculture. Advertised as "unused" or "underdeveloped" land, much of it was, in fact, commons used by pastoralists and smallholders (The Guardian 2014). The logic was classic development-cum-commodification: monetize the land to generate GDP growth and foreign exchange, assuming this will eventually benefit everyone. In practice, these deals often undermined local food security and provoked conflict. As The Guardian report described, it became a "'nightmare' for Ethiopian pastoralists as foreign investors buy up land," leading to human rights violations and displacement (The Guardian 2014).

Escobar's notion of epistemic erasure is helpful to understand what happens in these scenarios. Development's commodifying gaze not only dispossesses people of land but also erases the ways those people know and value the land. Local ecological knowledge, spiritual attachments, and customary tenure systems are rendered invisible or invalid. The land is reframed as empty or uncared for until capital arrives. In the words, development plans justify emptying the lowlands for investments by casting pastoral lands as blank frontiers awaiting civilization. This is exactly what Asebe Regassa et al. (2019) found in Ethiopia's South Omo Valley. Examining two sedentarization schemes imposed on agro-pastoral communities, they show how the state invoked a "civilizing impetus of 'improvement schemes' for 'backward' pastoralists" in order to appropriate grazing lands for corporate agriculture (Regassa, Hizekiel, and Korf 2019). Under the banner of modernization, pastoralists were coerced to abandon their migratory lifestyle; the land was fenced and transferred to investor control. As Regassa et al. put it, "frontier imaginations of pastoral lowlands" have legitimated violent practices of eviction, effectively development-by-dispossession (Regassa, Hizekiel, and Korf 2019). Private investors, often aligned with political elites, are "enrolled" in a project to extend state power; thus, "thickening state presence" into these margins, making land and resources more extractable for the regime's benefit (Regassa, Hizekiel, and Korf 2019). In short, the development paradigm in its various guises, colonial, socialist, neoliberal, consistently reduces land to an economic commodity, overriding indigenous and peasant conceptions of land as life, heritage, and sacred trust.

Yet, as post-development scholars argue, this paradigm is neither total nor uncontested. Arturo Escobar (1995) concluded his study by calling for alternatives to development, envisioning a "post-development era" rooted in local cultural practices and autonomy (Escobar 1995). Such alternatives reject the idea that land must be integrated into a global market to have value. Instead, they affirm that land's value might lie in community sustenance, cultural identity, or ecological balance. The challenge, of course, is that the institutions of modern sovereignty (states, banks, corporations) have largely been built around the commodification of land. As we turn to Ethiopia, we will see in microcosm how difficult it is to break from this paradigm; and yet also how communities have continually asserted different relationships to land, keeping alive the possibility of another way.

23.4 Sovereignty and Land in Ethiopia's Modernity

Ethiopia provides a rich and complex case study of how modern sovereignty has managed land, illuminating the shifts and continuities across imperial, socialist, and neoliberal regimes. Each era in Ethiopia's modern history, the imperial rule of Menelik II and Haile Selassie, the Derg military socialist regime, and the post-1991 neoliberal-oriented government, had its own ideological justification, yet all three approached land in ways that ultimately commodified it and dispossessed certain populations, notably the Oromo and other southern peoples, including pastoralists. Understanding this trajectory will ground our theoretical critique in concrete history.

23.4.1 Imperial Era (ca. 1880s–1974)

Ethiopia's imperial expansion under Emperor Menelik II in the late 19th century brought vast Oromo, Somali, Afar, and other lands in the south and east under Abyssinian control. This process, sometimes called "internal colonization," was driven by a quest for resources, tribute, and strategic territory. Conquered lands were often distributed as gult (fiefs) to northern soldiers and nobility, entrenching a system of semi-feudal land ownership. Indigenous systems of land tenure and use were supplanted or subordinated. For instance, among the Oromo, land had traditionally been managed communally under the Gadaa system, with complex rules (*safuu*) governing human-earth relations. The imperial regime, however, regarded itself as the sovereign owner of all land, free to allocate it in reward of loyalty or for plantation concessions. In effect, Menelik and later Haile Selassie treated land as both an instrument of power and a source of revenue, a commodity that could be taxed, leased, or gifted. Peasants became tenants on their own ancestral land, owing heavy rents to absentee landlords, especially in the fertile highlands. Meanwhile, in the lowland peripheries, imperial authority was less direct but increasingly felt through the establishment of towns, garrisons, and extraction of ivory, gold, and slaves (Marcus 2002, 85–92).

The theopolitical dimension was evident: the Emperor was often figured as having a quasi-sacred authority (Haile Selassie was literally called the "Elect of God, Lion of Judah"), yet his government simultaneously secularized land into imperial property. The Church played a role too, as the Orthodox Church was granted huge tracts of land (the ríst system), again consolidating the notion that land's control lay in elite hands. The imperial worldview largely dismissed the idea that pastoralists or shifting cultivators had legitimate claims to land; such people were often labeled as "backward", foreshadowing the civilizing discourse noted in the Asebe study (Regassa, Hizekiel, and Korf 2019). Thus, the imperial state set the pattern: land would be harnessed to serve state-building and elite accumulation, with little regard for local autonomy or sacred attachments; nonetheless, except when those could be co-opted to legitimize power.

23.4.2 Socialist Derg Era (1974–1991)

In 1974 a revolution ended the monarchy and brought to power the Derg, a Marxist military junta. The Derg's signature policy was the 1975 Land Reform Proclamation: "Land to the tiller!" was the rallying cry. Overnight, Ethiopia apparently abolished feudal land tenure; all land was nationalized and private ownership abolished. Peasants received use-rights to plots (in their home areas) and landlord exploitation was terminated. On paper, this was a radical decommodification of land, land could no longer be bought or sold. However, the state now became the ultimate owner of all land, and with that came new forms of control. The Derg pursued villagization and resettlement programs in the 1980s: it forcibly relocated millions of peasants (especially from drought-prone north) into "new towns" or to sparsely populated areas (often in the lowlands). These programs, done in the name of efficient service delivery and socialist modernization, often ignored the needs and wishes of local communities and led to great hardship and death. Pastoralists were pressured to sedentarize and adopt farming. The principle that the state knew best remained. The Derg's approach to land can be seen as high-modernist (Scott 1998), it sought to impose an order on the landscape (grid-like villages, state farms, production quotas) according to an abstract ideal. While they removed the land market (for a time), they continued the pattern of viewing land as a resource to be centrally planned for production.

Indeed, state farms and collectivization schemes treated land like a state-owned commodity, shifting it among uses by fiat. Traditional grazing routes or sacred groves held no weight against the barrel of the gun and the planner's pen. Those who resisted, or who simply did not fit the plan, whether nomadic herders, for example, were seen as obstacles to progress. In conclusion, the socialist period reinforced state sovereignty over land in a totalizing manner: if previously landlords were little kings over land, now the revolutionary state was the supreme lord. Importantly, the Derg also securitized land: insurgencies in Eritrea, Tigray, and Oromia made the regime suspect that allowing local autonomy over land could foster separatism. Therefore, tight control was justified in the name of national unity and security. Sovereignty's exception logic (à la Schmitt) was visible in the Derg's frequent declaration of emergencies and "red terror," during which land and life were at the mercy of military diktat.

23.4.3 Neoliberal/Developmental State Era (1991–present)

After 1991, a coalition of rebel forces overthrew the Derg and established a new government under the Ethiopian People's Revolutionary Democratic Front (EPRDF). This regime, led effectively by Prime Minister Meles Zenawi until 2012, embraced a curious blend of market economics and state developmentalism. Private land ownership remained prohibited in the constitution, land is formally public property held by the state on behalf of the people. This was justified as a way to prevent landlessness and speculative buying-up of land. In practice, however, the state's ownership has enabled it to lease land to investors on long-term contracts. The 2000s saw Ethiopia touted as a development success story, with high GDP growth. A cornerstone of this growth was commercial agriculture and resource extraction. The government identified millions of hectares of "idle" land available for investment, much of it in lowland regions like Gambella, Benishangul-Gumuz, the Omo Valley, and Afar. Here we see a reprise of earlier patterns but in the language of global markets: the lowlands are frontiers to be "civilized" by large-scale projects, whether sugar plantations, cotton farms, biofuel crops, or hydroelectric dams.

Pastoralists and shifting cultivators are once again viewed as wasting the land's potential. The EPRDF's Growth and Transformation Plan explicitly aimed to modernize agriculture by transitioning from subsistence to commercial farming. Consequently, local communities in these regions experienced mass displacements. The case of the Lower Omo (in South Omo zone) is illustrative: the construction of the massive Gibe III dam, completed 2015, ended the natural flood cycle of the Omo River, which had sustained the flood-retreat cultivation of Indigenous groups like the Mursi and Bodi for centuries. Simultaneously, the government paved roads and established state-run sugar cane plantations on their lands. Thousands of Indigenous agro-pastoralists were coerced into resettlement sites. Researchers have described this as a form of coercive agrarian development that echoes both colonial and socialist interventions (Regassa, Hizekiel, and Korf 2019). As Asebe et al. (2019) note, the language might have shifted to investment and productivity, but underneath is the same project of emptying and controlling peripheries in the name of a national vision (Regassa, Hizekiel, and Korf 2019).

It is worth underlying continuity: across these regimes, the Ethiopian state, though changing in ideology, consistently asserted a sovereign right to re-order land relations. Whether invoking divine mandate, socialist doctrine, or developmental necessity, the state centralized decision-making about land. The humanist logic, placing the state or party as the supreme arbiter, remained, and so did the treatment of land as an instrument. The victims of this process were often the most land-dependent:

peasant farmers and pastoralists, especially those in marginalized ethnic groups. Their reciprocal relationship with the land was systematically undermined.

Thus, Ethiopia's example crystallizes the crux of our inquiry: A sovereign power operating under a human-centered, productivist logic will manage land in a way that commodifies and often militarizes it. Land becomes a means to an end (be it wealth, national integration, or political reward) rather than an end in itself. But this trajectory is not without resistance. Ethiopian history is also replete with peasant rebellions (the Bale Oromo revolt in the 1960s, for instance), land rights movements, and everyday acts of subversion against state plans. Even within the EPRDF era, the Oromia-wide protests of 2015–2016, triggered by a master plan to expand Addis Ababa's territorial control over surrounding Oromo towns, showed a popular pushback to commodification of land. Protesters sloganed that the land of Finfinne (Addis) and Oromia is "not for sale" and that people's connection to their land must be respected. This contributed to a political shake-up in Ethiopia by 2018. The lesson is clear: treating land as commodity and ignoring its sacred or identity dimensions courts instability. Sovereignty that fails to respect sacred limits eventually faces crises of legitimacy.

23.5 Peace, Justice, and the Violence of Commodification

What are the consequences of commodifying land for peace and social justice? Peace research offers a lens to see the harm caused by the processes described above. Johan Galtung's theory of structural violence is particularly illuminating. Galtung defines structural violence as an "avoidable impairment of fundamental human needs" that is built into social structures (Galtung, 1969). In the case of land, structural violence occurs when land commodification and dispossession leave communities unable to meet basic needs including for food, shelter, livelihood, cultural meaning, despite no one firing a bullet. If a pastoral community is forced off its ancestral range to make way for a plantation, and as a result they suffer malnutrition and cultural disintegration, that is structural violence. Galtung would point out that no identifiable actor may be flogging them, yet the structure, the land tenure and economic system, is causing harm (even death) by depriving them of essentials (Galtung, 1969). In Galtung's terms, this is violence because it is avoidable and it originates in human decisions (not in, say, natural disasters).

The commodification of land is a driver of structural violence globally. It tends to concentrate land in fewer hands, or in corporate hands, creating landlessness or land poverty for many. Those without secure access to land often cannot grow food or earn income, which leads to hunger and poverty. Galtung listed poverty and repression as forms of violence (Galtung, 1969); land deals frequently produce both for local people. Moreover, structural violence begets direct violence: land conflicts spark armed clashes, and state repression is deployed to enforce land grabs. We have seen this in Ethiopia, where federal forces cracked down on indigenous groups opposing plantations, and in many other countries where land defenders are criminalized or even killed.

John Paul Lederach further observes peace as a relational quality, founded on justice and the restoration of right relationships (Lederach 1997, 26). A society cannot be at peace if significant portions of its population live under structural injustices. The commodification of land represents a breach of that relationship: between people and state, people and people (e.g., investors vs locals), and people and nature. It creates a condition that Lederach would call negative peace at best, an absence of overt war that masks underlying tension. True positive peace, in contrast, requires addressing those injustices and healing the relationships. From a Lederachian perspective, land dispossession is precisely the kind of injustice that peacebuilding efforts must confront if they are to achieve "sustainable reconciliation" (Lederach 1997). One can think of peace here not just as a political state but as *nagaa*, the Oromo notion of peace that implies harmony in all relationships. When land is commodified and sacred ties are broken, *nagaa* is disturbed; there is disquiet in the community and a spiritual rupture with the earth. This can only be mended by restoring a balance or returning land to those who belong to it.

Vivienne Jabri's work adds a critical postcolonial dimension: she argues that what is often packaged as "peacebuilding" in the international arena can reproduce colonial patterns unless it challenges deep power imbalances (Jabri 2013). In the context of land, one might say that a liberal peace that simply stabilizes a country for investment without altering land inequalities is a false peace. Jabri and others call for transformative approaches that foreground justice and local agency. If we apply this to our discussion, it means that answering "is land just a commodity?" has direct implications for peace: if we say yes (implicitly or explicitly) and allow unfettered commodification, we entrench a source of conflict; if we say no, and act to protect land as a social and sacred good, we remove a major source of structural violence and open space for genuine peace.

We should also consider cultural violence, another concept by Galtung, which is the way ideologies and narratives justify or normalize violence (Galtung, 1969). The development discourse that labels pastoralists as backward and their lands as empty is a form of cultural violence (Regassa, Hizekiel, and Korf 2019). It paints the victims of dispossession as if they were beneficiaries of a noble mission ("civilizing the frontier" (Regassa, Hizekiel, and Korf 2019)) and thus masks the violence being done to them. Over time, such narratives can make society at large indifferent to the plight of displaced communities, seeing it as the price of progress. Undoing cultural violence requires changing the stories we tell about land and those who live on it. Peace researchers increasingly talk about recognition as an element of peace, recognizing the dignity and rights of all groups. In our context, that means recognizing that a pastoralist or indigenous way of life is fully legitimate and that their connection to land deserves respect, not erasure.

In essence, commodifying land without regard for its human and ecological implications undermines what Johan Galtung would term *positive peace*: the presence of social justice and harmony. It creates structural violence that may not be immediately visible but manifests in poverty, marginalization, and eventually open conflict.

It violates what Lederach calls *just-peace*, a peace with justice. And it is enabled by cultural violence, pernicious narratives that rationalize the displacements and disconnections. If peace research teaches us anything, it is that peace is more than the silencing of guns; it is the cultivation of right relationships. By that measure, the global economy's treatment of land as commodity is profoundly unpeaceful. It drives a wedge between humanity and nature, as if we were not part of nature, and between the powerful and the powerless.

To turn this around requires structural changes, land reform, stronger community land rights, legal protections for indigenous territories, but also a different ethic. Here is where ethical and theological frameworks become crucial: they can re-sacralize what has been profaned, re-value what has been devalued. They can provide the moral arguments to buttress the legal and political changes needed. We now turn to such frameworks, looking at how integral ecology and indigenous wisdom offer pathways to heal the broken relationship between sovereignty, land, and people.

23.6 Land as Sacred Trust: Integral Ecology and Theopolitical Visions

In Catholic social teaching and eco-theology, the idea of integral ecology has emerged as a framework that unites concern for social justice with concern for the earth. Pope Francis's encyclical Laudato Si': On Care for Our Common Home (2015) is a landmark in this regard. It argues that we cannot separate the cry of the earth from the cry of the poor; environmental degradation and social injustice are interlinked. A key element of Laudato Si' is the critique of a technocratic, profit-driven paradigm that treats nature as something to dominate. Francis explicitly critiques the commodification of creation: "Nowadays, for example, we are conscious of the disproportionate and unruly growth of many cities... [and] social exclusion" (Francis 2015, §44). He laments that the earth has come to be seen as "an immense pile of filth" (§21) in the eyes of consumerist society. Against this, Francis lifts up the notion of the earth as sister and mother, echoing Saint Francis of Assisi, deserving of love and care, not exploitation.

One of the most salient passages of Laudato Si' for our inquiry is paragraph 146, which we cited earlier: "For [indigenous communities], land is not a commodity but rather a gift from God... a sacred space with which they need to interact if they are to maintain their identity and values" (Francis 2015, §146). The encyclical here validates indigenous cosmologies as teachers of a more truthful worldview. It implies that the rest of the world has much to learn from those who see land as gift and sacred. The word "gift" is crucial: a gift is given freely and entails responsibility and gratitude, unlike a commodity which is bought and entails only a transactional obligation. If land is a gift from the Creator, then humans are stewards or trustees, not absolute owners. This aligns with a long Judeo-Christian tradition: in Leviticus 25:23, God says, "The land shall not be sold in perpetuity, for the land is Mine; for you are strangers and sojourners with Me." The biblical Jubilee laws, which Laudato Si' references, mandated periodic return of lands and rest for the earth (sabbath for the land), reflecting the idea that the earth should not be exhausted by human greed (Francis 2015, §71). Thus, Catholic integral ecology provides a theological basis for sacred limits on land use: limits on how much can be taken, limits on private acquisition, and duties to care for creation and for the poor.

Matthew Fox, a theologian of creation spirituality, amplifies this by calling for a recovery of sacred Earth archetypes across religious traditions (Fox 2018). Fox (2018) argues that the split between humanity and nature is a modern illusion, one that has led to rapacious exploitation. He identifies spiritual concepts like the Cosmic Christ or the presence of divine wisdom in creation as ways to rekindle reverence for the natural world (Fox 2018). In his view, scientists and environmentalists can benefit from a spiritual consciousness that sees nature as Thou, not It. He writes that the loss of Earth's sacredness "lies at the heart of the current worldview" driving extractive capitalism (Fox 2018). By reclaiming a sense of the sacred, we in turn challenge the moral permissibility of treating land (and water, forests, animals) as mere instruments. One practical implication is the concept of rights of nature, which has gained traction in some legal systems (like in Ecuador's constitution or New Zealand's recognition of a river as a legal person). Such developments, though not explicitly religious, resonate with the idea of land as belonging to God or itself having value beyond human use.

Meanwhile, Jürgen Moltmann, in works like God and Creation and Ethics of Hope, has advanced a theology where God's covenant extends to all creatures, not just humans. Moltmann (1985) speaks of the Sabbath as the celebration of creation's inherent worth, when humans cease their labor and let the land rest, acknowledging that the world belongs to God. He asserts that humanity's role is not lordship in the sense of domination, but a priesthood of creation, to mediate blessing and care. In a 1984 essay "Political Theology and the Rights of the Earth," Moltmann argued that human political arrangements must be answerable to the Creator and the rights of other creatures (Moltmann 1984, 98). Though writing decades before Laudato Si', Moltmann anticipated its themes by declaring that a society is unjust if it exploits nature in a way that harms future generations or annihilates other life forms. He would frame the commodification of land as a sin of hubris, man putting himself in the place of God, owning what ultimately belongs to God.

An intriguing thread in new political theology is the suggestion that sovereignty itself could be reimagined not as solely a human prerogative but as something shared or limited by the sovereignty of nature or the divine. Lovász and Pető (2025) argue that "instead of abolishing sovereignty, a twenty-first century eco-theopolitics should affirm the nonhuman sovereignty of nature," effectively institutionalizing "nature's self-organizing capacities" (Lovász and Pető 2025). This radical idea proposes that, for example, rivers, mountains, or ecosystems could have their own authority in decision-making processes. While this is still largely theoretical, we see early glimmers in practices like granting legal personhood to ecosystems or including indigenous elders in legislatures to represent the land's interests. It is a direct affront to the

Westphalian model, but perhaps necessary in the Anthropocene age where human sovereignty unchecked has caused planetary crises.

Integrating these theological and ethical insights, we can articulate a vision of land as sacred trust. In this vision, sovereignty is not ownership but custodianship. The state or community holds land in trust for the common good, for future generations, and for the Creator (or, in secular terms, for the Earth community itself). This concept has concrete policy implications: it would prioritize tenure security for those who actually live on and care for the land (since they are the trust's beneficiaries and caretakers), it would entail strict environmental limits (you cannot destroy the trust asset), and it would demand inclusive decision-making (all stakeholders, including the marginalized and possibly non-humans via proxies, have a say). It is notable that some of these elements exist in customary systems around the world. For instance, many African customary land laws treat land as inalienable lineage property, you can't sell it outside the community because it belongs to past, present, and unborn members. Colonial regimes attacked these customs as obstacles to development (read: commodification), but today we see their wisdom anew.

Catholic integral ecology also speaks of structural conversion: not only personal change of heart but transforming the economic structures that cause environmental and social harm. In Laudato Si' (§139), Francis calls for "the economy and politics [to] be placed at the service of peoples and the common good, not the other way around." If we apply that to land, it means land governance should serve people's sustenance and cultural life, and the health of ecosystems, rather than serving abstract growth metrics. This could mean agrarian reform, support for small-scale agroecology, recognition of indigenous land rights, and curbs on speculation, all policies that treat land not as an ordinary commodity but as a special category. Indeed, some countries already label land and water as part of the "common heritage" (for example, Bolivia's law of Mother Earth). The challenge is immense because it asks for a reorientation of value: away from immediate profit, toward long-term flourishing.

Yet, as daunting as it is, this reorientation might be inevitable. Environmental ethicists often speak of planetary boundaries, ecological limits we transgress at our peril. One could say these boundaries are nature's sacred limits. Climate change, desertification, biodiversity loss, these crises are indicators that the commodification paradigm is reaching a dead end. We are being forced by reality to respect limits that our economic system tried to deny. Eco-theologians would add that this is not merely a biophysical necessity but a spiritual opportunity: a call to humility and conversion. It is reminiscent of the biblical call in 2 Chronicles 7:14: "If my people…turn from their wicked ways, then I will heal their land." Healing the land and healing our societies go hand in hand.

Before moving to the final section, let us synthesize: Integral ecology and related theopolitical visions do not propose a rejection of all human use of land. Rather, they propose a regulated, humble, and reverent use. Land can yield fruit, fibre, minerals, but our extraction must be bounded by ethical constraints, by *safuu* (moral limits) one might say, and guided towards *nagaa* (harmonious peace). The earth is abundant, but not an infinite warehouse; it is generous, but not voiceless. If sovereignty means anything in a redeemed sense, it is the exercise of care for this inheritance. In the next section, we turn to Oromo indigenous cosmology, which strikingly parallels many of these notions, illustrating a convergence between religious ecology and indigenous knowledge. This convergence can light a path forward for Ethiopia and the world.

23.7 Indigenous Visions: Oromo Concepts of *Safuu* and *Nagaa*

Long before the advent of modernity or environmental science, Indigenous peoples around the world developed their own frameworks for living in peace with the land. Among the Oromo people of the Horn of Africa, Ethiopia's largest ethnic group, the concepts of *safuu* and *nagaa* encapsulate an eco-theopolitical ethic that could hardly be more relevant today. These Afan Oromo terms are difficult to translate neatly, but roughly, *safuu* refers to a moral order or code of right relations, a sense of limits and respect, while *nagaa* means peace, not just as absence of conflict but as a state of wholeness and right relationship, with others and with *Waaqa* (God) and with nature that naturally emanates when *safuu* is enacted.

In Oromo cosmology, similar to most African Traditional Religions, the world is animate and ensouled. *Waaqa*, the creator, is immanent in creation. Mountains, rivers, trees, and animals are considered part of a divinely ordained network of life in which humans are participants, not masters. *Safuu* is essentially the principle that governs this network. It tells people what is permitted and what is forbidden in order to maintain cosmic balance. Breaking *safuu*, for example, by arrogantly taking more from the land than one should, or by disrespecting an elder, or by harming a defenseless creature, is believed to bring about misfortune or disharmony. It's not just a social contract; it's a cosmic and spiritual one.

Now, we need to ask: how does this translate to land? For one, land among traditional Oromo was not alienable property. It was often communal, managed through the Gadaa system (the Oromo democratic socio-political system). Clans had territories (*ardaa*) which were used according to agreed upon rules. The idea of selling land to outsiders was generally taboo as land was seen as belonging to the community, and ultimately to *Waaqa*. There were certainly power dynamics and inequalities in practice, but the underlying worldview did not see land as a commodity. Land was the source of life (*lafti nafa*, "land is a living body [life]" is an Oromo expression). In prayers and songs, Oromo elders refer to *horii fi lafa* (cattle and land) as blessings that need *nagaa* in order to thrive. If conflict or injustice arises over land, it is seen as a tear in the fabric of *nagaa*. The Gadaa system provided mechanisms, such as councils of elders and rituals like araara (reconciliation), to restore *nagaa* when it was broken.

There is a fascinating resonance between Oromo cosmology and the notion of integral ecology. Both assert an interconnection of social and natural realms and insist that moral-spiritual principles must guide material use. In Oromo myths, there are often cautionary tales of individuals who violated *safuu* and met with calamity, for instance, a farmer who farmed too close to a sacred tree or a hunter who killed more animals than he could use. These stories encode the principle of sustainable use and respect for limits. One could analogize *safuu* to the concept of "the rights of Mother Earth" in other indigenous traditions or to the idea of dharma (cosmic law) in Eastern philosophies.

Perhaps most relevant to our discussion is how *safuu* and *nagaa* could inform an alternative development path in places like Ethiopia. Imagine a development policy guided by *safuu*. It would ask of any project: does this maintain the moral balance? Does a dam or a farm preserve respect for people and nature, or does it transgress the boundary of greed and disrespect? If it transgresses, it violates *safuu* and thus jeopardizes *nagaa*. In practical terms, that might mean some projects should need substantial redesign to be community-led, and reciprocal.

Crucially, the Oromo concept of *nagaa* includes *nagaa* namaa (individual peace) and, *nagaa* olaanaa (peace of the community), and *nagaa* biyya (peace of the country); these are all nested and linked. But one also hears of *nagaa* with *Waaqa* (peace with God) which is contingent on *safuu*, and implicitly *nagaa* with the environment. The peace with God and nature is maintained by things like ceremonies at water springs (*Irreecha*), sacred groves, often led by spiritual leaders like *qaallus* (in parallel with a priest in Christian world) and by ethical conduct toward wildlife (certain species are totemic and not to be harmed, for example). What this indicates is a lived sense that human peace is impossible without peace with the natural world.

Bringing Oromo wisdom into dialogue with political theology and development studies yields a vision of what we might term eco-theopolitical reconciliation. It suggests that to heal the wounds caused by sovereignty's commodification of land, we must rehabilitate both the spiritual relationship to land and the political-economic structures. Practically speaking, in Ethiopia, it would mean empowering local communities, especially those like the pastoralists of Oromia, Afar, and others, with genuine land rights and voice in governance. It would mean the state accepting limits on its projects: for instance, if a community like the Borana says a certain wetland is sacred and essential for their culture, then even a lucrative investment should not override that. It would also mean incorporating traditional authorities and knowledge into environmental management, e.g., using the Gadaa time cycles and leadership to manage resources sustainably. Interestingly, there have been attempts to do this in some conservation projects in Ethiopia, where elders are involved in range management decisions.

One can anticipate the criticism: isn't this romanticizing indigenous culture? After all, not all was golden in the past, and population growth and climate pressures make some traditional ways harder to maintain. That is true; the goal is not to freeze anyone in time. Rather, it is to learn from these principles (*safuu* and *nagaa*) and adapt them to today's needs. Indeed, Oromo scholars and activists have been doing just that, advocating for a "peaceful development" paradigm rooted in Gadaa values and human rights (Megersa 2014). The Oromo protests mentioned earlier often included cultural displays and invocations of *safuu*, protesters would carry grass and wear traditional clothes as symbols of peace and identity, implicitly saying: development that tramples *safuu* and *nagaa* is not acceptable.

Furthermore, these concepts are not exclusive to Oromo. Many other Ethiopian ethnic groups have similar notions. The Afar speak of inaf (water/land as mother). The Konso have sacred groves and a moral order around farming. As such, there is a broader Ethiopian indigenous knowledge base that can be tapped to redefine the relationship to land. It is about re-indigenizing development, or one might say decolonizing it, taking it out of the narrowly economic frame and placing it in a cultural-ecological frame.

To the extent that Ethiopia, and other countries, seek to build sustainable peace domestically, integrating such perspectives could be invaluable. Imagine constitutional or legal provisions that recognize concepts like *safuu*. South American countries have adopted "Pachamama" (Mother Earth) language in constitutions; Ethiopia could likewise acknowledge that land is a communal gift that must be honored according to principles of respect and peace. This might sound abstract, but legal principles shape policy. For example, if *safuu* were recognized, an environmental impact assessment for a new project might have to include cultural/spiritual impact, not just economic benefit. It would elevate the conversation beyond cost-benefit calculus to questions of identity and meaning.

In closing this section, let us paint the vision that emerges: A nation where *nagaa* (peace) is the aim, must ensure *nagaa* with the land. That means no group feels alienated or dispossessed; everyone has a home and a stake. The state's sovereignty is exercised with a self-restraint born of understanding that it is under a higher law, call it God, call it natural law, call it *safuu*. Development projects are carried out, yes, but carefully, consultatively, and only in forms that do not violate the community's moral economy. Land might still produce wealth, but that wealth is shared and not at the cost of future fertility or social cohesion. In the Oromo tradition, before undertaking a significant action, elders invoke *Waaqa* for guidance and swear not to violate *safuu*. What if our modern politicians did the same, metaphorically at least, pausing to consider the sanctity of what they might disrupt? The outcome could be truly transformative.

23.7.1 Conclusion: Towards an Eco-Theopolitical Future

Land is not just a commodity: it is a living nexus of sovereignty, community, and sacred value. We have seen that the modern humanist regime of sovereignty, exemplified in global development paradigms and the Ethiopian state across eras, answered the question "Is land just a commodity?" in the affirmative through its actions, with destructive consequences. Treating land as an object for unlimited use and exchange has led to dispossession of the poor, cultural annihilation of indigenous peoples, and ravaging of ecosystems. It has undermined structural peace by building violence into the very structure of society, the violence of hunger, exclusion, and ecological disaster. It has also left us spiritually impoverished, estranged from the very ground that sustains us.

However, alternative answers have always existed. Political theology reminds us that earthly sovereignty must answer to higher sovereignties: to God or the moral law; and, thus cannot rightfully convert everything, and of course, everyone, under its domain into property. Peace research insists that true peace requires justice and right relationship, which means land must be allocated and cared for in ways that uphold human dignity and environmental integrity. Catholic integral ecology articulates a holistic vision in which land, and nature broadly, is part of the common good, a gift we hold in trust, calling for gratitude and restraint rather than exploitation. And Indigenous wisdom, exemplified by Oromo *safuu* and *nagaa*, offers time-tested principles for living within limits and honoring the sacredness of place.

Bringing these threads together, we can outline an eco-theopolitics of land as a framework for the future. Its key tenets would be: (1) Land belongs ultimately to the Creator, or, in secular terms, to the Earth itself and the community of life; hence no political authority or market actor has absolute dominion over it. (2) Humans hold land in stewardship; we have use rights that come with responsibilities to care for soil, water, and biodiversity, and to share the land justly among ourselves. (3) There are sacred limits to land use, certain areas must remain inviolate, for cultural, ecological reasons, and there are limits on how much can be extracted or how much inequality can be tolerated before peace is broken. (4) The voices of the land's traditional custodians must guide decision-making, whether indigenous peoples, local communities, or other vulnerable groups, their knowledge and attachment to the land are invaluable and merit legal protection. (5) Development must mean enhancement of life, not merely economic growth: projects should be evaluated by whether they improve the real wellbeing of people and the land. If a project increases GDP but causes social upheaval or ecological damage, it is not development in this new paradigm. (6) Sovereignty redefined as service: governments exist not to own land but to ensure it serves the common good. A theopolitically informed state would see itself as a guarantor of both social justice and the integrity of creation, mediating between human needs and ecological limits.

For Ethiopia, embracing this paradigm could help resolve some of the persistent challenges that have fueled conflict. Ethnic tensions often have a resource and land dimension; recognizing customary land rights and giving regions more say in land matters, as opposed to top-down allocation to investors as it mitigates grievances. An eco-theopolitical approach would also strengthen environmental resilience, crucial as climate change brings harsher droughts and floods to the region. Imagine a policy shift where pasture lands are not seen as "empty" but recognized for the carbon sequestration and climate adaptation services they provide when managed by pastoralists' indigenous techniques. Such recognition could open up international climate finance to support pastoral communities rather than displace them. In essence, aligning modern governance with the old wisdom that land cannot be simply owned and traded might be key to a more stable and sustainable Horn of Africa.

Globally, the stakes are just as high. The commodification of land is tied to global issues like deforestation in the Amazon, palm oil expansion in Southeast Asia, or the privatization of communal lands in many countries. These drive biodiversity loss and can even contribute to pandemics through habitat destruction. Movements like agroecology, food sovereignty, and land back such as indigenous land reclamation are already embodying elements of the alternative vision we have discussed. They often carry a spiritual or moral fervor, even if expressed in secular language: they essentially assert the sacredness of the relationship to land and community.

Of course, transitioning to treating land as sacred trust will not be simple. It challenges interests. It requires legal innovation, such as new forms of collective tenure or rights of nature laws. It may face ideological resistance from those who equate development with unfettered capitalism. Yet, the interdisciplinary insight from this research is that our diverse traditions both secular and religious, scientific and indigenous are converging on the truth that a commodified earth leads to dead ends. By contrast, a reverent and restrained approach offers hope. The concept of Sacred Limits invoked in our title is not a call for stasis or anti-modernism; it is a call for self-restraint guided by wisdom. It is akin to what the philosopher Hans Jonas called the imperative of responsibility, or what economists might frame as the precautionary principle. It's saying: just because we can do something with land, plow this forest, dam this river, evict these villagers, does not mean we ought to: there are higher considerations.

In closing, let us return to the fundamental question: Is land just a commodity? Our exploration leads to a resounding no. Land is community, land is spirit, land is life. If we acknowledge this, the praxis of sovereignty and development must change. A new covenant with the land can be forged, that mirrors peace with each other. To cite Laudato Si', "Peace, justice and the preservation of creation are three absolutely interconnected themes" (Francis 2015, §92). A future of peace demands an economics and politics of respect for creation. Ethiopia's own term for peace, *nagaa*, encapsulates this trinity: peace with others, with God, with nature. By relearning *safuu*, the art of living within sacred limits, Ethiopia and the world may find a path out of the ecological and social crises of our time.

References

Agamben, Giorgio. 2004. *State of Exception*. Chicago: University of Chicago Press.

Escobar, Arturo. 1995. *Encountering Development: The Making and Unmaking of the Third World*. Princeton, NJ: Princeton University Press.

Ferguson, James. 1994. *The Anti-Politics Machine: "Development," Depoliticization, and Bureaucratic Power in Lesotho*. Minneapolis: University of Minnesota Press.

Fox, Matthew. 2018. "Climate Change, *Laudato Si'*, Creation Spirituality, and the Nobility of the Scientist's Vocation." *Zygon: Journal of Religion & Science* 53(2): 586–605.

Francis (Pope). 2015. *Laudato Si': On Care for Our Common Home*. Vatican City: Libreria Editrice Vaticana.

Galtung, Johan. 1969. "Violence, Peace, and Peace Research." *Journal of Peace Research* 6(3): 167–191.

Jabri, Vivienne. 2013. *The Postcolonial Subject: Claiming Politics/Governing Others in Late Modernity*. London: Routledge.

Langford, Peter, and Saul Newman. 2023. *Order, Crisis, and Redemption: Political Theology after Schmitt*. Albany: State University of New York Press.

Lederach, John Paul. 1997. *Building Peace: Sustainable Reconciliation in Divided Societies.* Washington, DC: United States Institute of Peace Press.

Li, Tania Murray. 2007. *The Will to Improve: Governmentality, Development, and the Practice of Politics.* Durham, NC: Duke University Press.

Lovász, Ádám, and Zoltán Pető. 2025. "Political Theology After Humanism: Eco-Theopolitics for the Twenty-first Century." *Political Theology* (online Feb 17, 2025).

Marcus, Harold G. *A History of Ethiopia.* Berkeley: University of California Press, 2002.

Megersa, Gemetchu. 2014. "The Impact of Gadaa on Oromian Culture and Identity." In *Proceedings of the 4th Oromo Studies Conference*, 33–47. Oromia: Oromia Culture and Tourism Bureau.

Moltmann, Jürgen. "Political Theology and the Rights of the Earth." *Cross Currents* 34, no. 1 (Spring 1984): 97–109.

Moltmann, Jürgen. 1985. *God in Creation: An Ecological Doctrine of Creation.* San Francisco: Harper & Row.

Moltmann, Jürgen. 1999. *God for a Secular Society: The Public Relevance of Theology.* Minneapolis: Fortress Press.

Regassa, Asebe, Yetebarek Hizekiel, and Benedikt Korf. 2019. "'Civilizing' the Pastoral Frontier: Land Grabbing, Dispossession and Coercive Agrarian Development in Ethiopia." *Journal of Peasant Studies* 46(5): 935–955.

David Smith, "'Nightmare' for Ethiopian pastoralists as foreign investors buy up land," *The Guardian*, November 10, 2014, updated October 19, 2022, September 26, 2025, https://www.theguardian.com/global-development/2014/nov/10/ethiopia-pastoralists-foreign-investors-land

Chapter 24
Concluding

Richard Marcantonio, Ellis Adjei Adams, Emmanuel Ojeifo, and Emmanuel Katongole

The array of contributions in this volume has laid bare that integral ecology is a woven tapestry of lived experience, commitment, and feedbacks from the natural world and beyond. There is no singular method or way of living integral ecology.

In our own framing of integral ecology, we describe with six ecologies. The ecology of nature, the beautiful and complex biotic and abiotic flows of material and energy throughout the Earth system. The ecology of knowledge, understandings derived from experience, experimentation, and engagement that is as rich and varied as the means by which it can be transmitted between people and communities. The ecology of culture, the shared identity between members of a community that is dynamic and ever evolving. A social ecology of interactions between human and non-human systems, often referred to as socioecological systems or couple natural-human systems, again with an unfolding dynamism that is mutually evolving and shaping. There is the ecology of the spirit, an innate potential of all humans to feel called to and believing in a power greater than the individual and at the source of all life. And intergenerational solidarity recognizing the durational and persistent nature of ecological changes today that shift the trajectory and potential opportunities of future generations. These six ecologies emerge in many and overlapping forms throughout the volume.

R. Marcantonio (✉) · E. A. Adams · E. Ojeifo · E. Katongole
University Notre Dame, Notre Dame, Indiana, United States
e-mail: rmarcant@nd.edu

E. A. Adams
e-mail: eadams7@nd.edu

E. Ojeifo
e-mail: eojeifo@nd.edu

E. Katongole
e-mail: ekatongo@nd.edu

R. Marcantonio et al. (eds.), *Integral Ecology in the Anthropocene: Africa and Beyond*,
The Anthropocene: Politik—Economics—Society—Science 49,
https://doi.org/10.1007/978-3-032-32125-1_24

Distilling and summarizing the entire volume is not the purpose of concluding here. Rather the aim is to draw out key threads and highlighting what is needed to continue to weave these threads into broader understanding of the possibilities and potential of integral ecology.

Across the chapters, integral ecology as a lived ethic emerges as central. Whether as the motivation for risking transitioning from conventional agricultural production to regenerative agroecology or the basis for promoting non-violence in communities via comprehensive environmental and social justice training, the value proposition of integral ecology extends well beyond meeting material needs. In fact, it emerges as originating in a spiritual and/or moral commitment first and then secondarily as a means of meeting material needs. All authors are clear that meeting material needs is essential for integral ecology to durably take hold. But it is a different motivation that seems to be the point of origin and catalyst, and ultimately the real value, for integral ecology in the contexts presented.

It is also clear that the verdict on integral ecology as lived practice is still undetermined. Cursory evidence of the ecological impact of integral ecology suggests that in fact the promise sustaining the environment and human needs is possible. For example, soil profiles are improving where integral ecology via agroecology is deployed, such as at the Bethany Land Institute. But what the appropriate and most effective practices for operationalizing integral ecology via agroecology is still undetermined, with a hodgepodge of techniques and methods being used throughout the cases and limited longitudinal data to bear out their ultimate effect.

The volume makes clear that the general aims of many of the approaches and practices being deployed in the name of integral ecology are similar: sustain the land and human needs, a simultaneous care for creation and livelihoods, generative rather than extractive practice, a pursuit of social and ecological justice, and so on. But while the aims are similar, how it is being undertaken and then, correspondingly, what the resulting effects are is still unclear. The result is threefold. First, again, the evidence presented suggests that the promise of integral ecology, particularly for agriculturalists, as a lifeway and livelihood practice is plausible. Second, how integral ecology may extend beyond agricultural lifeways to the varied livelihood options humans pursue while holding true to these general aims is unclear and in need of much more experimentation and evaluation. And third, all of the above needs more uniform and longitudinal evaluation.

A question that emerges from this discussion as well is what about regenerative practice that does not have 'integral ecology' as its core? Would this be discounted or lesser in any way? For example, several of the case studies throughout the volume, while aligned with integral ecology and part of the formation of the various interventions and programs, were not explicitly or comprehensively motivated by integral ecology. In fact, much regenerative practice work exists that includes no mention of integral ecology, such as the numerous practices and projects collated by Project Regeneration (Hawken 2021). It has to be recognized that integral ecology may be nowhere in the purview of practioner's and believers of many of its tenants that are not motivated in anyway by a spiritual or religious origin. Even more, there are many for who such a conviction would in fact devalue the practice or make it off-putting. Fortunately, the key to considering this question is intentionally and explicitly stated at the core of Pope Francis's articulation of integral ecology. Care for Our Common Home, while a Catholic Social Teaching document, was not written only for Catholics (Francis, 2015). Rather, he writes for "all people of good will" to move towards more just, equitable, and sustainable practices and lifeways. As such, whether a practice that is aligned with integral ecology emerges from an intention of living integral ecology is not necessary. But what does emerge from our volume is that for those who are approaching it from this intention find an additional, even primary value from it that sustains them. It is what helps them persist even when the going gets tough or to find solidarity with others and the land. This value is not exclusive to integral ecologists but emerges as an essential characteristic of living integral ecology.

As the Anthropocene continues to unfold and evolve, more integral ecologists, whether self-identifying or not, are needed to shape a future characterized by hope and flourishing. Most characterizations of the Anthropocene are bleak (e.g., Dalby 2017; Folke et al. 2021)—and not without good reason. The term and framing of 'polycrisis' has come to the fore as an apt conceptualization of the interconnected and cumulative disturbances advancing around the globe (Lawrence et al. 2024; Scheffran 2025; World Bank 2025). The stories, reflections, experiments and interventions contained in this volume from Africa and beyond make clear these realities. But, like Pope Francis's evocation with integral ecology, so too do they offer hope. Through community, cooperation, and collective action the promise of integral ecology offers a message and, now evidenced, plausible pathway to a flourishing future. Integral ecology has the potential to be a, even the, essential instrument for realizing hope and reconciling the Anthropocene.

References

Dalby, S., 2017. Anthropocene Formations: Environmental Security, Geopolitics and Disaster. *Theory, Culture & Society* 34, 233–252. https://doi.org/10.1177/0263276415598629

Folke, C., Polasky, S., Rockström, J., Galaz, V., Westley, F., Lamont, M., Scheffer, M., Österblom, H., Carpenter, S.R., Chapin, F.S., Seto, K.C., Weber, E.U., Crona, B.I., Daily, G.C., Dasgupta, P., Gaffney, O., Gordon, L.J., Hoff, H., Levin, S.A., Lubchenco, J., Steffen, W., Walker, B.H., 2021. Our future in the Anthropocene biosphere. *Ambio* 50, 834–869. https://doi.org/10.1007/s13280-021-01544-8

Francis, P., 2015. Laudato si: On care for our common home. *Our Sunday Visitor*.

Hawken, P., 2021. *Regeneration: Ending the Climate Crisis in One Generation*. Penguin.

Lawrence, M., Homer-Dixon, T., Janzwood, S., Rockstöm, J., Renn, O., Donges, J.F., 2024. Global polycrisis: the causal mechanisms of crisis entanglement. *Global Sustainability* 7, e6. https://doi.org/10.1017/sus.2024.1

Scheffran, J., 2025. *Systemic Risks and Governance of the Global Polycrisis in the Anthropocene: Stability of the Climate-Conflict-Migration-Pandemic Nexus. Global Sustainability* 1–18. https://doi.org/10.1017/sus.2025.10026

World Bank, 2025. *Pathways Out of the Polycrisis*. World Bank, Washington, D.C.

Appendix
The Hans Günter Brauch Foundation on Peace and Ecology in the Anthropocene

The Hans Günter Brauch Foundation for Peace and Ecology in the Anthropocene (HGBS) is a foundation under civil law which has its registered office in 74821 Mosbach in Baden-Württemberg.

Aims and Purposes of this Foundation

The purpose of the Foundation is to promote science and research, education and upbringing, international peace and environmental protection, in particular peace research, ecology and the interrelationships of peace and ecology (peace ecology) in the Anthropocene, the new epoch in the history of the Earth.

The Purpose of the Foundation Shall be Realised Through

(a) Promoting local competitions on questions of the transition to sustainability and peace at general-education grammar schools in the Neckar-Odenwald County, which will be awarded annually to each school from 2022 onwards.
(b) An international science prize for peace and ecology that will be awarded from 2023 onwards with the prize for students.
(c) Maintenance of a foundation house with archives, library and the foundation office (after the founder's death).
(d) Conducting scientific events, research projects and their publication on peace research and ecology (currently not possible).

R. Marcantonio et al. (eds.), *Integral Ecology in the Anthropocene: Africa and Beyond*, The Anthropocene: Politik—Economics—Society—Science 49,
https://doi.org/10.1007/978-3-032-32125-1

Members of the Governing Board

- **Dr Hans Günter Brauch (1947), Mosbach, Germany,** Chairman of the Board Political Scientist,
- **Prof. Dr Úrsula Oswald Spring (1946), Mexico, Switzerland),** Deputy Board Chair, UNAM, CRIM, Cuernavaca, Mexico
- **Dr Hans Friedrich Happes, MA (1953),** Deputy Board Chair, former headmaster, Nicolaus-Kistner Grammar School in Mosbach (Ret.),
- **Georg Nelius (1949),** Member of the Board, former SPD member of the State Parliament (2007-2021), Deputy Mayor of Mosbach, Chair of the SPD Faction in the Town Council. Member of the County Council, Neckar-Odenwald Council.
- **Wolfgang Roth (1962), Mosbach-Lohrbach, lawyer, member of the Board and of the Town Council of Mosbach**

International Science Prize for Peace and Ecology in the Anthropocene

Starting in July 2023, an international science prize has been awarded annually by the *Hans Günter Brauch Foundation for Peace and Ecology in the Anthropocene* (HGBF) in Mosbach, Baden-Württemberg in South Germany that combines:

- *Peace and Ecology* as two distinct research programs in the social sciences that evolved since the 1950s and 1970s and that have coexisted without a joint research perspective.
- *Peace Research* emerged between World War I and World War II in the idealist or Kantian tradition of international relations that offered critiques of the new realist tradition in international law and politics that prevailed after World War II.
- *Peace Research* started in North America and Scandinavia and spread in the 1970s to other parts of the world after the *International Peace Research Association* (IPRA) had been established in 1964. J. W. Burton was one of the founding fathers.
- The *'ecology concept'* was coined in the 1860s by Ernst Häckel, a German biologist. Since World War II the concept widened as 'deep ecology', 'human ecology', 'political ecology', 'social ecology', 'geoecology' and 'ecofeminism'.
- In political science and international relations *ecological perspectives, environmental studies* started in the 1960s and *research on global environmental change* followed in the 1980s.
- *Peace Research and Environmental Studies* have coexisted as two research programs with little conceptual, theoretical and empirical interchange.
- *Paul J. Crutzen*, Nobel Laureate in Chemistry, propagated the term *Anthropocene* in February 2000 as a *New Epoch in Earth's History.*
- *The Anthropocene Working Group (AWG)* proposed in 2019 as its start the years 1945 (first nuclear weapon test in Alomogordo) and 1950 (start of the Great Acceleration).

The *Hans Günter Brauch Foundation* (HGBF) *on Peace and Ecology in the Anthropocene* (PEA) was established on 8 May 2020 on the occasion of the 75th anniversary of the end of World War II and of the liberation from Fascism to combine a joint peace research and ecology perspective on the new challenges facing humankind's survival in the Anthropocene.

The *HGBF on Peace and Ecology in the Anthropocene* supports *globally scientific research* and *locally education* in secondary schools in the county where the founder was born and has lived. The HGBF funds *local educational prizes* for grammar or high school students and globally this *international science prize*. Both prizes will be awarded jointly to enhance the exchange of scientific results to secondary schools and the cooperation between scientists and pupils.

Between peace and security issues on the one hand and environment and sustainable development problems on the other, various relationships and linkages have been addressed in the social science literature since the 1960s in the Anthropocene Epoch:

- early conceptual debate on linkages between peace and the environment (since 1980s);
- conceptual discussion on environmental and ecological security (since 1989);
- research on the impacts of weapons and wars on the environment (Westing et al.);
- impact of environmental degradation and stress on environmental conflict (Baechler, Homer-Dixon);
- environmental peace-making (Conca) and role of the environment in post-conflict peacebuilding;
- emerging discourse and policy debate on climate change and conflicts (since 1989);
- early approaches to peace ecology and their shortcomings (Kyrou, Amster, Oswald/ Brauch);
- theoretical and empirical work on sustainability transition, transformation of the economy, society and political realm and climate neutrality.

These manifold research themes have not yet resulted in a peace ecology research program in the Anthropocene Epoch. The theme of *the first international Science Prize of the HGB-Foundation was:* Climate Change and Conflicts in 2023. The first recipient is PD Prof. Dr Tobias Ide (senior lecturer, Murdoch University, Perth (Australia) who also teaches at the University of Hiroshima (Japan) and at the TH in Braunschweig (Germany). International Recognitions were awarded to Dr Tom Deligiannis, Laurier University (Canada) and Dr Richard Marcantonio, Notre Dame University (USA).

The theme of *the second international Science Prize of the HGB-Foundation* in 2024 *was*: E**cocide: Impacts of Wars and/or Climate Change on Food Security since 1945.** The second recipients were Prof. Dr Gregory Thaler, Oxford University, Oxford, UK, and Dr Victor Tsilonis, international lawyer, Greece. International Recognitions were awarded to Dr Annet Adong (Uganda), Dr Melissa Roxana Quispe-Zuniga (Peru), and Dr Christo Idowu Odeymi (Nigeria).

The third international science prize in 2025 will be awarded to Dr Angela J. Lederach, Assistant, Chapman University, California, USA for her book: *Feel the Grass Grow: Ecologies of Slow Peace* (Stanford University Press, 2023) and to Dr Matthew Gillett, Senior Lecturer (Associate Professor), *University of Essex* (NZ, UK) for his book: *Prosecuting Environmental Harm before the International Criminal Court* (Cambridge University Press, 2022). **An international scientific recognition will be awarded *to*** *Dr Anselm Vogler for his PhD thesis* (2025): *Navigating External and Existential Threats: National-Level Policy Responses to Climate Security*, Hamburg University, Germany; Dr Barbara Magalhães Teixeira (2024) for her thesis at Lund University on: *The Nature of Peace and the Continuum of Violence in Environmental Conflicts*, and Dr ASM Mostafizur Rahman, Heidelberg University, for his thesis (2024) on *Garments and Growth—State, Ideas, Policy Learning, and Globalization in Bangladesh.*

Chairman of the Board of Trustees (*Kuratorium*) of the HGB-Foundation: Lord Mayor of the town of Mosbach, Julian Stipp. Mosbach has been a member of Mayors for Peace since 2021.

See at: http://hgb-stiftung.org/index.html and http://hgb-stiftung.de/index.html.

Global Jury Members of the HGBS Science Prize

- **PD Dr. Hans Günter Brauch**, Board Chair of the HGB Foundation, Germany;
- **Prof. Dr. Úrsula Oswald Spring**, Deputy Board Chair of the HGB Foundation, Mexico;
- **Assoc. Prof. Dr. Jean Chrysostome Kimbuku Kiyala**, DUT, Durban, South Africa, born in DRC;
- **Prof. Dr. Richard Matthew**, Director, Blum Center, Univ. of California, Irvine, USA;
- **Prof. Dr. Antonina Ivanova Berchova**, Universidad Autónoma de Baja California Sur, Mexico;
- **Prof. Dr. Simon Dalby**, Wilfrid Laurier University, Waterloo, Canada;
- **Assoc. Prof. Dr. Dahlia Simangan**, Hiroshima University, Japan;
- **Prof. Dr. Imtiaz Ahmed**, Centre for Alternatives, Dhaka, Bangladesh;
- **Prof. Dr. Susan Park**, Sydney University, Sydney, NSW, Australa;
- **Prof. Dr. Jürgen Scheffran**, Department of Geography, Hamburg University, Germany;

- **Dr. Judith Nora Hardt,** Researcher, Centre Marc Bloch, Humboldt Univ., Berlin, Germany (on leave).

List of Books edited by Hans Günter Brauch in the Hexagon Series

<https://www.springer.com/series/8090 > and
<http://www.afes-press-books.de/html/hexagon.htm >

List of Books in the APESS Series

< https://www.springer.com/series/15232> and
< http://www.afes-press-books.de/html/APESS.htm>

List of Books in the ESDP Series

<https://www.springer.com/series/10357> and
<http://www.afes-press-books.de/html/SpringerBriefs_ESDP.htm >

List of Books in the PSP Series

< https://www.springer.com/series/10970> and
<http://www.afes-press-books.de/html/SpringerBriefs_PSP.htm >

List of Books in the PAHSEP Series

<https://www.springer.com/series/15230 > and
http://www.afes-press-books.de/html/PAHSEP.htm

Open Access Books Published in these five Series

List of Open Access Books in the Hexagon Series

- **HESP 1:** Truong, Thanh-Dam; Gasper, Des; Handmaker, J., Bergh, S.I. (Eds.): Migration, Gender and Social Justice—Perspectives on Human Insecurity. Hexagon Series on Human and Environmental Security and Peace, vol. 9 (Heidelberg—New York—Dordrecht—London: Springer, 2013); at: < https://link.springer.com/book/10.1007/978-3-642-28012-2>
- **HESP 2:** Jeroen Kool, Saeb Bamya, Samer Talozi, EcoPeace, Tel Aviv, Israel—Bethlehem, Palestine—Amman, Jordan: *Sustainable Development in the Jordan River Valley: Final Report of the Regional NGO Master Plan.* Hexagon Series on Human and Environmental Security and Peace, vol. 13 (Berlin—Heidelberg—New York: Springer, 2016); at: <https://link.springer.com/book/10.1007/978-3-319-30036-8>.

List of Open Access Books in the APESS Series

- **APESS 1:** Maja Göpel: *The Great Mindshift: How a New Economic Paradigm and Sustainability Transformations go Hand in Hand* (Cham—Heidelberg—New York—Dordrecht—London: Springer International Publishing, 2017); at: < https://link.springer.com/ book/10.1007/978-3-319-43766-8>.
- **APESS 2:** Carl Middleton, Vanessa Lamb (Eds.) (Eds.): *Knowing the Salween River: Resource Politics of a Contested Transboundary River* (Cham: Springer International Publishing, 2019); at: <https://link.springer.com/book/10.1007/978-3-319-77440-4>.
- **APESS 3: S.** Benner, G. Lax, G., P.J. Crutzen, , U. Pöschl,, J. Lelieveld, H.G. Brauch **(Eds.).** *Paul J. Crutzen and the Anthropocene: A New Epoch in Earth's History* (Cham: Springer International Publishing, 2021); at: <https://link.springer.com/book/10.1007/978-3-030-82202-6>.
- **APESS 4:** H.G. Brauch **(Eds.):** *Towards Rethinking Politics, Policy and Polity in the Anthropocene—Multidisciplinary Perspectives* (Cham: Springer International Publishing, 2021); at: <https://link.springer.com/book/10.1007/978-3-031-71807-6>.

List of Open Access Books in the ESDP Series

- **ESDP 1:** So far no Open Access Volume is available in this book series.

List of Open Access Books in the PSP Series

- **PSP 1:** Paul J. Crutzen, Hans Günter Brauch (Eds.): *Paul J. Crutzen: A Pioneer on Atmospheric Chemistry and Climate Change in the Anthropocene*. Springer Briefs on Pioneers in Science and Practice No. 50 (Cham—Heidelberg—New York—Dordrecht—London: Springer International Publishing, 2016; at: <https://link.springer.com/book/10.1007/978-3-319-27460-7>.
- **PSP 2:** Nils Petter Gleditsch: *Pioneer in the Analysis of War and Peace*—Presented by Peter Wallensteen. Springer Briefs on Pioneers in Science and Practice No. 29 (Cham—Heidelberg—New York—Dordrecht—London: Springer, 2016); at: < https://link.springer.com/ book/10.1007/978-3-319-03820-9>.
- **PSP 3:** Gleditsch, Nils Petter (Ed.): *R. J. Rummel: An Assessment of His Many Contributions*. Springer Briefs on Pioneers in Science and Practice, vol. 37 (Cham—Heidelberg—New York—Dordrecht—London: Springer-Verlag, 2017); at: <https://link.springer.com/book/ 10.1007/978-3-319-54463-2>.

List of Open Access Books in the PAHSEP Series

PAHSEP 1: Nils Petter Gleditsch (Ed.): *Lewis Fry Richardson: His Intellectual Legacy and Influence in the Social Sciences.* Pioneers in Arts, Humanities, Science, Engineering, Practice (PAHSEP), No. 27 (Cham: Springer International Publishing, 2020); at: <https://link.springer.com/book/10.1007/978-3-030-31589-4>.

University of Notre Dame

UNIVERSITY OF NOTRE DAME

The Keough School of Global Affairs at the University of Notre Dame has been a primary supporter of bringing this project to fruition, housing several of the Institutes involved in this work. Founded in 2014, the Donald R. Keough School of Global Affairs is a vibrant community committed to serving human dignity through research and scholarship, teaching and learning, and policies and practices designed to advance the development of the whole person and of each person in their specific socio-cultural context.

Through our commitment to integral human development, we seek solutions in ways that promote the common good and prioritize the needs of the most vulnerable and marginalized. Our commitment to advancing human dignity requires that we meet fundamental demands with respect to how policies, practices and processes of development, poverty-alleviation, sustainability, peacebuilding and governance are designed, implemented and evaluated. As a policy school that emphasizes integral human development, we seek solutions to today's challenges by working with policymakers and influencers throughout the world and through our presence in our Washington D.C. Office.

Our faculty and students conduct rigorous research that addresses some of the world's most pressing, and complex challenges, with an emphasis on sustainability and environmental justice; poverty, inequality and marginalization; democracy, governance, institutions and rights and systems and structures of violence and peace. By building long-term partnerships within the academy, business, government and civil society, the Keough School integrates disciplines and best practices to identify solutions responsive to the interconnected nature of global challenges.

R. Marcantonio et al. (eds.), *Integral Ecology in the Anthropocene: Africa and Beyond*, The Anthropocene: Politik—Economics—Society—Science 49,
https://doi.org/10.1007/978-3-032-32125-1

Because contemporary challenges are not bound by the perspective of any one country, language, culture, religion, or discipline, we teach our students to address these issues in a holistic way through a wide array of courses, experiential and cross-cultural learning opportunities and interactions with international scholars, policymakers and cultural leaders.

The University of Notre Dame's Kroc Institute for International Peace Studies, part of the Keough School of Global Affairs, is one of the world's leading centers for the study of the causes of violent conflict and strategies for sustainable peace. Kroc Institute faculty and fellows conduct interdisciplinary research on a wide range of topics related to peace and justice.

The Kroc Institute for International Peace Studies was inspired by the vision of Rev. Theodore M. Hesburgh, C.S.C., and philanthropist Joan B. Kroc of a world free from the threat of nuclear holocaust. During his 35 years as president of Notre Dame, Father Ted was a leading national voice for civil and human rights and the abolition of nuclear weapons. Joan Kroc was deeply concerned about the nuclear arms race and committed to world peace.

Founded at the University of Notre Dame in 1986, the Kroc Institute focuses on research on a wide range of issues relating to peace and justice and the causes of violent conflict.

Prize Winners since 2023 of the International Science Prizes of the Hans Günter Brauch Award on Peace and Ecology in the Anthropocene

< http://hgb-stiftung.org/index.html>

2023

2025

The Editors

Marcantonio is a researcher, teacher, and practitioner focused on regenerative and durable livelihoods, environmental management and policy, environmental and other violence, and peacebuilding. He has published numerous peer-reviewed, policy, and public facing articles in periodicals ranging from Peacebuilding to Environmental Science and Policy to the popular periodical the Bulletin of Atomic Scientists. He has conducted research and practice on these critical issues on five continents working with communities, elected officials, regulators, corporations, and NGOs alike. Marcantonio holds a B.A. in geography and the environment from the University of Texas—Austin, an M.P.A. in environmental management from Indiana University—Bloomington, and an M.A. in anthropology and a Ph.D. in peace studies from the University of Notre Dame.

1. Marcantonio, R. (2026) The Environment in Conflict: Implications and pathways. Cambridge University Press.
2. Marcantonio, R. and Field, S. 2025. Environmental vulnerability and conflict occurrence are tightly related. Nature Communications Earth and Environment. 6(1), 1-10.
3. Marcantonio, R., Fuentes, A., and Lederach, J.P. eds. (2024) Exploring Environmental Violence: Perspectives, experience, expression, and engagement. Cambridge University Press.
4. Marcantonio, R. 2024. The Glaring Gaps: Environmental Violence and Peace Research and Practice. Peacebuilding. 13(2), 145-167.
5. Marcantonio, R. (2022) Environmental Violence: In the Earth System and the Human Niche. Cambridge University Press.

Address:0318D Hesburgh Center for International Studies, Notre Dame, IN, 46556. USA
Email: rmarcant@nd.edu
Website: https://keough.nd.edu/about/faculty-staff-directory/richard-marcantonio/

(continued)

R. Marcantonio et al. (eds.), *Integral Ecology in the Anthropocene: Africa and Beyond*, The Anthropocene: Politik—Economics—Society—Science 49,
https://doi.org/10.1007/978-3-032-32125-1

(continued)

Ellis Adjei Adams is an Associate Professor of Geography and Environmental Policy at the Keough School of Global Affairs, University of Notre Dame. He holds a PhD in Geography, Environment, and Spatial Sciences from Michigan State University, M.S in Environmental Policy from Michigan Tech. University, and B.S in Natural Resources Management from the Kwame Nkrumah University of Science and Technology, Ghana. His work examines the social, political, institutional, and governance dimensions of environmental and natural resources. He conducts research in Ghana, Malawi, Kenya, Uganda, and in the United States, and his work has been funded by numerous agencies including the US National Science Foundation. Adams was the recipient of the 2020 Distinguished Emerging Scholar Award in African Geography from the Association of American Geographers.

1. Beresford, M., Adams, E., Budds, J., Harris, L. M., Jepson, W., Marley, T., ... & Wutich, A. (2025). Can household water sharing advance water security? An integrative review of water entitlements and entitlement failures. Environmental Research Letters, 20(1), 013003.
2. Adams, E. A. (2024). "Why Should a Married Man Fetch Water?" Masculinities, gender relations, and the embodied political ecology of urban water insecurity in Malawi. Social & Cultural Geography, 25(4), 582-600.
3. Adams, E. A., Thill, A., Kuusaana, E. D., & Mittag, A. (2023). Farmer–herder conflicts in sub-Saharan Africa: drivers, impacts, and resolution and peacebuilding strategies. Environmental Research Letters, 18(12), 123001.
4. Adams, E. A., Byrns, S., Kumwenda, S., Quilliam, R., Mkandawire, T., & Price, H. (2022). Water journeys: Household water insecurity, health risks, and embodiment in slums and informal settlements. Social Science & Medicine, 313, 115394.
5. Stoler, J., Guzmán, D. B., & Adams, E. A. (2023). Revisiting transformative WASH: measuring impact. The Lancet Global Health, 11(4), e493-e494.

Address: O304 Hesburgh Center for International Studies, Notre Dame, IN, 46556, USA
Email: eadams7@nd.edu
Website: https://keough.nd.edu/about/faculty-staff-directory/ellis-adams/

Emmanuel Ojeifo is a doctoral candidate in the Department of Theology and graduate coordinator of the Kellogg Integral Ecology Working Group at the University of Notre Dame. His research interests include ecological and agrarian violence, eco-martyrdom, black sanctity, and human flourishing in Africa. His dissertation research examines the question of land violence, identity, and belonging in sub-Saharan Africa through the lens of Christian social ethics. His works have been published in *Journal of Catholic Social Thought* and *Religions*.

1. Ojeifo, E.O. (2025). Christian Social Ethics and Land Reform in Africa: Insight from Oscar Romero. Journal of Catholic Social Thought 22(2), 286-310. Doi.
2. Ojeifo, E.O. (2025). Ecology and Religion in Chinua Achebe's Things Fall Apart. Religions 16 (1132), 1-16.
3. Ojeifo, E.O. (2025). Mineral Extraction, Human Rights Violation, and the Church's Social Responsibility in the Democratic Republic of Congo. African Theological Journal for Church and Society 6(1), 26-52.
4. Ojeifo, E.O. (2025) Integral Ecology as Theology and Ethics of "Place." Religions 16(659), 1-14.
5. Ojeifo, E.O. (2025). African and Christian: Black Sanctity and the Future of African Catholicism. In M. Agbaw-Ebai & M. Levering (Eds.), Continuing the Quest for Morality Truly African, Truly Christian: The Achievement of Paulinus Odozor C.S.Sp. (pp. 256-276). St Augustine's Press.

Address: Dept. of Theology, Notre Dame, IN, 46556, USA
Email: eojeifo@nd.edu
Website: https://theology.nd.edu/people/emmanuel-ojeifo/

(continued)

(continued)

Emmanuel Katongole is professor of theology and peace studies in the Keough School of Global Affairs at the University of Notre Dame. A Catholic priest ordained by the Archdiocese of Kampala, Uganda, he is a core faculty member of the Keough School's Kroc Institute for International Peace Studies. Katongole is the author of books on the Christian social imagination, the crisis of faith following the genocide in Rwanda, and Christian approaches to justice, peace, and reconciliation. His most recent book is "Who Are My People? Love, Violence, and Christianity in Sub-Saharan Africa" (Notre Dame Press, 2022) which won first place honors from the Catholic Media Association in the category of Gender Issues–Inclusion in the Church. Katongole earned a Ph.D. in philosophy from the Catholic University of Louvain, undergraduate degrees in philosophy and in theology (Urbaniana, Rome) and a diploma in theology and religious studies from Makerere University in Kampala, Uganda. He has served as associate professor of theology and world Christianity at Duke University, where he was the founding co-director of the Duke Divinity School's Center for Reconciliation

1. Katongole. E. (2022) Who Are My People? The Invention of Love in Sub Saharan Africa. *University of Notre Dame Press.*
2. Katongole. E. (2017) The Journey of Reconciliation: Groaning for New Creation in Africa. *Orbis Press.*
3. Katongole. E. (2017) Born from Lament: On the Theology and Politics of Hope in Africa. *Eedermans Press.*
4. Katongole. E. (2013) Stories from Bethany: On the Faces of the Church in Africa. *Paulines Publications.*
5. Katongole. E. (2010) The Sacrifice of Africa: A Political Theology for Africa. *Eerdmans Press.*

Address:
Email:
Website:

Heather Foucault-Camm is the Program Director for the Science & Religion Initiative (SRI) at the McGrath Institute for Church Life, where she develops the pedagogical components of SRI programming, and delivers presentations on Bioethics, Technology, and Chemistry. A veteran science teacher and curriculum designer, she has a BSc and MSc in the field of Chemistry (Chemical Physics), a PGCE, an M.A. in Theology (Notre Dame) and she continues to work on her Ph.D. in Moral Theology (Notre Dame). Heather's scholarly work focuses on the social implications of emerging technologies, with a particular influence on Artificial Intelligence. Heather is a Scholar Associate of the Society of Catholic Scientists, upon special invitation from its Board, and is a Lay Dominican of the Province of St. Albert the Great.

Address: 370 Geddes Hall, Notre Dame, IN, 46556, USA
Email: hfoucaul@nd.edu
Website: https://mcgrath.nd.edu/about/faculty-staff/heather-foucault-camm-pgce-m-sc-m-a/

The Contributors

Ikechukwu Alamezie—Pan-Atlantic University, Nigeria, ialamezie@pau.edu.ng;
Elsa Barron—University of Notre Dame, USA, ebarron3@nd.edu;
Brett Beasley—University of Notre Dame, USA, bbeasle1@nd.edu;
Geleta Tesfaye Berisso—University of Notre Dame, USA, gberisso@nd.edu;
Vincent Ferrer Bukenya—Uganda Martyrs University, Uganda, bukenya.vincent@stud.umu.ac.ug;
Portia Ozioma Chigbu—University of Notre Dame, USA, pchigbu@nd.edu;
Godfrey Ddungu—California State University Fresno, USA, goddungu@mail.fresnostate.edu;
Landing Diedhiou—Assane Seck University of Ziguinchor, Senegal, l.d7@zig.univ.sn;
Heather M. Foucault-Camm—University of Notre Dame, USA, hfoucaul@nd.edu;
Nicolaas Ahmad Heeren—Catholic Committee Against Hunger and for Development-Terre Solidaire, France, nicheeren@gmail.com;
Obinna Ikejimba—Pan-Atlantic University, Nigeria, oikejimba@pau.edu.ng;
Benard Islanda—Justice, Peace, Development and Integrity of Creation Commission, Vatican City, benardisandacmm@gmail.com;
Ryan Juskus—University of Notre Dame, rjuskus@nd.edu;
Richard Kachungu—Young Emerging Farmers Initiative, Zambia, richard.kachungu@gmail.com;
Denis Katusiime—Uganda Martyrs University, Uganda, dkatusiime@unik.ac.ug;
Sister Sheila Kinsey—Integrity for Nonviolence, Vatican City, skinsey.fcjm@gmail.com;
Mitchell Liddick—University of Notre Dame, mliddick@nd.edu;
Alonso de Llanes—Laudato Si' Action Platform, Dicastery for Promoting Integral Human Development, Vatican City, alonsodellanes@gmail.com;
Richard Marcantonio—University of Notre Dame, rmarcant@nd.edu;
David Mugambe Mpiima—Makerere University, Uganda, davidmpiima@gmail.com;

R. Marcantonio et al. (eds.), *Integral Ecology in the Anthropocene: Africa and Beyond*, The Anthropocene: Politik—Economics—Society—Science 49,
https://doi.org/10.1007/978-3-032-32125-1

Nicholas Mugabi—Makerere University, Uganda, mugabinicoxs@gmail.com;
Jeremiah Anene Nwankwegu—Ebonyi State University, Nigeria, jeremiah.nwankwegu@ebsu.edu.ng;
Raymond Offenheiser—University of Notre Dame, USA, roffenhe@nd.edu;
Father Emmanuel Ojeifo—University of Notre Dame, USA, eojeifo@nd.edu;
Father Ugochukwu Okoye—University of Notre Dame, USA, uokoye@nd.edu;
Sister Damien Marie Savino—University of Notre Dame, USA, ksavino@nd.edu;
Deena Shelton—LeTourneau University, USA, deenashelton@letu.edu;
Mark J. Shimave—Boston College, USA, shimave@bc.edu;
Jimmy Spire Ssentongo—Uganda Martyrs University, Uganda, jsssentongo@gmail.com;
David Ssekamatte—Uganda Management Institute, Uganda, dmssekamatte@gmail.com;
JosephSsekandi—Uganda Martyrs University, Uganda, sekjoseph@gmail.com;
Shawn T. Storer—Holy Cross College, USA, sstorer@hcc-nd.edu;
Jennifer Tank—University of Notre Dame, jtank@nd.edu;
Charles L. Tumuhe—Uganda Martyrs University, Uganda, tumchaz34@gmail.com.

Index

R. Marcantonio et al. (eds.), *Integral Ecology in the Anthropocene: Africa and Beyond*, The Anthropocene: Politik—Economics—Society—Science 49,
https://doi.org/10.1007/978-3-032-32125-1

H

I

J

L

M

N

W

FSC
www.fsc.org
MIX
Papier | Fördert gute Waldnutzung
FSC® C083411

Kolibri 360 GmbH
Ferdinand-Jühlke-Straße 7
99095 Erfurt, Deutschland
produktsicherheit@kolibri360.de